Embodied Intelligence

Embodied Intelligence

Multidisciplinary Perspectives on Natural, Artificial, and Hybrid Systems

edited by Sheila L. Macrine, Jennifer M. B. Fugate, Arsen Abdulali, and Josie Hughes

The MIT Press
Cambridge, Massachusetts
London, England

The MIT Press
Massachusetts Institute of Technology
77 Massachusetts Avenue
Cambridge, MA 02139
mitpress.mit.edu

The MIT Press would like to thank the anonymous peer reviewers who provided comments on drafts of this book. The generous work of academic experts is essential for establishing the authority and quality of our publications. We acknowledge with gratitude the contributions of these otherwise uncredited readers.

This book was set in Stone Serif and Stone Sans by Westchester Publishing Services. Printed and bound in the United States of America.

Library of Congress Cataloging-in-Publication Data is available.

ISBN: 978-0-262-05349-5

EU Authorised Representative: Easy Access System Europe, Mustamäe tee 50, 10621 Tallinn, Estonia | Email: gpsr.requests@easproject.com

Collectively, we wish to thank our families, our mentors for their inspiration, and the growing embodied community.

Sheila Macrine dedicates this book to her son, Gavin, whose intellectual curiosity keeps her constantly engaged, and to her husband, Nicholas, for his support.

Jennifer Fugate would like to dedicate this book to her husband, Matthew, for keeping his humor throughout the process. She also dedicates this book to her son, Connor, whose intelligence has grown tremendously throughout the last year through all forms of interaction with the natural and artificial worlds.

Josie Hughes would like to thank her family for their continuous understanding and support and her lab members and colleagues with whom she gets to share the journey of continuing to grow her understanding of Embodied Intelligence.

Arsen Abdulali dedicates this book to his parents, Ahmed and Ramilya, for their unwavering support, and to his nephew, Rinal, for reminding him that perspectives may shift in an evolving world, but fundamental questions endure and drive science.

Contents

Foreword

The study of embodiment and embodied intelligence (EI) has deep roots, dating back to philosophical and religious debates about mind-body dualism. Over time, these investigations have persisted, not because we have reached full understanding but because embodiment touches on the very core of life, intelligence, and creativity. The phenomena explored in this field are vast and complex and go significantly beyond what traditional science has managed to explain.

In the last decades, the study of EI has reshaped how we think about intelligence. No longer can we view the mind as independent from the body or the environment. EI reveals that intelligence emerges from the interaction between the physical body and the world. This perspective, which integrates ideas from biology, neuroscience, psychology, computer science, robotics, and philosophy, offers a more comprehensive view of how intelligent behavior arises from bodily experiences.

For too long, the study of intelligence has been confined to the realms of abstract thought and disembodied computation. This volume tackles those limitations, embracing a holistic perspective that recognizes the profound interconnectedness of mind, body, and environment. By exploring the embodied nature of intelligence across a diverse spectrum of agents, this book opens up exciting new avenues for research and discovery.

This evolving reconceptualization of EI was recently explored with the organization of annual online conferences on EI,[1] which fostered a vibrant community of researchers dedicated to exploring this fascinating field. This volume builds on that momentum, bringing together leading voices to offer a comprehensive overview of the current state of the art and a glimpse into the future of EI.

What truly sets this volume apart is its commitment to inclusivity. By embracing a broad definition of EI, it transcends the traditional divide between the biological and the artificial, recognizing the potential for intelligence to emerge in unexpected forms. This perspective challenges us to reconsider our assumptions about the nature of intelligence and to appreciate the remarkable diversity of intelligent behavior in the world around us.

The editors of this volume have assembled an impressive roster of world-class contributors, each bringing their unique expertise to bear on this fascinating topic. This transdisciplinary approach is essential for unlocking the full potential of EI research.

As a researcher who has dedicated their career to the intersection of robotics and biology at the Bio-Inspired Robotics Lab at the University of Cambridge, I am delighted to endorse this groundbreaking volume, *Embodied Intelligence*. This book is not simply a collection of scholarly chapters; it is a testament to the power of interdisciplinary collaboration and a bold manifesto for a paradigm shift in how we understand intelligence. Emerging from the very heart of this exciting field, where the boundaries between the living, artificial, and hybrid entities, in the case of living robots, are constantly being challenged, this volume promises a uniquely insightful perspective.

I am particularly excited by the book's emphasis on practical applications. By exploring the use of EI in fields such as robotics, prosthetics, and education, this volume demonstrates the transformative potential of this paradigm shift. It is a testament to the power of EI to not only deepen our understanding of ourselves and the world around us but also to improve our lives in tangible ways.

Embodied Intelligence is a must-read for anyone interested in the future of intelligence research. With its roots in the cutting edge, this book offers a unique and invaluable perspective on the embodied nature of intelligence. It is a bold and visionary work that will challenge your assumptions, expand your horizons, and inspire you to explore the fascinating world of EI. I wholeheartedly recommend it to scholars, students, and anyone curious about the nature of intelligence and its potential to shape our future.

I hope this volume will inspire further exploration into the embodied nature of intelligence, guiding future research and innovation in fields as diverse as biology, cognitive science, artificial intelligence, robotics, and beyond. By embracing a holistic view of intelligence—one that integrates the body, the environment, and the processes that connect them—we stand to make significant strides in understanding what it means to be truly intelligent in an interconnected world.

—Fumiya Iida
Bio-Inspired Robotics Laboratory, Department of Engineering,
University of Cambridge, UK. October 17, 2024

Note

1. EI 2025: Embodied Intelligence Conference, www.embodied-intelligence.org.

Preface

As two cognitive psychologists immersed in the world of embodied cognition (EC), Jennifer and I have witnessed a burgeoning interest in the concept of embodiment across diverse fields—neuroscience, neurobiology, psychology, computer science, cognitive science, robotics, engineering, artificial intelligence (AI), evolutionary psychology, and biology, to name a few. It became increasingly apparent that there were significant overlaps in terminology, labels, and concepts across these disciplines, and sometimes even a misuse of the term "embodied" itself (i.e., a robot with a head labeled as embodied). We began to encounter phrases like "embodied robotics" and "embodied AI," which sparked our curiosity and led us to embark on a journey to explore the depths of this fascinating area.

This exploration required expertise beyond our own. Recognizing this, we invited two robotics scientists, Josie Hughes (École Polytechnique Fédérale de Lausanne) and Arsen Abdulali (Queen Mary University of London) to join us. Their profound understanding of robotics, AI, and soft robotics proved invaluable as we embarked on the ambitious project of editing this volume. Our goal was to bring together leading scholars across disciplines to provide clarity and depth to a reconceptualized and broadened understanding of embodied intelligence (EI). This new perspective expands on the foundational ideas about EI pioneered by Rodney Brooks—specifically, his work with the insect-like robot Genghis, which demonstrated the power of situatedness and interaction in generating intelligent behavior. It also draws inspiration from the work of Rolf Pfeifer and Josh Bongard, as well as Shaun Gallagher, Andy Clark, and phenomenologist Maurice Merleau-Ponty, further enriching our understanding of embodiment. To achieve this, we approached the MIT Press with a proposal for a volume that would explore this expanding understanding of the embodiment of intelligence across a wide range of disciplines and entities, from humans and other biological organisms to traditional robots, hybrid systems, and even entirely novel life forms.

This volume represents the culmination of that effort. It aims to capture the current state of the field, distill its core principles and concepts, and chart potential future directions. It offers a rich variety of perspectives on EI and its implications for the future of intelligent systems across diverse entities.

Under the guidance of our talented MIT Press acquisition editor, Susan Buckley, and with advice from her colleague Philip Laughlin, our initial focus, which tended to be human-centric, contrasting human and robotic intelligence (as explored in chapter 2), expanded significantly, leading us to investigate a vast new world. This deep dive led us to consider EI across a wider array of entities. With the insights from our fellow editors, we moved beyond simply comparing human and robotic intelligence to consider embodiment across a broader spectrum, encompassing biological organisms, traditional and hybrid robots, virtual entities, and even novel life forms like the Xenobots, created by Professor Josh Bongard and Professor Michael Levin. These "living robots" opened up a completely new realm of possibilities, challenging our preconceptions about the nature of EI and inspiring us to embrace a more inclusive perspective.

This Volume

Embodied Intelligence represents a pivotal shift, emphasizing the synergistic relationships among biological and artificial embodiments and beyond. It incorporates the integration of body/mechanical exoskeletons, sensorimotor/sensory systems, and mind/computational elements. A broader approach fosters significant research advancements, allowing for transdisciplinary exploration of the parallels and synergies between biological and artificial systems, leading to a more holistic and inclusive definition of EI. The added expertise of robotics scientists, Josie Hughes and Arsen Abdulali, enriches this interdisciplinary discussion with their unique perspectives.

This volume serves as a demonstration of the growing interest in EI. By emphasizing the interconnectedness of biological and artificial systems, this volume has the potential to reveal novel insights and advancements in fields ranging from health care and regenerative medicine to human-computer interaction. Promoting interdisciplinary dialogue fosters transformative understanding and utilization of EI, making this volume a catalyst for reshaping how we conceptualize and leverage its power to inform the future of intelligence research. By uniting researchers across disciplines, this volume encourages idea exchange, promotes interdisciplinary learning, and breaks down knowledge barriers. This, in turn, fosters a collaborative spirit and the sharing of knowledge, which combined with the groundbreaking insights presented within, makes it an invaluable resource for anyone seeking to understand and shape the future of this inclusive view of intelligence in a world increasingly intertwined with technology. We believe this work will inspire further exploration of EI and its potential to reshape our world.

Acknowledgments

This volume would not have been possible without the collaborative efforts and generous support of many individuals. We express our sincere gratitude to Susan Buckley, acquisitions editor at the MIT Press, whose invaluable insights, guidance, and partnership were instrumental in shaping and refining this project from conception to completion. Her contributions to reimagining the volume and refining our framework proved essential. We also extend our heartfelt thanks to the dedicated team at the MIT Press for their expertise and commitment and to the external reviewers for their thoughtful and constructive feedback.

We are deeply grateful for the contributions of several key individuals who have been integral to this project. We thank Fumiya Iida for his masterful foreword, which eloquently captures the passion and potential of embodied intelligence (EI). Our sincere appreciation goes to Distinguished Professor Karl Friston, whose unwavering support from the project's inception, including being the first scholar to contribute and directing us to emerging scholars and researchers in the field, has been invaluable. We also acknowledge Emeritus Professor Rolf Pfeifer for his encouragement and endorsement and Andy Clark for his continued support since our last volume with the MIT Press.

We are particularly indebted to Josh Bongard and Michael Levin, whose cutting-edge research on living robots and related areas has profoundly impacted the field of EI, pushing the boundaries of what's possible and inspiring new avenues of inquiry. We also extend our gratitude to Shaun Gallagher for his insightful philosophical work, which has been instrumental in clarifying the complex interplay of embodied cognition in humans and artificial agents.

Finally, we extend our deepest gratitude to all the contributors to this volume. We are incredibly thankful for these leading EI scholars and emerging researchers who came together in this unique interdisciplinary effort to work toward a transdisciplinary and inclusive reconceptualization of EI. Their expertise, dedication, and collaborative spirit have made this volume a reality.

Introduction: Toward Embodied Intelligence

Sheila L. Macrine, Jennifer M. B. Fugate, Arsen Abdulali, and Josie Hughes

Nihil est in intellectu quod non sit prius in sensu.
There is nothing in the understanding that is not first in the senses.
—Thomas Aquinas, De Veritate, q. 2 a. 3 arg. 19

Introduction

The frontier of intelligence research is evolving, moving beyond the traditional confines of brains and code to explore the profound and multifaceted role of embodiment. This volume explores the expanding field of embodied intelligence (EI) across disciplines, advocating for a broadened definition that encompasses a diverse range of entities. We present a collection of transdisciplinary perspectives on EI, contributed by world-class scholars and emerging researchers, intending to enrich this nascent transdisciplinary field with fresh insights and analyses.

Transcending purely biological or computational views, EI offers a powerful interdisciplinary approach to understanding intelligence as an emergent property. This perspective emphasizes the continuous and dynamic interplay between an agent's physical or virtual body—living, artificial, hybrid, and living-robotic—and its cognitive apparatus, enabling continuous adjustments based on environmental stimuli.

Integrating knowledge from cognitive psychology, neuroscience, biology, robotics, bioengineering, computer science, and related fields, EI posits that an entity's goals, capabilities, and environmental engagement collectively define its intelligence, significantly shaped by the intricate interactions between its morphology and environment (Pfeifer & Bongard, 2006).

Furthermore, EI is characterized by inherent adaptive processes enabling dynamic responses to environmental changes (Pfeifer & Scheier, 1999). Recognizing that intelligence is not confined to brains or algorithms, this volume seeks to foster a vibrant ecosystem of cross-disciplinary research sharing and dialogue.

To further explore the landscape of EI, we begin by examining its manifestations in human cognition (HC) and artifical intelligence (AI)/robotics. This initial exploration highlights conceptual parallels and traces similarities in how these distinct systems learn through environmental interaction. Subsequent chapters examine how EI is being taken up and characterized across various disciplines.

A central goal of this volume is to forge epistemic alignment regarding EI across diverse fields. This means establishing a shared understanding for discussing and researching EI and moving beyond disparate terminologies and methodologies to create a common ground for future research. To this end, we aim to promote shared conceptual frameworks and catalyze synergistic research initiatives by rigorously examining contemporary advancements and proactively identifying future challenges across a spectrum of entities.

Building on this foundational aim, our explicit objective is to demonstrate, through concrete examples and insightful exploration, the diverse ways in which EI is currently being embraced and uniquely interpreted across disciplines, thereby highlighting its pervasive influence and potential for transformative impact.

We also hope to instigate transdisciplinary dialogues, strategically leveraging the foundational concept of embodiment—a principle already holding significant weight across numerous disciplines—to catalyze a fundamental reconceptualization of EI. While this project does not articulate a comprehensive framework for EI, it represents a crucial initial step in exploring its multifaceted nature across disciplines. Further research and analysis, building on the diverse perspectives presented here, will be necessary to fully develop foundational principles and robust theoretical frameworks for this evolving field.

By exploring current trends and anticipating future challenges across diverse disciplines, our collective endeavors to foster shared concepts and drive collaborative research, explicitly demonstrating how EI is being adopted and interpreted across these varied fields. This emergent paradigm promises a dynamic and significantly broadened understanding of intelligence as it manifests across a spectrum of biological and artificial substrates, holding immense promise to profoundly reshape our understanding of ourselves, our technologies, and the very essence of intelligence itself.

The Philosophical and Phenomenological Roots of Embodied Intelligence

Historically, the dominant view of cognition was rooted in Cartesian dualism, rigidly separating mind from body and treating thought as abstract computation. However, this long-held perspective, which has roots in the philosophical views of thinkers like Plato, was fundamentally transformed by the phenomenological movement. Key figures like Edmund Husserl (1913/1982), Martin Heidegger (1927/1962), and Maurice Merleau-Ponty (1968) provided paradigmatic insights.

Edmund Husserl, the founder of phenomenology, introduced the concept of intentionality, highlighting the essential directedness of consciousness toward an object. He explored the "life-world" (Lebenswelt), the everyday world of lived experience, underscoring the significance of subjective experience and how meaning arises from our interactions within this world. Heidegger's related concept of "being-in-the-world" (*in-der-Welt-sein*) (1927/1962) challenged the Cartesian view by asserting that human existence is fundamentally situated and relational, with us actively engaged in the world, not detached from it.

Merleau-Ponty (1968), deeply influenced by both Husserl and Heidegger, provided a powerful account of embodied perception and the "lived body" (*le corps vécu*), emphasizing the crucial role of bodily movement, sensation, and perception in shaping our experience of the world. The collective work of Husserl, Heidegger, and Merleau-Ponty had a profound impact, laying the groundwork for future developments by establishing the inseparability of the body and cognition and moving the field away from abstract, disembodied views of the mind.

Contemporary Perspectives on Embodiment

Maturana and Varela's (1973/1980) concept of "autopoiesis"—the self-production defining living systems—was central to contemporary perspectives on embodiment that understand the brain is deeply intertwined with the body and the environment. They proposed a nonobjective view of reality, suggesting that we create our world through embodied interaction. Di Paolo (2005, 2009) expanded on autopoiesis, exploring adaptivity and agency, and with De Jaegher, investigated participatory sense-making (De Jaegher & Di Paolo, 2007).

This emphasis on interaction and the organism's role in shaping its environment was further elaborated in Varela, Thompson, and Rosch's (1991) theory of enactivism. "Enaction" refers to thinking that involves processes outside the brain through an active engagement with an agent's environment, arising from how we creatively match actions to situations, not internal representations. This perspective challenged the traditional view of the brain as the sole locus of cognition, suggesting that the mind extended beyond the physical boundaries of the brain, encompassing the body and the environment.

While enactivism and the ecological perspectives (influenced by Gibson [1966, 1977, 1979]) both rejected traditional views, they differed in perception and meaning-making. Enactivists emphasized sensorimotor interactions, while ecological theorists advocated for direct perception of environmental affordances. Within enactivism, Noë (2004) highlighted action in perception, and Hutto and Myin (2017) emphasized narrative. Baggs and Chemero (2018, 2021) reconciled these by suggesting that ecological psychology focus on the objective environment (habitat) and enactivism on the organism's sense-making activity.

Phenomenological and philosophical insights from Gallagher (2005), Zahavi (2005), and Thompson (2007) underscored the mind-body-intersubjectivity relationship. Whereas Gallagher linked bodily experience to self and Zahavi emphasized relational embodiment, Thompson connected embodiment to biological and ecological contexts. Phenomenologist Sheets-Johnstone (1982) highlighted the role of movement in cognition, and philosophers like Clark and Chalmers (Chalmers, 1996; Clark & Chalmers, 1998) argued for extended cognition and raised questions about consciousness and embodied processes.

This concept of embodiment is also crucial for understanding how intelligence evolved (Scheier & Pfeifer, 2001). Evolution's constant interplay between learning and adaptation within complex environments led to the vast diversity of body forms seen across living organisms. This process has continuously refined bodies from the simple structures of ancient bilaterian worms to the intricate morphologies of humans. Thus, organisms with well-adapted bodies exhibit remarkable embodied intelligence. In this view, intelligence emerges from the deep connection between physical structures and behaviors, highlighting how the body laid the foundation for the evolution of sophisticated cognitive abilities across varied entities.

Building on this evolutionary understanding, a broader perspective suggests that EI goes beyond cerebral structures, intertwined with the intrinsic physicality of organisms and agents. This holistic view posits that intelligence may be more fundamental than traditionally perceived, extending even to organisms lacking complex nervous systems. Essentially, EI is not solely a function of the brain but is intricately woven into how living entities and certain artificial systems engage with and understand the world. Levin's (2022b) Technological Approach to Mind Everywhere (TAME) framework further supports this view, emphasizing the evolutionary and adaptive nature of embodied cognitive processes across diverse entities. Accordingly, intelligence can be described as all activities undertaken by a "self" that involves gathering, processing, and acting on information for adaptive action (Levin, 2022). This encompasses basic goal-directed activity, learning, predicting, active inference, and more complex skills, such as symbolic reasoning and metacognition.

Having explored the philosophical and phenomenological foundations of EI and its contemporary theoretical developments, the following section examines the historical context and paradigm shifts that further shaped our understanding of mind and computation.

Shifting Landscapes: Origin Stories

The mid-twentieth century witnessed a period of profound intellectual upheaval, marked by transformative paradigm shifts in our understanding of both the mind and computation. Driven in part by wartime exigencies, the computational advancements of this era, exemplified by the codebreaking techniques at Bletchley Park, which culminated in the 1944 development of Colossus (Sale, 2000), provided a crucial backdrop for a parallel intellectual revolution in psychology—the "cognitive revolution." This "revolution" marked a decisive departure from the prevailing behaviorist focus on observable behaviors and external stimuli, a perspective that had largely sidelined the study of internal mental processes (e.g., Gardner, 1985). This intellectual movement brought core cognitive functions—including memory, attention, perception, language, and problem-solving—back to the forefront of psychological inquiry (Neisser, 1967).

Following WWII, several pivotal paradigm shifts further transformed our understanding of intelligence. First, in 1950, Alan Turing introduced the Turing test, a foundational concept in the burgeoning field of computational sciences (Turing, 1950). Shortly after, the Dartmouth Summer Research Project on Artificial Intelligence officially coined the term "Artificial Intelligence," establishing a new field dedicated to creating intelligent machines (McCarthy et al., 1955/ 2006). That same year, Newell and Simon (1956) developed the logic machine, another milestone in computational development.

Crucially, the emergence of the computational model in computer science significantly influenced this shift (Miller, 2003). The development of computers and the conceptualization of information processing provided a powerful new framework for understanding the complexities of the human mind. This fostered the widespread adoption of the "information processing" metaphor, conceptualizing the mind as an active system that receives, processes, stores, and retrieves information, drawing a direct analogy to the operations of a computer (e.g., Broadbent, 1958). Furthermore, George Miller's (1956) work explored the limitations of information processing capacity, and Ulric Neisser's (1967) seminal work, *Cognitive Psychology*, provided a foundational text for the field. These developments, alongside the formulation of numerous specific models, aimed to elucidate the intricacies of human cognition. This shift was further formalized by the computational theory of the mind, which posited that neural computations constitute the fundamental basis of cognitive processes, thus revolutionizing our understanding of learning and thought (Fodor, 1975, 1981; Putnam, 1975). This framework profoundly shaped the conceptualization and investigation of mental processes by cognitive scientists aligning with the broader view of the brain as a computational system akin to software running on neural "hardware" (Fodor, 1975, 1981; Putnam, 1975). In addition, Fodor (1975, 1981) proposed a modular organization of cognitive functions, in which sensory input is transformed into amodal representations (abstract symbols) and manipulated by rules to produce cognitive outputs. This symbol-based, rule-driven approach significantly influenced models of information processing, resulting in a reliance on representational structures such as feature lists, schemas, and semantic networks (Fodor, 1975; Haugeland, 1985; Marr, 1982; Newell & Simon, 1972; Pylyshyn, 1984).

A particularly influential contribution to this computational perspective came from David Marr (1982), whose three levels of analysis provided a crucial framework for understanding information processing systems. Marr proposed that any such system, including the brain, could be analyzed at the computational level: defining the problem being solved (e.g., visual object recognition); the algorithmic level: specifying the representations and processes used to solve that problem (e.g., edge detection algorithms); and the implementational level: describing how those algorithms are physically realized (e.g., in neural circuits).

As experimental psychologists redefined their field through the lens of the cognitive revolution, the computational sciences were simultaneously undergoing parallel advancements. Consequently, the historical relationship between HC and AI was deeply intertwined (chapter 1). Early AI research was significantly driven by the quest to unravel the mechanisms of human brain function in areas like learning, memory, and decision-making, with the explicit goal of developing intelligent machines. However, with the rapid evolution of both HC and AI in recent years, their respective paths have, to some extent, diverged, leading to differing perspectives on the mind and intelligence. Nevertheless, both HC and AI retain the potential to complement each other, especially when viewed through the unifying lens of embodiment.

The Embodied Turn in Psychology, Evolutionary Biology, and Computer Science

Embodied cognition (EC) represented a significant shift from traditional computational frameworks. EC challenged the compartmentalization of information and distinct cognitive modalities, emphasizing the inextricable link between our physical bodies (biological or engineered), the environment, and our cognitive processes. This perspective posits that algorithms of thought are not confined solely to the brain but rather arise from the body's dynamic interactions with its environment, exploring how the body's form, sensory input, and environmental interactions influence various cognitive functions.

Psychology

Barsalou's perceptual symbol systems theory (1999) offered an alternative perspective to cognitivism, emphasizing grounded cognition, which suggested that mental representations were often rooted in our sensory and motor experiences. Here, cognitive representations were made possible through perceptual and sensory systems by simulation of these initial systems (Barsalou, 1999, 2008). This contrasting viewpoint challenged the primacy of amodal symbols in understanding how our minds represent and process information. Lakoff and Johnson's (1981) work on metaphors, which emphasized the role of language in reactivating sensorimotor information, further challenged traditional views of cognition. Finally, Eleanor Rosch's (2002) work on categorization highlighted the importance of the "basic level" in how we classify and understand the world, influencing language processing and communication.

These embodied perspectives were further supported by work in dynamical systems. The dynamical hypothesis (van Gelder, 1998) proposed that intelligence should be understood more as a process rather than primarily viewed as a computational phenomenon. In other words, it suggested that the dynamics of interactions and relationships within systems played a crucial role in shaping intelligent behavior, emphasizing a more holistic and dynamic approach to understanding cognition and intelligence. Damasio's work (1994, 1999) argued that emotions were integral to rational thought and decision-making and that consciousness arose from the neural state in the present moment. He highlighted emotions as guides, driven by bodily states and past experiences, and emphasized interoception's role in self-awareness and consciousness, creating a "proto-self" (Damasio, 2012). Importantly, this aligns with the concept of niche construction, which highlights the bi-directional feedback between the agent and the environment (Pio-Lopez et al., 2025).

British philosopher Andy Clark (1997, 2008) explained how external artifacts significantly guide our actions, such that environmental changes induce behavioral reactions in the cognitive agent. Clark (1997, 2008) argued for an extended mind theory, such that mental processes were not phenomena restricted to the brain but encompassed their connection to the body and the environment. The "extended" mind thesis proposed that cognitive processes were "off-loaded" to external devices and the environment. Accordingly, he wrote that human cognition involved the "inextricable tangles of feedback, feed-forward, and feed-around loops: loops that promiscuously crisscross the boundaries of brain, body, and world" (Clark, 2008, p. xxviii). These ideas were exemplified by "information pickup" and aligned with Gibson's affordances (see Chemero, 2009) and situatedness (Clancey, 1997; Suchman, 1987).

Specifically, this work paved the way for a new synthesis, where predictive brains served as entry points, with embodiment and action inherently extending into the environment to make meaningful connections with the world (Clark, 2013). Clark and Friston's work connected early human adaptive behavior studies to contemporary theories of predictive processing, active inference, and EC. The predictive processing framework (Clark, 2013; Friston, 2010) emphasizes how our brains minimize "surprises" through continuous updates, following the laws of the free energy principle (FEP) (Friston et al., 2006, 2017). Both became pivotal in shifting from classical cognitivist views to an ecologically grounded understanding of intelligence as emerging from real-time environmental interaction.

In contrast to the extended mind theory, another influential perspective, radical embodied cognition (Chemero, 2009), emphasized "embeddedness" and focused on causal dependency, in which the agent and the environment exerted mutual influence on each other. This perspective emphasized that cognition was more than being "extended" or constitutive of each other, but it was, rather, causal (see also Chalmers, 2008; Seifert et al., 2020). Kiverstein and Miller (2015) suggested that the agent interacts with the environment through dynamical systems and ecological psychology (see Chemero, 2009; Hutto & Myrin, 2013). These ideas harkened back to Vygotsky's original argument (Moll, 1997) in that representations exist publicly externally first and are then "internalized". Thus, the evolution and maintenance of these public sociocultural practices explain both the initial emergence and the continued maintenance of content-involving minds (Hutto & Satne, 2017).

Evolutionary Biology

The concept of embodiment typically refers to the human mind and body; though, it can be extended to other biological systems like animals, plants, and even single-celled organisms in certain contexts. Each species possesses a unique constellation of embodied cognitive capacities, finely adapted to its specific environment, or *umwelt* (Uexküll, 1936/2001), enabling diverse animals within the same ecosystem to detect and respond to different environmental cues. Consequently, variations in sensorimotor abilities across species directly shape their cognitive capabilities (Bräuer et al., 2020).

Animals across the spectrum provide compelling examples of EI—the idea that intelligence emerges from the interaction of an animal's body and its environment—in which they utilize their diverse senses and physical bodies to navigate complex environments to efficiently locate food sources and effectively avoid potential predators. This is evident in examples such as the echolocation of bats that allows them to "see" in darkness, the electroreception of sharks to detect hidden prey, the magnetoreception of migratory birds for navigation, the polarized light detection in bees, and the infrared vision of snakes. Challenging traditional, anthropocentric classifications of senses, Ed Yong (2022) argues that these abilities show that animals can perceive the world in ways that extend far beyond the limitations of human sensory experience, sometimes even with forms of perception that do not rely on conventional sensory organs, blurring the traditional distinctions between different senses and creating unique and often unimaginable sensory experiences.

The careful study of behavior and cognitive abilities in nonhuman animals has revealed a shared fundamental capacity for intelligent action across diverse species. This understanding demonstrates that intelligence is a far more widespread biological phenomenon than previously appreciated. To further illustrate this breadth, recent studies reveal that plants may exhibit rudimentary yet significant forms of intelligent behavior, demonstrating measurable responses to stimuli such as physical touch and sound vibrations (Parise et al., 2020). For example, plants actively utilize their root systems and leaves to sense and dynamically respond to various aspects of their surrounding environment. Specifically, cells within tree roots possess the ability to navigate their growth patterns to actively seek out essential resources like water and mineral nutrients, and they can even direct root growth to "escape" from stressful or harmful environments, such as areas with toxic soil or other hazardous conditions (Burbach et al., 2012).

Expanding on this perspective, plants actively participate in shaping their immediate surroundings, even potentially offloading certain elements of cognitive processing into the complex network of the soil and associated fungal networks. This suggests a sophisticated form of intelligence that significantly challenges traditional and limited notions of what constitutes "thinking" (Segundo-Ortin & Calvo, 2023). Similarly, at the most fundamental level of biological organization, single-celled organisms utilize their inherent chemical and physical properties to actively interact with and make crucial decisions within their immediate microscopic environment (Polychronidou et al., 2023).

This understanding of embodiment has been further enriched by the work of biologists such as Michael Levin (2023), who highlights the active role of living systems in shaping not only their own physical structures but also the environments they inhabit through bioelectric processes, strongly reinforcing the crucial role of form and function in biology. Intriguingly, even before the evolution of brains in animals for goal-directed behaviors, individual cells demonstrate sophisticated and seemingly "intelligent" mechanisms to coordinate their activities for survival via essential processes such as regeneration, embryogenesis, and metamorphosis (Baluška & Levin, 2016).

To further contextualize these perspectives, it is important to note that evolutionary biology also provides a crucial context for understanding regenerative medicine and morphogenesis by situating these embodied processes within the framework of biological adaptation and historical constraint. Morphogenesis, the development of an organism's shape, is not an unconstrained process but rather is influenced by the evolutionary history of that organism. Inherited structural plans, such as the basic limb structure in vertebrates, limit and guide the possibilities of developmental change and regenerative capacity (Murugan et al., 2022). Furthermore, the capacity for regeneration itself is an evolved trait that varies significantly across species.

Computer Science

Traditional Good Old-Fashioned Artificial Intelligence (GOFAI) predominantly conceived of intelligence as a form of disembodied symbolic manipulation and abstract reasoning, akin to computation occurring independently of a physical substrate or real-world interaction (Ciocarlie, 2022). In stark contrast, the embodied perspective posits that intelligence is inextricably linked to an agent's physical embodiment and its continuous, dynamic engagement with the environment (Aguilar et al., 2014). This perspective emphasizes the indispensable role of sensorimotor experiences—encompassing an agent's movement, perception, actions, and situatedness within a specific context—in the very emergence and manifestation of intelligent behavior, drawing profound inspiration from the principles observed in the intricate workings of natural biological systems.

The early conceptual groundwork that paved the way for embodied approaches to machine intelligence can be attributed to the insightful work of Valentino Braitenberg (1986). Through his elegant thought experiments involving simple, hypothetical "vehicles," Braitenberg demonstrated how even rudimentary connections between an agent's sensory inputs and its motor outputs could generate behaviors that appeared surprisingly complex and even goal-directed. These abstract vehicles, devoid of internal symbolic representations or sophisticated reasoning mechanisms, served as a foundational conceptual stepping-stone for the field of behavior-based robotics. They compellingly illustrated the inherent potential for a form of "intelligence" to arise directly from the interaction between a physical form and its surrounding world, effectively bypassing the need for explicit, internal models or intricate computational processes.

Building on these foundational conceptualizations, the late 1980s and the 1990s witnessed tangible and impactful advancements in the realm of embodied robotics, largely spearheaded by the pioneering research of Raibert's (1986) work on dynamically stable-legged robots. His robots provided compelling empirical evidence for the critical role of physical embodiment in achieving complex motor skills and demonstrated how sophisticated behaviors, such as running and maintaining balance, emerged from the intricate and real-time dynamic interplay of a body and the physical constraints and affordances of its environment. These advancements underscored the importance of direct physical interaction for achieving robustness and adaptability in real-world scenarios.

Further solidifying the embodied turn and profoundly influencing the trajectory of AI research was Rodney Brooks's seminal introduction of the subsumption architecture (Brooks, 1991). His influential Genghis robot, designed and controlled using this novel architecture, served as a powerful demonstration of the efficacy of situated and embodied action. Brooks's work directly challenged GOFAI, which heavily emphasized symbolic representations and centralized planning as the cornerstones of intelligent systems. His highly influential 1991 paper, "Intelligence without Representation," articulated a compelling vision of EI, which argued that complex, adaptive behaviors arose from a decentralized collection of simple, parallel behavioral modules that directly sensed and acted on the environment. Brooks's work sparked significant research interest in the development of various embodied approaches, including behavior-based robotics, evolutionary robotics (which employs evolutionary algorithms to design both the physical morphology and the control systems of robots), and developmental robotics (which seeks to understand how intelligence emerges and develops through an agent's interaction with its environment over time). Embodied paradigm shifts have marked the evolution of AI.

The growing emphasis on embodied interaction also brought the philosophical concept of multiple realizability (Rouleau & Levin, 2023) into sharper focus within the AI community. Drawing parallels with the vast diversity of biological intelligences observed in nature, the realization that similar cognitive functions or behavioral outcomes can be achieved by vastly different physical substrates and organizational principles presented a significant challenge to purely computational or brain-centric perspectives on intelligence. This understanding highlighted the complex and context-dependent interplay between an agent's physical form (its morphology), the specific challenges and opportunities presented by its environmental context, and the particular strategies and processes it employs to achieve its goals.

Recognizing this dynamic relationship has become crucial for both understanding existing forms of intelligence and for designing novel intelligent systems capable of operating in unprecedented and diverse environments, pushing the boundaries beyond the limitations of traditional computational models. The convergence of these influential ideas, alongside the burgeoning field of artificial life (Langton, 1989), explored the fundamental principles of living systems through the creation of artificial ones. Neuroscience also revealed diverse implementations of EI, including the mechanisms of predictive processing and error minimization in both biological and artificial neural networks (ANNs), providing insights into how embodied agents generate intelligent behavior. Finally, evolutionary robotics (Nolfi & Floreano, 2000), applied evolutionary computation to the design of autonomous robots, collectively laying the essential groundwork for the formal emergence and widespread acceptance of embodied artificial intelligence (EAI) as a distinct and highly influential paradigm within the broader field.

The Rise of Embodied Artificial Intelligence

EAI represented a significant paradigm shift in AI, moving beyond purely computational models to emphasize the crucial role of a physical body and its dynamic interaction with the environment in the emergence of intelligence. EAI recognizes the unique opportunities for learning and adaptation that arise from physical embodiment and views the body as both a constraint and an enabler. EAI agents can perceive and act on their surroundings purposefully and adaptively with diverse sensors and actuators, enabling effective navigation, object manipulation, and task completion. EAI aims to distribute intelligence beyond traditional controllers, underscoring the importance of intelligent materials and structures within. The goal is to enable self-organizing and self-learning processes within the physical form itself (Pfeifer & Iida, 2006), marking a fundamental shift in how we conceptualize and engineer intelligent machines. This fundamentally contrasts with disembodied computation, highlighting the inherent connection between intelligence and an agent's physical presence and interaction within a specific environment. At a more immediate level, EAI considers factors such as the material properties of exoskeletal systems, recognizing that perceptions and movements are shaped not only by physical or virtual bodies but also by the environment. This shift implies that the agents' morphologies and materials, rather than just their processors, can be partly assumed through self-organizing and self-learning processes, distributing intelligence beyond centralized control (Pfeifer & Iida, 2006).

EAI's foundation was laid by Brooks (1991), who argued for intelligence arising from direct physical interaction without complex internal representations. Additionally, the integration of predictive processing theories and active inference (Clark, 2013; Friston et al., 2006) highlighted the crucial role of error-minimizing predictions and the adaptation of morphology to the environment for achieving robust and efficient intelligent behavior, which offered powerful frameworks for understanding how agents actively engaged with and made sense of their environments. Such theories also provided valuable insights into the neural basis of understanding and enacting agency, informing the development of more sophisticated EAI systems. These early works established three principles foundational to EAI: avoiding predefined logic, incorporating evolutionary learning, and recognizing the environment's role in shaping behavior and cognition (Liu & Wu, 2024).

Recent advancements have integrated powerful foundation models, such as large language models (LLMs) and visual language models (VLMs) (OpenAI & Figure AI, 2024; Hu et al., 2023), building on the significant AI advancements of the last decade driven by deep learning, powerful hardware, and large data sets. This integration is particularly important for developing robust embodied reasoning in robots—a set of capabilities including object recognition, spatial reasoning, and trajectory prediction—enabling them to ground their understanding in the physical constraints and affordances of their surroundings, a key tenet of EAI (Gemini Robotics, 2025). This suggests a potential path toward artificial general intelligence (AGI) through the continued scaling and refinement of such models. These principles are exemplified by successes in diverse fields like protein structure prediction, robotic control, and content generation (Di Paolo et al., 2024). The Deep Evolutionary Reinforcement Learning (DERL) framework (Gupta et al., 2021) serves as a concrete example of incorporating evolutionary learning to enable robots to adapt both their morphology and behavior in response to environmental challenges, demonstrating key principles of EI. Gupta and colleagues' (2021) "Unimals"—virtual stick figures featuring a head and multiple limbs—illustrate this, as they can predict, problem-solve, and self-learn by adapting their morphology through evolutionary mutations. Initially trained for tasks like locomotion or object manipulation across diverse terrains, the most successful Unimals are chosen for reproduction, with mutations introduced to their offspring. These new generations are then retrained from the beginning within the environment. This iterative evolutionary learning process, repeated extensively, has resulted in successful adaptations, including limb addition or removal and length and flexibility modifications. These advancements have enhanced the perceptual abilities of robots, allowing them to move beyond digital tasks and achieve effective interaction within the physical world (Haddadin, 2025).

Current research is exploring training AI through simulated physics videos (Piloto et al., 2022) and comparing the learning trajectories of neural networks with those of children (Buschoff et al., 2023), all with the underlying goal of grounding intelligence in physical experience. Despite this progress, problems and hurdles remain. For example, how can we endow robots with a fundamental understanding of the physical world, including concepts such as gravity? Another challenge is developing efficient virtual environments like Habitat (Savva et al., 2019) to provide scalable platforms to generate crucial interaction data (Liu et al., 2012). While progress is being made, achieving a fully integrated system possessing a deep, intuitive understanding of physical laws remains a key challenge (Torresen, 2018). Ultimately, EAI—with its emphasis on physical intelligence, self-regulation, and the emergence of complex behaviors from designed structures (Hughes et al., 2022; Pfeifer & Iida, 2006)—promises a transformative approach to creating machines that can interact with and learn from the world more naturally and independently, fully embracing the tenets of the embodiment hypothesis and the guiding principles of the field.

Bio-Inspired Soft Robotics: A Synergy for Embodied AI

A foundational principle driving advancements in robotics and AI is biologically inspired robotics, which champions adopting biological strategies to imbue artificial systems with enhanced adaptability and efficiency. Methodologies such as ANNs modeled on biological nervous systems and evolutionary algorithms inspired by natural selection perfectly illustrate this paradigm. These approaches empower machines to interact more naturally, learn autonomously, and acquire new skills, highlighting the crucial connection between intelligence, physical embodiment, and environmental engagement.

The synergy between EAI and soft robotics, a prominent branch of bio-inspired robotics, exemplifies this trend. Soft robotics, characterized by flexible structures enabling safe interaction and delicate manipulation ideal for EAI (Floreano & Keller, 2010), utilize adaptable, often biomimetic materials mimicking biological systems like muscles and tissues. Soft robotics exemplifies embodied bio-inspiration (figure 0.1) by translating core biological principles into novel robotic designs. This design allows for natural environmental interaction, enabling safer and more nuanced operations through bending, squeezing, and conforming—a key advantage for achieving EAI's goal of autonomous learning and operation.

The perspective of evolutionary biology further underscores the significance of embodiment in the emergence of intelligence. The fundamental principles governing this relationship are likely conserved across species, regardless of brain complexity (Levin, 2022). The imperative for survival in diverse environments generates selective pressures that sculpt physical forms and refine sensorimotor systems through adaptation and morphological change. Successful physical actions and subsequent sensory

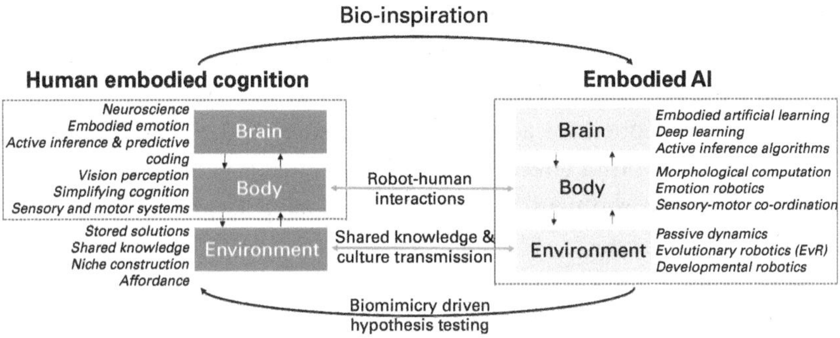

Figure 0.1
Bio-inspiration. *Source*: the author.

feedback are hypothesized as key drivers in the evolution of sophisticated nervous systems, facilitating nuanced movement and enhanced resource exploitation, even in simple organisms lacking centralized brains.

Contemporary research (Floreano & Mattiussi, 2023) highlights a shift in understanding intelligence as arising not just from the brain, but from the interplay of cells, bodies, social interactions, evolution, development, and learning. While traditional AI often aims to replicate human cognition, current innovative inquiries explore a broader spectrum of biological systems exhibiting autonomous self-organization as inspiration. This includes evolutionary computation, ANNs, artificial immune systems, bio-robotics, and swarm intelligence. This research underscores the potential of these bio-inspired computational models to advance the field by leveraging biological self-organization for more robust, adaptable, and intelligent artificial systems (Floreano & Mattiussi, 2023).

By studying nature's intelligent behaviors, scientists gain valuable insights for designing and programming embodied machines. Data from soft robots is analyzed by AI, facilitating learning and adaptation through trial and error, leading to the construction of internal spatial and temporal world representations for enhanced task proficiency (Firoozi et al., 2024). Josie Hughes's team at the École Polytechnique Fédérale de Lausanne epitomized this approach with their development of a soft robot arm, inspired by elephant trunks and octopus tentacles (Mitchell, 2023). By ingeniously incorporating a spring-like spiral structure, termed a "trimmed helicoid" it demonstrates unprecedented dexterity and adaptability for flexible and robust environmental interaction, it mirrors natural limbs (Guan et al.,

2023). This shift toward embodied systems has the potential to revolutionize human-robot interaction (HRI), fostering collaboration and understanding.

Bio-inspiration plays a pivotal role in the ongoing shift toward EAI. Biological systems honed for adaptability, resilience, and resource efficiency offer a vast reservoir of effective intelligent behaviors across diverse physical forms and environments and provide invaluable lessons for truly intelligent embodied systems.

Biomimicry

Biomimicry, the practice of emulating nature's designs, unlocks the potential to enhance AI capabilities by drawing inspiration from the remarkable adaptations found in living organisms. By replicating human senses like touch, temperature, and pressure, these systems can interact more effectively with the physical world. By creating intelligent systems without brains, such as bioengineered microbes that adapt to their environment (Bongard & Levin, 2021; Rouleau & Levin, 2023), biomimicry demonstrates that intelligence can manifest in diverse forms beyond the human model. This field highlights the crucial interplay between an organism's physical form, its environment, and its cognitive abilities, allowing researchers to explore how different body designs influence behavior and problem-solving (McMillen & Levin, 2024; Muhsen et al., 2024).

Biomimicry transcends the imitation of human capabilities, unlocking profound potential and drawing inspiration from the extraordinary sensory systems prevalent in nature. To illustrate, researchers are modeling swarm behaviors in robots that emphasize the importance of interdependence and cooperation—showing a deliberate application of EC that mirrors nature's inherent mind-body unity (Ishige et al., 2019). By harnessing insights from the exceptional visual perception of dragonflies, the electroreception of eels, resources used in ant colonies, the resilient fault-tolerance of beehives, potent regenerative abilities in flatworms, and the echolocation of whales, researchers are paving the way for new applications (e.g., Davies et al., 2017).

Much like ANNs replicate brain structures, evolutionary algorithms mirror natural selection to potentially guide the development of agile robots (Doncieux et al., 2015). This approach can guide us in designing technologies that work in harmony with—rather than in opposition to—the environment, with applications spanning health care, robotics, and virtual reality. Through the lens of these diverse biological examples, we see how the principles of animal sensory systems can also inform the creation of advanced artificial sensors. By studying these natural patterns and abilities, we are likely to find the key to developing smarter cities and intelligent systems capable of harnessing collective intelligence for effective environmental monitoring (Muhsen et al., 2024).

Bioengineering: Pioneering Embodied Intelligence

Bioengineering is at the forefront of challenging anthropocentric views by demonstrating intelligence inherent in diverse, nonhuman biological systems (Levin, 2023). The very existence of bioengineered systems capable of goal-directed behavior, independent of human-like consciousness, compellingly questions the long-held belief that intentionality is solely a product of the human brain. This paradigm shift is amplified by bioengineering's vital role in regenerative medicine, where biochemical, molecular, and cellular insights are expertly integrated with engineering principles to create innovative therapies for tissue repair, replacement, rejuvenation, and regeneration.

Evolutionary principles, particularly the concept of evolutionary optimization of form and function, are also increasingly relevant to bioengineering. By studying how natural selection has shaped tissues and organs for efficiency and resilience, researchers can gain valuable insights into the design of artificial biological systems. Wang et al. (2025) demonstrate how this convergence of biology and engineering is further exemplified by the development of neuromorphic interfaces, such as the peripheral sensory-computer interface (PSCI). PSCIs employ an ultrasensitive, low-power, and atomically thin two-dimensional transistor array designed to mimic the exquisite sensitivity and real-time processing capabilities of the biological nervous system (Wang et al., 2025).

The remarkable emergence of hybrid agents, including Xenobots (self-organizing entities constructed from frog cells [Kriegman et al., 2020]) and Anthrobots (human tracheal cell–based constructs with promising biomedical applications [Gumuskaya et al., 2024]) provides compelling evidence for the versatility of biological building blocks in realizing novel forms of EI. These groundbreaking developments, alongside the progress in brain-computer interfaces and bio-inspired robotics, underscore the accelerating and significant convergence of living and nonliving systems around EC's fundamental principles. Proactively investigating the diverse strategies by which varied physical systems, whether biological or artificial, achieve cognitive outcomes through their specific embodied forms is no longer just an academic pursuit but rather is a crucial imperative for the future development of truly intelligent AI and a deeper understanding of the foundations of HC. This embodied perspective offers a transformative framework for all future inquiries into the nature and diversity of intelligence.

Embodied Intelligence: Toward a Unified View

The past century has witnessed a significant evolution in our understanding of intelligence across diverse life forms, spanning fields including psychology, biology, and computer science. Pioneers like Merleau-Ponty, Barsalou, Brooks, Pfeifer and Bongard, Clark, Friston, and Levin, all challenged traditional notions of disembodied symbolic reasoning, highlighting the fundamental connection between intelligence and our physical presence and interaction with the world. This shift is leading to the emergence of the expanded concept of EI, a rapidly evolving framework that explores the interconnectedness of intelligence, physical form (or its virtual equivalent), and the environment. EI acknowledges that intelligence can manifest in biological, artificial, and hybrid forms, with an agent's body playing a crucial role in its environmental interaction and learning.

This broader perspective aligns with the principle of teleonomy—goal-directed behavior—which blurs the traditional lines between engineered and naturally occurring systems (Clawson & Levin, 2023). This convergence opens the possibility of rich inner lives existing beyond humans, extending to animals, plants, microorganisms, and potentially synthetic beings. Unlocking the secrets of basal cognition, the intelligence found across life forms even without traditional brains, promises a deeper comprehension of life and intelligence itself.

Bongard's and Levin's (2021) Xenobots are a compelling example of this developing perspective on embodiment. These programmable organisms challenge traditional categories of "machine" and "biological organism" by suggesting that cognition is not confined to complex neural networks but arises from physical embodiment and dynamic environmental interaction. Ultimately, this paradigm shift toward EI promises to fundamentally reshape our understanding of intelligence across diverse systems, from biological organisms to artificial agents.

Remaining Challenges

A significant challenge confronting the field of EI lies in critically evaluating the necessity of traditional physical bodies for artificial agents to exhibit embodiment. While the conventional understanding of EC strongly posits a crucial role for a physical body, analogous to biological organisms in facilitating meaningful interaction with the environment, recent advancements in EAI raise fundamental questions about this prerequisite.

For instance, the remarkable reasoning and planning abilities demonstrated by LLMs, entities lacking a conventional physical form, prompt inquiry into whether EI can manifest through alternative means. The capacity of AI agents to learn, adapt, and make decisions based on their engagement with diverse environments—be they physical, simulated, or even abstract linguistic spaces—underscores the need to investigate whether the nature of this interaction, rather than the specific materiality of a body, constitutes a sufficient basis for a functional embodiment. The increasing sophistication of simulations and robotic interactions further exemplifies how AI agents can effectively sense and act on their surroundings, leveraging reinforcement learning techniques to develop complex behavioral repertoires beyond traditional physical instantiations.

This challenge to traditional views of embodiment is directly addressed by Michael Levin's TAME—a framework for understanding and manipulating cognition in unconventional substrates (2022b). TAME fundamentally questions the conventional life/machine dichotomy and the exclusive link between cognition and biological embodiment. It proposes a continuum of agency and intelligence, emphasizing the primacy of functional organization and goal-directedness as defining characteristics of intelligent systems, irrespective of their material instantiation (Levin, 2022; chapter 17). Here, Levin introduces the concept of the "cognitive light cone" to illustrate the scope of an agent's attainable goals, suggesting that while current AI might lack intrinsically emergent goals comparable to biological systems, their capacity to operate effectively within a defined problem space and exhibit functional predictability remains a critical aspect of their agency (Levin, 2022).

Furthermore, Levin's research on morphogenesis and regeneration in biological systems that lack centralized nervous systems provides significant empirical evidence of how sophisticated problem-solving and goal-directed behavior can emerge in unconventional biological "embodiments" (Levin, 2019). This view offers a tangible biological precedent for the potential realization of EI in nontraditional artificial systems. His vision of future technology potentially birthing novel forms of intelligence would blend evolved and designed components across multiple levels of organization within their own unique *umwelt* (Levin, 2022).

Embracing this broader perspective necessitates a more dynamic and flexible understanding of embodiment itself. This perspective challenges static, biocentric views of embodiment, emphasizing the constant adaptive adjustments and the crucial role of the embedded "body" in shaping both cognitive processes and behavioral outputs as agents actively interact with and learn from their environment. While the direct application of principles derived from biological embodiment to nonbiological entities presents significant conceptual and practical hurdles, ongoing research endeavors are crucial for exploring the diverse ways in which different forms of embodiment can enhance capabilities and deepen the fundamental understanding of EI. Addressing these remaining challenges through a critical deconstruction of traditional assumptions offers a glimpse into a future populated by a diverse array of "intelligences" embodied in potentially unconventional hardware and software substrates, thereby significantly enriching our broader comprehension of the very nature of cognition.

Conclusion

This collection represents a crucial departure point in the study of EI, moving beyond preliminary observations to demand a rigorous and nuanced analysis of how embodiment fundamentally shapes both the nature and mechanisms of intelligence and learning across the diverse spectrum of biological and artificial systems. A central and vital goal has been to forge epistemic alignment across disciplines, establishing a shared vocabulary, conceptual framework, and methodological foundation from the varied lenses, perspectives, and insights of multiple fields. By rigorously examining the constraints, affordances, and emergent properties inherent in the rich diversity of physical and computational architectures of embodied agents, we lay the groundwork for transformative comparative studies. These studies are essential to progress beyond simplistic analogies and establish a genuinely integrated and comprehensive perspective on EI.

The inherent disciplinary breadth of these explorations compels a profound reconsideration of our foundational preconceptions: What truly constitutes EI? Where, precisely, do the boundaries of "self" reside within the embodied and environmentally coupled system? The implications of this evolving understanding are far-reaching, poised to revolutionize not only technological fields such as robotics and AI, but also fundamental scientific disciplines including philosophy, psychology, biology, and beyond. By studying EI across a spectrum of entities, we are driven to radically rethink the intricate interconnectedness of mind, body, and environment, dismantling traditional scientific silos.

Ultimately, the ongoing exploration of an expanded and transdisciplinary view of EI charts a course toward the future. This view transcends the limitations of traditional physical forms by integrating diverse disciplinary perspectives, envisioning a future where an ever-expanding landscape of intelligence (biological, artificial, hybrid, and yet unconceived) dynamically interacts with and shapes our world in ways we are only beginning to glimpse. Achieving this ambitious vision requires a sustained commitment to building a robust shared foundational knowledge base across disciplines and substrates.

This collection contributes a vital first step, establishing the groundwork for future research that will not only compare but also synthesize the unique insights and methodologies of each field into a unified and comprehensive understanding of how embodiment fundamentally informs the diverse expressions of intelligence we observe and are yet to discover.

References

Aguilar, W., Santamaría-Bonfil, G., Froese, T., & Gershenson, C. (2014). The past, present, and future of artificial life. *Frontiers in Robotics and AI, 1*(8), 1–15. https://doi.org/10.3389/frobt.2014.00008

Baggs, E., & Chemero, A. (2018). Radical embodiment. In A. Newen, L. De Bruin, & S. Gallagher (Eds.), *The Oxford Handbook of 4E cognition* (pp. 17–27). Oxford University Press. https://doi.org/10.1093/oxfordhb/9780198735410.013.1

Baggs, E., & Chemero, A. (2021). Radical embodiment in two directions. *Synthese, 198*(Suppl 9), 2175–2190. https://doi.org/10.1007/s11229-018-02020-9

Baluška, F., & Levin, M. (2016). On having no head: Cognition throughout biological systems. *Frontiers in Psychology, 7*, 902. https://doi.org/10.3389/FPSYG.2016.00902

Barsalou, L. W. (1999). Perceptual symbol systems. *Behavioral Brain Science, 22*(4), 577–609. https://doi.org/10.1017/s0140525x99002149

Barsalou, L. W. (2008). Grounded cognition. *Annual Review of Psychology, 59*, 617–645. https://doi.org/10.1146/annurev.psych.59.103006.093639

Bongard, J., & Levin, M. (2021). Living things are not (20th century) machines: Updating mechanism metaphors in light of the modern science of machine behavior. *Frontiers in Ecology and Evolution, 4*, 9. https://doi.org/10.3389/fevo.2021.650726

Braitenberg, V. (1986). Vehicles: Experiments in synthetic psychology. *Philosophical Review, 95*(1), 137–139.

Bräuer, J., Hanus, D., Pika, S., Gray, R., & Uomini, N. (2020). Old and new approaches to animal cognition: There is not "one cognition." *Journal of Intelligence, 8*(3), 5–28. https://doi.org/10.3390/jintelligence8030028

Broadbent, D. E. (1958). *Perception and communication*. Pergamon Press. https://doi.org/10.1037/10037-000

Brooks, R. A. (1991). Intelligence without representation. *Artificial Intelligence, 47*(1–3), 139–159. https://doi.org/10.1016/0004-3702(91)90053-M

Burbach, C., Markus, K., Yin, Z., Schlicht, M., & Baluška, F. (2012). Photophobic behaviour of maize roots. *Plant Signaling & Behavior, 7*, 876–880. https://doi.org/10.4161/psb.21012

Buschoff, L. M., Schulz, E., & Binz, M. (2023). *The acquisition of physical knowledge in generative neural networks*. In A. Krause, E. Brunskill, K. Cho, B. Engelhardt, S. Sabato, & J. Scarlett (Eds.), *Proceedings of the 40th International Conference on Machine Learning (ICML),202* (pp. 30321–30341). PMLR. https://icml.cc/Conferences/2023

Chalmers, D. J. (1996). *The conscious mind: In search of a fundamental theory*. Oxford University Press.

Chalmers, D. J. (2008). Foreword. In A. Clark, *Supersizing the mind: Embodiment, action, and cognitive extension* (pp. vii–xii). Oxford University Press.

Chemero, A. P. (2009). *Radical embodied cognitive science*. MIT Press. https://doi.org/10.7551/mitpress/8367.001.0001

Ciocarlie, M. (2022, March 23). What is embodied intelligence? Artificial intelligence @ Columbia. Columbia University. https://ai.columbia.edu/news/what-embodied-intelligence

Clancey, W. J. (1997). *Situated cognition: On human knowledge and computer representations*. Cambridge University Press.

Clark, A. (1997). *Being there: Putting brain, body, and world together again.* MIT Press.

Clark, A. (2008). *Supersizing the mind: Embodiment, action, and cognitive extension.* Oxford University Press.

Clark, A. (2013). Whatever next? Predictive brains, situated agents, and the future of cognitive science. *Behavioral and Brain Sciences, 36*(3), 181–204. https://doi.org/10.1017/S0140525X12000477

Clark, A., & Chalmers, D. (1998). The extended mind. *Analysis, 58*(1), 7–19. http://www.jstor.org/stable/3328150

Clawson, W. P., & Levin, M. (2023). Endless forms most beautiful 2.0: Teleonomy and the bioengineering of chimaeric and synthetic organisms. *Biological Journal of the Linnean Society, 139*(4), 342–356. https://academic.oup.com/biolinnean/article/139/4/457/6643577

Damasio, A. (1994). *Descartes' error: Emotion, reason, and the human brain.* Putnam.

Damasio, A. (1999). *The feeling of what happens: Body and emotion in the making of consciousness.* Harcourt Brace and Co.

Damasio, A. (2012). *Self comes to mind: Constructing the conscious brain.* Pantheon Books.

Davies, E. L., Lei, K., Seidel, C. W., Kroesen, A. E., McKinney, S. A., Guo, L., Robb, S. M., Ross, E. J., Gotting. K., & Alvarado, A. S. (2017). Embryonic origin of adult stem cells required for tissue homeostasis and regeneration. *Elife, 6,* e21052. https://doi.org/10.7554/eLife.21052

De Jaegher, H., & Di Paolo, E. (2007). Participatory sense-making: An enactive approach to social cognition. *Phenomenology and the Cognitive Sciences, 6*(4), 485–507. https://doi.org/10.1007/s11097-007-9076-9

Di Paolo, E. A. (2005). Autopoiesis, adaptivity, teleology, agency. *Phenomenology and the Cognitive Sciences, 4,* 429–452. https://doi.org/10.1007/s11097-005-9002-y

Di Paolo, E. A. (2009). Editorial: The social and enactive mind. *Phenomenology and the Cognitive Sciences, 8*(4), 409–415. https://doi.org/10.1007/s11097-009-9143-5.

Di Paolo, L., White, B., Guénin-Carlut, A., Constant, A., & Clark, A. (2024). Active inference goes to school: The importance of active learning in the age of large language models. *Philosophical Transactions of the Royal Society B,* 379(1911), 20230148. https://doi.org/10.1098/rstb.2023.0148

Doncieux, S., Bredeche, N., Mouret, J.-B., & Eiben, A. E. (Gusz). (2015). Evolutionary robotics: What, why, and where to. *Frontiers in Robotics and AI, 2,* 4. https://doi.org/10.3389/frobt.2015.00004

Firoozi, R., Tucker, J., Tian, S., Majumdar, A., Sun, J., Liu, W., Zhu, Y., Song, S., Kapoor, A., Hausman, K., Ichter, B., Driess, D., Wu, J., Lu, C., & Schwager, M. (2024).

Foundation models in robotics: Applications, challenges, and the future. *The International Journal of Robotics Research, 43*(14), 1612–1662. https://doi.org/10.1177/02783649241281508

Floreano, D., & Keller, L. (2010). Evolution of adaptive behaviour in robots by means of Darwinian selection. *PLoS Biology, 8*(1), e1000292. https://doi.org/10.1371/journal.pbio.1000292

Floreano, D., & Mattiussi, C. (2023). *Bio-inspired artificial intelligence: Theories, methods, and technologies*. MIT Press.

Fodor, J. A. (1975). *The language of thought*. Harvard University Press.

Fodor, J. A. (1981). *Representations: Philosophical essays on cognition*. MIT Press. https://doi.org/10.7551/mitpress/2355.001.0001.

Friston, K. (2010). The free-energy principle: A unified brain theory? *Nature Reviews Neuroscience, 11*, 127–138. https://doi.org/10.1038/nrn2787

Friston, K., FitzGerald, T., Rigoli, F., Schwartenbeck, P., & Pezzulo, G. (2017). Active Inference: A Process Theory. *Neural Computation, 29*(1), 1–49. https://doi.org/10.1162/NECO_a_00912

Friston, K., Kilner, J., & Harrison, L. (2006). A free energy principle for the brain. *Journal of Physiology, Paris, 100*(1–3), 70–87. https://doi.org/10.1016/j.jphysparis.2006.10.001

Gallagher, S. (2005). *How the body shapes the mind*. Oxford University Press.

Gardner, H. (1985). *The mind's new science: A history of the cognitive revolution*. Basic Books.

Gemini Robotics. (2025). Gemini robotics: Bringing AI into the physical world. *arXiv*. https://arxiv.org/abs/2503.20020

Gibson, J. J. (1966). *The senses considered as perceptual systems*. Houghton Mifflin.

Gibson, J. J. (1977). *The theory of affordances*. Erlbaum Associates.

Gibson, J. J. (1979). *The ecological approach to visual perception*. Houghton Mifflin.

Gumuskaya, G., Srivastava, P., Cooper, B. G., Lesser, H., Semegran, B., Garnier, S., & Levin, M. (2024). Motile living biobots self-construct from adult human somatic progenitor seed cells. *Advanced Science, 11*(4), 2303575. https://doi.org/https://doi.org/10.1002/advs.202303575

Guan, Q., Stella, F., Della Santina, C., Leng, J., & Hughes, J. (2023). Trimmed helicoids: An architectured soft structure yielding soft robots with high precision, large workspace, and compliant interactions. *NPJ Robotics, 1*. https://doi.org/10.1038/s44182-023-00004-7

Gupta, A., Savarese, S., Ganguli, S., & Fei-Fei, L. (2021). Embodied intelligence via learning and evolution. *Nature Communications, 12*(1), 5721. https://doi.org/10.1038/s41467-021-25874-z

Haddadin, S. (2025). What embodied AI means for discovery and humanity. Columbia Engineering, March 13, 2025. https://www.engineering.columbia.edu/about/news/what-embodied-ai-means-discovery-and-humanity

Haugeland, J. (1985). *Artificial intelligence: The very idea of it.* MIT Press.

Heidegger, M. (1962). *Being and time* (J. Macquarrie & E. Robinson, Trans.). Harper & Row. (Original work published 1927)

Hu, N., Mitchell, E., Manning, C. D., & Finn, C. (2023). Meta-Learning online adaptation of language models. *arXiv.* https://arxiv.org/abs/2305.15076

Hughes, J., Abdulali, A., Hashem, R., & Iida, F. (2022). Embodied artificial intelligence: Enabling the next intelligence revolution. In *Proceedings of the International Workshop on Embodied Intelligence 2021* (IOP Conference Series: Materials Science and Engineering, *1261*, 012001). IOP Publishing. https://doi.org/10.1088/1757-899X/1261/1/012001

Husserl, E. (1982). *Ideas pertaining to a pure phenomenology and to a phenomenological philosophy—first book: General introduction to pure phenomenology.* (F. Kersten, Trans.). Nijhoff. (Original work published 1913).

Hutto, D. D., & Myin, E. (2013). Radicalizing enactivism: Basic minds without content. MIT Press.

Hutto, D. D., & Myin, E. (2017). *Evolving enactivism: Basic minds meet content.* MIT Press.

Hutto, D. D., & Satne, G. (2017). Continuity skepticism in doubt: A radically enactive take. In C. Durt, T. Fuchs, & C. Tewes (Eds.), *Embodiment, enaction, and culture: Investigating the constitution of the shared world* (pp. 107–127). MIT Press.

Ishige, M., Umedachi, T., Taniguchi, T., & Kawahara, Y. (2019). Exploring behaviors of caterpillar-like soft robots with a central pattern generator-based controller and reinforcement learning. *Soft Robotics, 6*(5), 579–594. https://doi.org/10.1089/soro.2018.0126

Kiverstein, J., & Miller, M. (2015). The embodied brain: Towards a radical embodied cognitive neuroscience. *Frontiers in Human Neuroscience, 9,* 237. https://doi.org/10.1007/s11229-017-1583-9

Kriegman, S., Blackiston, D., Levin, M., & Bongard, J. (2020). A scalable pipeline for designing reconfigurable organisms. *Proceedings of the National Academy of Sciences of the United States of America, 117*(4), 1853–1859. https://doi.org/10.1073/pnas.1910837117

Langton, C. G. (1989). Artificial life. In C. G. Langton (Ed.), *The proceedings of an interdisciplinary workshop on the synthesis and simulation of living systems* (pp. 1–47). Addison-Wesley Publishing Company.

Levin, M. (2019). The computational boundary of a "self": Developmental bioelectricity drives multicellularity and scale-free cognition. *Frontiers in Psychology, 14*, 10. https://doi.org/10.3389/fpsyg.2019.02688

Levin, M. (2022). Technological approach to mind everywhere: An experimentally-grounded framework for understanding diverse bodies and minds. *Frontiers in Systems Neuroscience, 16*, 768201. https://doi.org/10.3389/fnsys.2022.768201

Levin, M. (2023). Bioelectric networks: The cognitive glue enabling evolutionary scaling from physiology to mind. *Animal Cognition, 26*(6), 1865–1891. https://doi.org/10.1007/s10071-023-01780-3

Levin, M. (2024). Self-Improvising memory: A perspective on memories as agential, dynamically reinterpreting cognitive glue. *Entropy, 26*(6), 481. https://doi.org/10.3390/e26060481

Liu, S., & Wu, S. (2024, April 29). *A brief history of embodied artificial intelligence, and its outlook*. blog@CACM. https://cacm.acm.org/blogcacm/a-brief-history-of-embodied-artificial-intelligence-and-its-future-outlook/

Liu, W., Wang, J., & Chang, S. F. (2012). Robust and scalable graph-based semisupervised learning. *Proceedings of the IEEE, 100*(9), 2624–2638. https://doi.org/10.1109/JPROC.2012.2197809

Macrine S. L., & Fugate, J. M. B. (2022). *Movement matters: How embodied cognition informs teaching and learning*. MIT Press.

Marr, D. (1982). *Vision: A computational investigation into the human representation and processing of visual information*. W. H. Freeman and Company.

Maturana, H. R., & Varela, F. J. (1973/1980). *Autopoiesis and cognition: The realization of the living*. Springer. https://doi.org/10.1007/978-94-009-8947-4

McCarthy, J., Minsky, M. L., Rochester, N., & Shannon, C. E. (2006). A proposal for the Dartmouth summer research project on artificial intelligence. *AI Magazine, 27*(4), 12–14. https://doi.org/10.1609/aimag.v27i4.1904 (Original work published 1955)

McMillen, P., & Levin, M. (2024). Collective intelligence: A unifying concept for integrating biology across scales and substrates. *Communications Biology, 7*, 378. https://doi.org/10.1038/s42003-024-06037-4

Merleau-Ponty, M. (1968). *The visible and the invisible* (A. Lingis, Trans.). Northwestern University Press.

Miller, G. A. (1956). The magical number seven, plus or minus two: Some limits on our capacity for processing information. *Psychological Review, 63*(2), 81. https://doi.org/10.1037/h0043158

Miller, G. A. (2003). The cognitive revolution: A historical perspective. *Trends in Cognitive Sciences, 7*(3), 141–144. https://doi.org/10.1016/S1364-6613(03)00029-9

Mitchell, M. D. (2023, October 26). Soft, elephant trunk-like robot for close interaction with humans. https://actu.epfl.ch/news/soft-elephant-trunk-like-robot-for-close-interacti/

Moll, L. (1997). Vygotsky, la educación y la cultura en acción. In A. Álvarez (Ed.), *Hacia un currículum cultural: La vigencia de Vygotski en la educación* (pp. 39–53). Fundación Infancia y Aprendizaje.

Muhsen, D. K., Sadiq, A. T., & Raheem, F. A. (2024). A survey on swarm robotics for area coverage problem. *Algorithms, 17*(1), 3. https://doi.org/10.3390/a17010003

Murugan, N. J., Vigran, H. J., Miller, K. A., Golding, A., Pham, Q. L., Sperry, M. M., Rasmussen-Ivey, C., Kane, A. W., Kaplan, D. L., & Levin, M. (2022). Acute multidrug delivery via a wearable bioreactor facilitates long-term limb regeneration and functional recovery in adult *Xenopus laevis. Science Advances, 8*(4), eabj2164. https://doi.org/10.1126/sciadv.abj2164

Neisser, U. (1967). *Cognitive psychology*. Appleton-Century-Crofts.

Newell, A., & Simon, H. A. (1956). The logic theory machine: A complex information processing system. *IRE Transactions on Information Theory, 2*(3), 61–79.

Newell, A., & Simon, H. A. (1972). *Human problem solving*. Prentice Hall.

Noë, A. (2004). *Action in perception*. MIT Press.

Nolfi, S., & Floreano, D. (2000). *Evolutionary robotics. The biology, intelligence, and technology of self-organizing machines*. MIT Press.

Paolo, G., Gonzalez-Billandon, J., & Kegl, B. (2024). A call for embodied AI. *arXiv*. 2402.03824v4. https://arxiv.org/pdf/2402.03824

Parise, A. G., Gagliano, M., & Souza, G. M. (2020). Extended cognition in plants: Is it possible? *Plant Signaling & Behavior, 15*(2), 1710661. https://doi.org/10.1080/15592324.2019.1710661

Pfeifer, R., & Bongard, J. C. (2006). *How the body shapes the way we think: A new view of intelligence*. MIT Press.

Pfeifer, R., & Iida, F. (2006). Morphological computation: Connecting body, brain, and environment. In F. Iida, R. Pfeifer, L. Steels, & Y. Kuniyoshi (Eds.), *Lecture notes in computer science: Vol. 3853. Embodied artificial intelligence* (pp. 66–90). Springer. https://doi.org/10.1007/978-3-642-00616-6_5

Pfeifer, R., & Scheier, C. (1999). *Understanding intelligence.* MIT Press. https://doi.org/10.7551/mitpress/6979.001.0001

Piloto, L. S., Weinstein, A., Battaglia, P., & Botvinick, M. (2022). Intuitive physics learning in a deep-learning model inspired by developmental psychology. *Nature Human Behavior, 6*(9), 1257–1267. https://doi.org/10.1038/s41562-022-01394-8

Pio-Lopez, L., Pezzulo, G., & Levin, M. (2025, March 11). Scale-free niche construction: Expanding agent-microenvironment co-development to unconventional substrates. OSF Preprints. https://doi.org/10.31219/osf.io/ytg35_v1

Polychronidou, M., Hou, J., Babu, M. M., Liberali, P., Amit, I., Deplancke, B., Lahav, G., Itzkovitz, S., Mann, M., Saez-Rodriguez, J., Theis, F., & Eils, R. (2023). Single-cell biology: What does the future hold? *Molecular Systems Biology, 19*(7), e11799. https://doi.org/10.15252/msb.202311799

Putnam, H. (Ed.). (1975). *Philosophical papers I: Mind, language, and reality.* Cambridge University Press.

Pylyshyn, Z. W. (1984). *Computation and cognition.* MIT Press. https://doi.org/10.7551/mitpress/2004.001.0001

Raibert, M. H. (1986). *Legged robots that balance.* MIT Press.

Rosch, E. (2002). Principles of categorization. In D. J. Levitin (Ed.), *Foundations of cognitive psychology: Core readings* (pp. 251–270). MIT Press.

Rouleau, N., & Levin, M. (2023). The multiple realizability of sentience in living systems and beyond. *eNeuro, 10*(11), ENEURO.0375-23.2023. https://doi.org/10.1523/ENEURO.0375-23.2023

Sale, T. (2000). The colossus of Bletchley Park: The German cipher system. In R. H. Rojas (Ed.), *The first computers: History and architecture* (pp. 351–64). MIT Press.

Savva, M., Kadian, A., Maksymets, O., Gordon, D., Jain, Y., Zamir, A., Collett, T., Batra, D., Feichtenhofer, C., Goswami, D., Greff, K., Malik, J., Mandyam, P., Murthy, V., Wijmans, E., Wu, E., & Yevseienkova, Y. (2019). Habitat: A platform for embodied AI research. In K. M. Lee, D. Forsyth, M. Pollefeys, & X. Tang (Eds.), *Proceedings of the 2019 IEEE/CVF International Conference on Computer Vision (ICCV)* (pp. 9332–9342). IEEE. https://doi.org/10.1109/ICCV.2019.00933

Scheier, C., & Pfeifer, R. (2001). *Understanding intelligence.* MIT Press. https://mitpress.mit.edu/9780262661256/understanding-intelligence/

Segundo-Ortin, M., & Calvo, P. (2023). Plant sentience? Between romanticism and denial: Science. *Animal Sentience*, 33(1), 1.

Seifert, L., Davids, K., Hauw, D., & McGann, M. (2020). Editorial: Radical embodied cognitive science of human behavior: Skill acquisition, expertise and talent development. *Frontiers in Psychology, 11,* 1376. https://doi.org/10.3389/fpsyg.2020.01376

Sheets-Johnstone, M. (1982). Thinking in movement. *Journal of Aesthetics and Art Criticism, 39*(4), 399. http://doi.org/10.2307/430239

Suchman, L. (1987). *Plans and situated actions: The problem of human/machine communication.* Cambridge University Press.

Thompson, E. (2007). *Mind in life: Biology, phenomenology, and the sciences of mind.* Harvard University Press.

Torresen, J. (2018). A review of future and ethical perspectives of robotics and AI. *Frontiers in Robotics and AI, 4,* 75.

Turing, A. M. (1950). Computing machinery and intelligence. *Mind, 49,* 433–460. https://doi.org/10.1093/mind/LIX.236.433

Von Uexküll, J. (2001). An introduction to umwelt. *Semiotica, 134*(1–4), 107–110. (Original work published 1936)

van Gelder, T. (1998). The dynamical hypothesis in cognitive science. *Behavioral and Brain Sciences, 21*(5), 615–665. http://doi.org/10.1017/s0140525x98001733

Varela, F. J., Thompson, E., & Rosch, E. (1991). *The embodied mind: Cognitive science and human experience.* MIT Press.

Wang, S., Chen, K., Zhang, Q., Ma, J., Zhu, L., Sun, Y., Liu, X., Wang, B., Xu, Z., You, X., Liu, C., Hu, B., Chen, H., & Zhou, P. (2025). Neuromorphic peripheral sensory-computer interface embodied by two-dimensional ultrasensitive circuits. *Materials Science and Engineering: R: Reports, 162,* 100884. https://doi.org/10.1016/j.mser.2024.100884

Yong, E. (2022). *An immense world: How animal senses reveal the hidden realms around us.* Random House.

Zahavi, D. (2005). *Subjectivity and selfhood: Investigating the first-person perspective.* MIT Press.

I Philosophical and Theoretical Foundations

Part I examines the philosophical and theoretical foundations of embodied intelligence (EI). It begins by highlighting a significant convergence between human-embodied cognition (HEC) and embodied artificial intelligence (EAI), emphasizing a paradigm shift that recognizes the crucial and reciprocal influence of the body and environment in shaping intelligent behavior, moving away from traditional brain-centric perspectives. Furthermore, it proposes a "pattern theory of self" as a dynamic and interconnected system for both humans and robots, introducing the concept of "robodiment" for robots, characterized by physical continuity, self-movement, and intelligible behavior, alongside outlining capacities for "ipse identity." This part also addresses the ongoing debate surrounding brain representation, introducing active inference with generative models as a framework for understanding representational processes like belief updating, perception, and action through approximate Bayesian inference. Finally, it expands on predictive processing and active inference in the context of higher-order cognition and evolutionary trajectories, suggesting an ontology of generative models that become increasingly complex in more evolved animals.

Common Themes

EI challenges traditional computational models by emphasizing the interconnection between form, environment, and cognitive processes (e.g., Varela et al., 1991; chapter 1). HEC and EAI share a rich history, whose similarities can be addressed by several themes: (1) Primitives (the fundamental building blocks of cognition, arising from the dynamic interplay of an agent's body and the environment), (2) priors (preexisting beliefs about the world that are shaped by innate factors and past experiences), (3) adaptive behaviors (behaviors that are shaped by both environmental pressures and new opportunities), (4) artificial neural networks (ANNs) and biological neural networks (BNNs) (ANNs are computational models consisting of interconnected nodes organized in layers to create distributed networks similar to BNNs) (Tian et al., 2020), and (5) recognizing emotions (emotions are fundamental to human [and many nonhuman] animals). Although large language models (LLMs) can accommodate large amounts of information quickly and have made increasingly better attempts to represent emotions, concerns remain about whether artificial systems can experience emotions or merely mimic them based on learned patterns (Casas et al., 2021; Serapio-Garcia et al., 2023; chapter 1).

Active inference provides a unifying self-organization framework linking perception, action, and cognition (Parr et al., 2022), in which action is explained as a solution to the inverse problem (chapter 3). It proposes that the brain maintains causal models of the world (generative models) that map onto neuroanatomy as well as the dynamics that realize that cognitive function. In a representation-free version of active inference, deontic values might can take the places of representations because beliefs about the states of the world are conditioned on associated sensory consequences (chapter 3).

The free energy principle (FEP) formalizes active inference by minimizing free energy caused by prediction errors—the difference between model predictions and actual sensory input (Friston et al., 2006; Friston, 2019; (chapter 4). An agent must *adaptively control* the surprise of free energy through their action-perception loops (Pezzulo et al., 2024). These "perception-action" loops facilitate self-learning and can be applied to artificial, as well as biological, agents (chapter 4).

Self-evidencing models unify perception, learning, and action through Bayesian approaches, in which actions are selected that maximize free energy and prediction errors by encoding preferences as prior beliefs (Friston, 2011; Ramstead et al., 2019). The complexity of a model can be measured in terms of the divergence of posterior inferences from priors, providing a natural form of regularization. When agents share generative models, they don't develop identical models but rather develop models with partial overlaps and complementary elements (see also part II, chapter 5).

Internal generative models can help explain how brains developed more complex behaviors (Pezzulo et al., 2022). More sophisticated generative models can duplicate predictive circuits, create multimodal links across distinct modalities, or capture environmental regularities and the external causes of sensations. This allows them to encode (beliefs about) hidden states of the world, including desired goal states, and the flexible sets of behaviors needed to achieve them (see also *world models*, e.g., Ha & Schmidhuber, 2018). Importantly, exhibiting sophisticated behaviors requires the ability to represent events unfolding at multiple levels of abstraction, across different time scales, and demonstrate some independence between factors (Friston et al., 2024) or conditionally independent causes (factorization) (chapter 4).

Social awareness can be explained by what is stable among a being (*idem* identity) and that which varies (*ipse* identity). A robot body that is physically continuous, capable of self-movement, and dynamically capable of intelligible behavior (performing a function that may involve deciding among alternatives but could also be a kind of preprogramed action) would have *idem* identity. A robot would have to have the capacity for a self-narrative to have *ipse* identity. This would include having (1) temporal integration, (2) first-person self-reference, (3) episodic autobiographical memory, and (4) reflective metacognition (chapter 2).

References

Casas, J., Senft, E., Gutierrez, L. F., Rincon-Rocancio, M., Munera, M., Belpaeme, T., & Cifuentes, C. A. (2021). Social assistive robots: Assessing the impact of a training assistant robot in cardiac rehabilitation. *International Journal of Social Robotics, 13*(6), 1189–1203. https://doi.org/10.1007/s12369-020-00708-y

Friston, K. (2011). What is optimal about motor control? *Neuron, 72*(3), 488–498. https://doi.org/10.1016/j.neuron.2011.10.018

Friston, K. J. (2019). Waves of prediction. *PLoS Biology, 17*(10), e3000426. https://doi.org/10.1371/journal.pbio.3000426

Friston, K. J., Kilner, J., & Harrison, L. (2006). A free energy principle for the brain. *Journal of Physiology-Paris, 100*(1–3), 70–87. https://doi.org/10.1016/J.JPHYSPARIS.2006.10.001

Friston, K. J., Parr, T., Heins, C., Constant, A., Friedman, D., Isomura, T., Fields, C., Verbelen, T., Ramstead, M., Clippinger, J., & Frith, C. D. (2024). Federated inference and belief sharing. *Neuroscience and Biobehavioral Reviews, 156*, 105500. https://doi.org/10.1016/j.neubiorev.2023.105500

Ha, D., & Schmidhuber, J. (2018). World models. *arXiv*. https://doi.org/10.5281/zenodo.1207631

Parr, T., Pezzulo, G., & Friston, K. J. (2022). *Active inference: The free energy principle in mind, brain, and behavior*. MIT Press.

Pezzulo, G., Parr, T., & Friston, K. (2022). The evolution of brain architectures for predictive coding and active inference. *Philosophical Transactions of the Royal Society B, 377*(1844), 20200531. https://doi.org/10.1098/rstb.2020.0531

Pezzulo, G., Parr, T., & Friston, K. (2024). Active inference as a theory of sentient behavior. *Biological Psychology, 186*, 108741. https://doi.org/https://doi.org/10.1016/j.biopsycho.2023.108741

Ramstead, M. J. D., Constant, A., Badcock, P. B., & Friston K. J. (2019). Variational ecology and the physics of sentient systems. *Physics of Life Reviews, 31*, 188–205. https://doi.org/10.1016/j.plrev.2018.12.002

Serapio-García, G., Safdari, M., Crepy, C., Sun, L., Fitz, S., Romero, P., Abdulhai, M., Faust, A., & Matarić, M. (2023). Personality traits in large language models. *arXiv*. https://arxiv.org/abs/2307.00184

Tian, C., Xu, Y., Li, Z., Zuo, W., Fei, L., & Liu, H. (2020). Attention-guided CNN for image denoising. *Neural Networks, 124*, 117–129. https://doi.org/10.1016/j.neunet.2019.12.024

Varela, F. J., Thompson, E., & Rosch, E. (1991). *The embodied mind: Cognitive science and human experience*. MIT Press.

1 Convergences of Human and Artificial Embodied Perspectives

Sheila L. Macrine and Jennifer M. B. Fugate

Introduction

The historical separation of the mind and body can be originally traced back to Plato (428–348 B.C.E.), who viewed the body as subservient and an obstacle to the rational mind. The separation of the mind from the body was later popularized by Rene Descartes (1596–1650) and became known as Cartesian dualism (Ryle, 1949). Descartes, in his First Meditation (Descartes, 1637/1998), argued that the senses can be deceptive. For example, the fact that distant objects appear small shows that the senses can be wrong (i.e., the distant object is not indeed smaller, although it looks that way). Rather, true knowledge derives from a mind uncorrupted by bodily or social influences, minimizing the body's role as an active agent in shaping understanding.

However, Cartesian dualism faced increasing challenges in the nineteenth and early twentieth centuries as philosophers advocated for an integrated view, recognizing the body's active role in experience and agency. Gilbert Ryle's (1949) critique of dualism, encapsulated in his "ghost-in-the-machine" metaphor, argued that mental phenomena are intrinsically intertwined with bodily experiences in the physical world, highlighting the flaw of independent mental and physical realms and implicitly suggesting the fundamentality of embodied agency.

Similarly, Maurice Merleau-Ponty's phenomenological approach (1945/1962) rejected Cartesian dualism by introducing the "embodied subject" and "intercorporeality," emphasizing the inseparable relationship between mind, body, and worldly experience. He argued that bodily engagement fundamentally shapes perception, rejecting the body as a mere object and highlighting its role as the subject of experience and a primary source of agency.

Although Ryle and Merleau-Ponty approached the mind-body relationship from different philosophical perspectives, both converged on a more holistic understanding of the human being, proposing intertwined mind and body as aspects of a unified lived experience characterized by embodied agency. Their contributions significantly shifted philosophical thought toward recognizing the importance of embodiment in understanding human consciousness and behavior, acknowledging the body as an active participant.

The purpose of this chapter is to examine the historical and philosophical context (see also the introduction to this volume) to explore the intersections and convergences between human-embodied cognition (HEC) and embodied artificial intelligence (EAI). Specifically, it examines the degree to which these disciplines share a similar understanding and justification of knowledge claims related to embodied cognition (EC), acknowledging potential variations in their epistemological frameworks, assumptions, and methods. This exploration contributes to a more comprehensive understanding of embodied intelligence (EI), a concept that challenges the traditional belief that intelligence is confined to the brain or a complex algorithm. Instead, EI offers a more inclusive perspective, recognizing the significant role that an agent's physical form, cognitive abilities, and interaction with the environment play in shaping intelligence and enabling agentic behavior—the capacity of an embodied agent for effective and purposeful environmental interaction. It also highlights how the physical structure of an agent influences its ability to learn and adapt through processes like evolution and morphological computation (MC) (Pfeifer & Bongard, 2006), ultimately impacting its capacity for effective agency.

The Embodied Turn: Reorienting the Study of Cognition

The study of cognition has undergone a significant transformation, often referred to as the "embodied turn," which offers a fundamentally different perspective on thinking and intelligence than its predecessors. This shift marks a departure from the traditional cognitivist paradigm, a view deeply influenced by foundational work in computation (Turing, 1950), abstract linguistics (Chomsky, 1959), and modular mind theory (Fodor, 1975, 1981). While these figures provided crucial insights to cognitive science, the embodied turn critically reevaluates their implicit downplaying of the body's role. Turing's computational metaphor, Chomsky's and Fodor's focus on abstract symbol manipulation, and their general tendency toward a Cartesian-inspired mind-body separation are key points of divergence for this emerging perspective.

The embodied turn is largely motivated by a critique of Cartesian dualism and its separation of mind and body. This view emphasizes the body's fundamental role in shaping cognitive processes and enabling meaningful agency. Consequently, understanding shifts from brain-centric computation to a phenomenon arising from the continuous interplay of brain, body, and environment.

This embodied approach has spurred extensive interdisciplinary investigations across psychology, computer science, biology, neuroscience, and cognitive science, exploring how our bodies constitutively influence various cognitive functions, including perception, action, language, emotion, and learning (Barsalou, 1999; Pfeifer & Bongard, 2006). These approaches have inspired novel theoretical frameworks and experimental models that prioritize the situated nature of cognition, studying it within ecological contexts where the body actively participates with and shapes the cognitive landscape. In psychology and cognitive science, HEC refers to the idea that cognition is not merely abstract symbolic processing but a dynamic and constitutive process deeply intertwined with our sensorimotor experiences (perception, action, interoception, proprioception) and our active engagement with the environment. For instance, conceptual understanding is grounded in sensorimotor actions (Glenberg & Kaschak, 2002), which highlights how our physical being informs our cognitive landscape and enables agency, including the development of bodily self-awareness and a more integrated sense of self (Friston, 2018, 2022; Matheson et al., 2015; Parma et al., 2024).

Beyond psychology and cognitive science, the embodied perspective has significantly impacted artificial intelligence (AI). For example, EAI aims to create physically situated intelligent agents capable of interacting with and learning from their surroundings, moving beyond purely disembodied algorithms and opening new avenues for understanding artificial agency. Neuroscience has further illuminated the neural mechanisms underlying EC, such as mirror neurons (Gallese, 1996; Keysers, 2009; Rizzolatti & Craighero, 2004; Shimada, 2022) and visuo-proprioceptive integration, providing insights for developing sophisticated EAI systems.

EAI emerged from a desire to overcome the limitations of traditional, disembodied AI, emphasizing the crucial role of physical embodiment in shaping intelligent behavior and enabling autonomous agency. Rodney Brooks (1991) challenged the need for complex internal models, arguing for intelligence arising from direct environmental interaction, exemplified by his insect-like robots. This resonated with researchers and spurred further EAI development. Pfeifer and Scheier (1999) further theorized that intelligence is an emergent property of the entire agent (body, sensors) and environment, challenging the purely computational view.

Embodied theories have also encouraged brain-inspired AI models like those utilizing predictive processing and active inference (Clark, 2013; Friston et al., 2006), offering frameworks for understanding agentic interaction. This embodied perspective, further enriched by theoretical neuroscience concepts like Friston's free energy principle (FEP) and active inference (Friston et al., 2006, 2019), provides powerful tools for understanding how embodied agents exercise agency through environmental interaction.

Free Energy Principle: Self-Learning, Predictive Processing, and Error Minimization

The FEP offers a compelling mathematical and philosophical foundation for understanding how embodied agents minimize surprise and engage in purposeful behavior, effectively framing agency as inherent to this minimization. Serving as a cornerstone of embodiment in both biological and artificial systems (Friston et al., 2006), the FEP posits an interplay between predictive processing, error minimization, and embodiment. It emphasizes that systems, from the human brain to advanced robots, strive to minimize "free energy"—a measure of surprise or discrepancy between internal predictions and actual sensory data (Friston, 2010; Friston et al., 2006, 2016). These prediction errors drive learning and adaptation by prompting the revision of internal models and guiding agentic actions to reduce them, enabling flexible and adaptive action (Friston, 2019; see also Zhang & Xu, 2024). Consequently, agents actively generate and refine predictions to maintain environmental equilibrium, driving their agentic behaviors.

Viewing intelligence as the capacity to accumulate evidence for a generative model of the world (Clark, 2013; Friston et al., 2019; chapter 5), the FEP highlights its essential role in informed and effective agency. Specifically, it proposes that neuronal networks function as predictive models, continuously updating expectations through active inference. This involves agents adaptively controlling the surprise of free energy via action-perception loops (Pezzulo et al., 2024; chapter 4), actively shaping sensory input to align with internal models. Pezzulo et al. (2024) further apply active inference to sentient behavior, suggesting that curiosity and uncertainty resolution drive individual and collective learning through belief sharing, enhancing sophisticated agency. They advocate for a shared modeling language and transaction protocol to facilitate collaboration within a "shared intelligence" ecosystem (Friston et al., 2024), underscoring active inference's foundational role in future AI and collaborative artificial agency.

Similarly, in robotics, the FEP guides the development of artificial agents with internal models that predict outcomes and minimize errors through active inference, facilitating self-learning and adaptation (Friston, 2010; see also Barp et al., 2022; Da Costa et al., 2020, 2023; Parr et al., 2022), enabling autonomous and agentic behaviors. This framework offers a powerful lens for understanding intelligent behavior across systems, providing insights into fundamental principles and paving the way for truly intelligent and autonomous machines.

The vision of a shared intelligence ecosystem (chapter 5) emphasizes the FEP's potential to advance our understanding of intelligence and shape future AI through human-machine collaboration and coevolution (see chapter 13), where the interplay of different forms of agency yields novel outcomes. This vision also highlights HEC's significance in shaping intelligent behavior through the dynamic interplay of an agent's physical form, cognitive capabilities, and environmental interaction (chapter 4).

Evolution and Evolutionary Robotics: Mirroring Nature's Design

Human evolution has been driven by genetic mutations and natural selection over millions of years, shaping us into complex beings capable of remarkable feats (Barfield, 2019). As a result, evolutionary processes are intrinsically linked to intelligence (H. C. Barrett, 2019; Lakatos & Janka, 2008), including the evolution of sophisticated forms of agency.

For example, human morphological variation is thought to be partially shaped by natural selection associated with environmental factors, particularly climate (Bell, 2023), influencing our physical capabilities. Evolutionary robotics (EvR) draws inspiration from these biological principles to create increasingly sophisticated machines (Gupta et al., 2021). Both biological and artificial evolution rely on variation, selection, adaptation, and heredity to optimize systems for specific tasks or environments (Pfeifer & Bongard, 2006; Saeb & Al-Naqeb, 2016; Savell et al., 2016). This key concept in EvR still mirrors these core evolutionary principles, even though it operates on a faster timescale driven by human ingenuity and technology. This includes selecting morphologies and control systems that exhibit more effective agency (chapter 10).

Expanding on these evolutionary approaches, evolutionary reinforcement learning (EvoRL) combines evolutionary algorithms with reinforcement learning to enhance AI adaptability and performance (Lin et al., 2024). It leverages the global search capabilities of evolutionary algorithms to optimize reinforcement learning policies, leading to improved adaptability and performance in complex scenarios, including goal-directed behavior. While promising, EvoRL faces challenges in scalability, adaptability, sample efficiency, robustness, and ethical considerations related to the increasing autonomy and potential agency of such systems (Lin et al., 2024).

The influence of morphology extends beyond the evolutionary selection of advantageous forms. MC specifically examines how an agent's physical structure (morphology) and its interaction with the environment can simplify control and information processing (Pfeifer & Bongard, 2006; also chapter 11). Instead of the brain or controller doing all the work, the body itself offloads some of the computational burden, directly influencing the agent's ability to act. For example, the shape of a bird's wing during flight emerges from muscular control and dynamic interactions with aerodynamic, inertial, and elastic forces (Ghazi-Zahedi et al., 2016). This can also be seen in the passive dynamics of walking, where the shape of the legs and the interaction with gravity contribute to the walking motion without requiring complex calculations (McGeer, 1990).

The interconnectedness of MC and human evolution underscores the importance of physical structure in shaping intelligence and behavior: MC shows how the body's design actively participates in information processing (McGeer, 1990; Pfeifer & Iida, 2006), whereas human evolution demonstrates how morphology has contributed to our cognitive abilities and the scope of our agency. This connection emphasizes the profound role of embodiment in our interaction with and understanding of the world and, consequently, our ability to act as agents within it.

Human and AI Emotions

William James once wrote, "Of two things concerning the emotions, one must be true. Either separate and special centers, affected by them alone, are their brain-seat, or else they correspond to processes occurring in the motor and sensory centers already assigned" (James 1890/1998, p. 473).

Human emotions, rooted in biological and evolutionary history, arise from complex embodied interactions intertwined with subjective experiences and conceptual knowledge (L. A. Barrett, 2015, 2017). Human emotion categories are built from the probabilistic patterns that develop from previous coordinated bodily activity (including sensorimotor and interoception) and sensory information (L. A. Barrett, 2017; Wilson-Mendenhall et al., 2011) within a given context, shaping how we perceive and act in the world.

Historically, "emotion detection AI models" were based on a rigid view of emotions, which—as W. James posited—could not be both embodied and reduced to dedicated circuitry. With this in mind, AI typically views emotions as computational constructs, lacking the depth and complexity of human emotions and relying on pattern recognition of discrete emotional signatures (Savery & Weinberg, 2020; Šimić et al., 2021), which may limit their capacity for truly human feelings. Although such emotion recognition algorithms are becoming increasingly sophisticated, including integrating sensors capable of capturing subtle cues (D'Mello & Graesser, 2015; Spezialetti et al., 2020), the general premise is that there are reliable and distinct emotional patterns that can be identified among humans. Indeed, such efforts have faced intense criticism for their reliance on universal prototypes and the problem of inference (for robust discussion, see L. A. Barrett, 2017; L. A. Barrett et al., 2019).

Alternatively, both HEC and EAI recognize that emotions are not merely abstract mental states but are rather deeply intertwined with bodily states and physiological responses (L. A. Barrett, 2017; Damasio, 1994), profoundly influencing our motivations and guiding our agentic behaviors. In the context of HEC, an embodied perspective on emotions emphasizes the crucial role of bodily sensations and physiological changes in shaping our emotional experiences (L. A. Barrett, 2017, L. A. Barrett et al., 2019; Niedenthal, 2007; Wilson-Mendenhall et al., 2011; see also Owens et al., 2018). Research in this field explores the intricate relationship between bodily states and emotions, along with the impact of bodily feedback on emotional regulation and decision-making.

Replicating the full complexity of human emotions in AI remains a challenge. Although large language models (LLMs) (e.g., BingAI, ChatGPT 4, GoogleBard, etc.) are now common and allow more flexibility (Liu et al., 2024), questions remain as to whether they genuinely experience emotions or merely mimic them based on learned patterns (Casas et al., 2021; Serapio-Garcia et al., 2023). The future of EAI will require developing a deeper understanding of human emotions (chapter 14).

Convergences between HEC and EAI

Many of the convergences between HEC and EAI, as highlighted in Table 1.1, point toward a growing shared understanding of intelligence across these disciplines. This evolving perspective signifies a significant departure from traditional, brain-centric views, with both fields increasingly recognizing the crucial role of embodiment and the active agency of intelligent systems (Staddon, 1983; Tolk et al., 2022). This shared understanding is reflected in their converging perspectives on the importance of dynamic interaction with the environment, the embodied grounding of neural cognition (primitives and priors rooted in physical structure and sensorimotor-linguistic links), and the emergent nature of psychological categories such as emotions and empathy (Clark, 2013, 2015; Gupta et al., 2021; Hoffman, 2021; Mazzaglia et al., 2023; Pfeifer & Iida, 2004). Below, we summarize these similarities into five key areas (see Table 1.1).

Primitives

Psychological primitives are the fundamental building blocks of cognition, arising from the dynamic interplay of an agent's body, core networks within the brain, and the agent's environment (L. A. Barrett, 2017; Bear et al., 2020; Franklin, 2019; Staddon, 1983; Tolk et al., 2022). Primitives continuously evolve as the agent learns and interacts with its environment (Clark, 2013, 2015). This adaptive nature is a cornerstone of HEC and EAI and underpins the flexibility of goal-directed action. Different bodies and sensorimotor capabilities lead to diverse ways of experiencing and interacting with the world, resulting in a variety of primitive building blocks (Li et al., 2022; Pfeifer & Iida, 2006; Shen et al., 2021), influencing the nature and scope of potential actions. These ideas are represented in MC, in which the interconnectedness of a physical structure is integral to intelligence and behavior and can impact how an agent can act on its environment (Cangelosi et al., 2015; Müller & Hoffman, 2017; Pfeifer & Iida, 2006). In addition to simple primitives, dynamic movement primitives, often utilized in EAI, represent more complex actions by modeling them as stable dynamical systems (Balconi & Crivelli, 2023; Saveriano et al., 2023).

Examples of psychological primitives include basic visual processes like edge detection (Medina et al., 2015; Sučević & Schapiro, 2023), sensorimotor actions like grasping (Bartolozzi et al., 2022; Chauhan et al., 2018; Ororbia & Kifer, 2022; Sučević & Schapiro, 2023; Zhong et al., 2014), and the building blocks that help establish more complex categories, such as emotion (L. A. Barrett, 2009, 2015). A key aspect of using psychological primitives to establish more complex categories is the ability to categorize objects by comparing them to mental prototypes, a process highlighted by Rosch's research on categorization (Rosch, 1978), and the ability to use language to anchor diverse sensorimotor experiences (see Lindquist & L. A. Barrett, 2012; Oosterwijk et al., 2015).

An excellent example that bridges human vision and categorization with EAI is Fei-Fei Li's creation of ImageNet (Li et al., 2007). This extensive and evolving data set of images has been instrumental in advancing visual recognition systems (Li & Krishna, 2022; Zhu et al., 2021) by providing a vast and diverse collection of images organized according to the WordNet hierarchy (Millier, 2021; Ng et al., 2021). This allows deep learning models to learn from a rich and comprehensive representation of the visual world, enabling them to achieve near-human accuracy in object recognition and categorization tasks, which is crucial for effective interaction with visual environments. ImageNet functions as both a training set and a benchmark, advancing computer vision models by mirroring human strategies for organizing and recognizing objects in complex environments (Deng et al., 2009; Zahra et al., 2022). This example underscores how HEC and EAI are converging to facilitate effective interaction with visual environments.

Priors

Priors—preexisting beliefs and expectations about the world—constitute a fundamental shared belief between HEC and EAI. Previous knowledge, shaped by innate factors and past experiences, guides learning and adaptation, enabling agents to use sensory input and choose appropriate actions, thereby directing agency. The FEP posits that error minimization and active inference are the core mechanisms by which agents maintain and update these priors (Friston, 2019), implying that our prior beliefs are continuously tested, refined, and potentially revised through our interactions with the world. Thus, the FEP views intelligence as the ability to gather and maintain a generative model of the world (Clark, 2013; Friston, 2019; chapter 5), which serves as the foundation for informed and effective agency. Priors function as a crucial filter, shaping an agent's perception and interaction with its environment by guiding the selection and utilization of affordances, thereby enabling cognitive enhancement and facilitating the extension of agency (Lakatos & Janka, 2008).

Understanding and beliefs about the world are built on how an agent interacts with the world and can ground their interactions with their sensorimotor systems that influence future agentic choices. The theory of affordances (Gibson, 1977, 1979) proposes that an agent shapes their identity from the environment, suggesting that the environment itself contributes to the development of priors that guide an agent's actions and sense of agency. Part of this shaping ability may also come from grounding diverse experiences not only within the body but also through language to anchor embodied simulations (see Lindquist & L. A. Barrett, 2012; Oosterwijk et al., 2015; Vong et al., 2024).

Natural language processing in situ (Nygaard & Tzeng, 2021) has been used to show that language can serve as a powerful prior, which can affect subsequent agentic behavior either directly by activating sensorimotor experiences or indirectly by anchoring embodied knowledge. For example, studies have shown that reading action words like "kick" activates the visuomotor cortex, the same brain region involved in planning and executing movements (e.g., Aziz-Zadeh et al., 2006; Glenberg & Kaschak, 2002; Pulvermuller & Hauk, 2006). This example shows how our understanding of action words is fundamentally grounded in embodied experience, further supporting the view that visual perception is not a passive reception of stimuli but rather an active and embodied process wherein prior experiences and knowledge shape our perception and inform our engagement with the world (Gordon et al., 2023).

Adaptive Behaviors

The process of learning and adaptation is driven by both environmental pressures ("push" factors) and new opportunities ("pull" factors) (Bell, 2023, p. 309), influencing the development of flexible and agentic responses to diverse situations. EAI aims to create intelligent agents that similarly learn and adapt through environmental interactions (Zahra et al., 2022), enabling them to exhibit autonomous and agentic behaviors in novel contexts. Levin's (2022) Technological Approach o Mind Everywhere (TAME) framework challenges traditional AI architectures, emphasizing the importance of an agent's ability to operate within a problem space and exhibit competency for evolution to function. Rouleau and Levin (2023) go on to suggest that even simple organisms possess a degree of inherent agency that AI should strive to emulate. This framework suggests a deeper cognitive complexity even in simple organisms, opening exciting research avenues to push the understanding of biological agency in simpler organisms.

Further, research in Deep Evolutionary Reinforcement Learning (DERL), demonstrates the interplay between environmental complexity, morphological intelligence, and learning ability (Gupta et al., 2021), showcasing how the physical form of an agent can facilitate more efficient learning and thus more effective agency. For example, the development of virtual creatures like "Unimals" (Gupta et al., 2021), which evolve and adapt their bodies to navigate their environments, supports the morphological Baldwin effect—how learned behaviors or adaptations that initially require effort or flexibility can become increasingly ingrained or "hardwired" into an agent's physical form or morphology over generations. This effect suggests that evolution favors morphologies that learn faster (Weber & Depew, 2003).

This shift has profound implications for the development of AI systems. By embracing the principles of EAI, we can create artificial agents that are not merely sophisticated computational machines but are also truly intelligent entities capable of learning, adapting, and interacting with the world meaningfully, exhibiting more autonomous and context-aware agency. These agents can seamlessly integrate with and complement human capabilities, leading to a future where humans and machines collaborate and coevolve within a shared cognitive ecosystem (chapter 5).

Networks

Both biological neural networks (BNNs) and artificial neural networks (ANNs) exhibit distributed and parallel processing capabilities, enabling complex computations and pattern recognition. However, while ANNs have achieved remarkable success in specific domains, they currently exhibit limitations compared to BNNs in terms of overall complexity, robust adaptability to novel situations, and the capacity for emergent phenomena such as consciousness and creativity (Adadi & Berrada, 2018; Hassabis et al., 2017; Mead, 1989; Nicolelis, 2001). Research is striving to bridge this gap through innovations like neuromorphic computing (Mead, 1989), brain-computer interfaces (Nicolelis, 2001), biologically inspired AI design (Hassabis et al., 2017), and explainable AI (Adadi & Berrada, 2018; Saeed & Omlin, 2023). Such advancements offer promising avenues for enhancing ANNs' efficiency, adaptability, and ethical use, potentially leading to more symbiotic human-AI systems and more human-like AI capabilities.

Emotion

While the full complexity of human emotion remains a significant challenge for AI, the development of emotionally aware AI systems holds the potential to enhance human-machine interaction, provide social and emotional support, and augment human emotional experiences, including agency and choice. Park and Whang (2022) define empathy in the context of human-robot interaction (HRI) as the robot's ability to recognize and understand a human's emotional state, thoughts, and situation and to generate appropriate responses that lead to a positive perception by the human. This emphasizes that for robots to interact and collaborate with humans successfully, they need to develop this capacity for empathy. This aligns with the broader understanding of empathy as crucial for social interactions and fostering cooperation and prosocial behavior (Hoffman, 2001). By incorporating this emotional awareness, robots and AI systems can better understand and adapt to human users, fostering a sense of empathy and connection and more harmonious and productive collaborations (Breazeal, 2003; chapter 6).

Toward an Inclusive Embodied Intelligence

Looking toward a future characterized by increased integration between HEC and EAI, there is a growing need for a more inclusive conceptualization of EI that encompasses all forms of intelligence—whether biological, artificial, hybrid, or something not yet conceived. This inclusive perspective emphasizes the crucial role that an agent's body, brain (or processor), and environment play in shaping learning and adaptation. By recognizing the deep interconnectedness of these elements, we challenge traditional, brain-centric views of cognition, moving beyond the limitations of cognitivism and Cartesian dualism. By recognizing the shared principles that govern intelligent behavior across these diverse forms, we can unlock the full potential of EI, paving the way for a future where humans and machines work together harmoniously to navigate the complexities of the twenty-first century.

An inclusive EI framework highlights how cognition is fundamentally intertwined with perception, action, the physical body, and the environment, offering a more holistic understanding of intelligence across various embodiments (Barsalou, 1999, 2008; Clark, 2009; Lakoff & Johnson, 1999; Pfeifer & Bongard, 2006; Shapiro, 2010, 2019). This approach can lead to more profound insights into AI, paving the way for developing systems that are more intelligent and capable systems that interact seamlessly with their environments and collaborate effectively with humans. Current robotic technologies often lack the EI necessary for effective interaction, adaptation, and appropriate responses within their environments (Putra et al., 2024). Neuromorphic computing, utilizing spiking neural networks, is one possible solution that works by replicating the brain's biological functions. However, accuracy, adaptability, efficiency, reliability, and security challenges persist. Overcoming these hurdles necessitates developing effective learning rules, energy-efficient optimizations, benchmarks, and reliability and security enhancements. This understanding is crucial for comprehending representation's role in action, especially in human motor control, which relies on planning and representing actions in terms of intended outcomes (chapter 12).

Therefore, EI is not just about brains/processors in bodies; It is about the deep intertwining of an agent's physical form, its environment, and its learning process. This applies whether that agent is a human, a robot, a purely digital AI, a hybrid system, or even a Xenobot (Blackiston et al., 2023; Bongard & Levin, 2021, 2022; chapters 10 and 17). EI challenges us to rethink intelligence as something that emerges from dynamic interactions, pushing us toward AI systems that can genuinely adapt, learn, and meaningfully engage with the world around them.

Still, while HEC and EAI share the embodied perspective, challenges remain. Teaching disembodied AI to grasp concepts humans learn through our physical experiences is difficult, as representing these deeply embodied concepts in systems without a physical form is a significant hurdle. There is also an ongoing debate as to whether representations are always embodied or can also originate from internal processes (Clark, 2013, 2015) and whether EAI represents "weak" or "strong" embodiment (chapter 8). Some scholars recommend distinguishing between concepts and conceptualizations and recognizing the interplay between form and function in conceptual understanding (chapter 7). Others question whether an artificial agent can have both narrative and bodily continuity, addressing whether reidentification and characterization can apply to a nonhuman agent (chapter 2).

Conclusion

The key convergences observed between HEC and EAI necessitate a more inclusive understanding of intelligent behaviors. The inclusive view of EI, on which this volume focuses, challenges traditional disembodied views of cognition, asserting that intelligence is inextricably linked to the dynamic interplay of brain/processor, body/morphology, and environment across diverse systems and offers a valuable lens for future inquiry across transdisciplinary perspectives. This approach, built on the foundational insights gained from comparing HEC and EAI, will facilitate the development of a more integrated and epistemically aligned framework for future research across a range of disciplines.

Our preliminary examination of some parallels, such as sensorimotor feedback loops and MC, serves as a crucial entry point, establishing a foundation for more in-depth comparative studies across biological and artificial agents. This includes investigating shared concepts and themes related to the embodied nature of psychological primitives and priors that shape agentic behavior, the role of adaptive behaviors and modeling of emotion, and the underlying neural networks (both biological and artificial) that implement these processes.

Here, we must be mindful of the risk of obscuring critical distinctions that arise from the fundamentally different substrates and evolutionary histories of biological and artificial systems. Ultimately, it is this expanded conceptualization of EI, encompassing different entities and substrates,

HEC	EAI
Adaptive Behaviors	
• Collective intelligence, evolution, and morphogenesis (Levin, 2023)	• DERL, evolutionary reinforcement learning (Lin et al., 2024), and evolutionary virtual robots (Gupta et al., 2021) • Self-supervised transfer learning (Eitel et al., 2019)
Networks	
• Mirror neurons (Aziz-Zadeh & Ivry, 2009; Gallese,1996; Keysers, 2009; Rizzolatti & Craighero, 2004; Shimada, 2022) • Biological neural network model (BNNs) (Bear et al., 2020; Cohen et al., 2022; Yeo et al., 2011) • Neural network for category learning (Sučević & Schapiro, 2023)	• Bio-inspired mirroring mechanism (Zahra et al., 2022) • Brain-inspired neural generative models (Ororbia & Kifer, 2022; Sučević & Schapiro, 2023) • Artificial neural networks (ANNs) (Chauhan et al., 2018; Ororbia & Kifer, 2022; Sučević & Schapiro, 2023) • Convolutional neural network (Li et al., 2007; Krizhevsky et al., 2017; Tian et al., 2020) • Neuromorphic-neural computation (Bartolozzi et al., 2022)
Emotions	
• Embodied emotion theories (L. F. Barrett, 2023; Cañamero 2021; Fugate & Wilson-Mendenhall, 2022; Sumer et al., 2023; Wilson-Mendenhall et al., 2013) • Emotion & active inference (L. A. Barrett, 2017; Smith et al., 2019) • Human empathy (Hoffman, 2001) • Interoceptive inference (altered interoception to psychopathology) (Owens et al., 2018)	• Emotion modeling in AI & in robots (Roemmich, et al., 2023; Savery & Weinberg, 2020) • Affective computing (Picard, 1997) • Artificial empathy or computational empathy (Yalçın & DiPaola, 2020) • Empathy in HRI (Park & Whang, 2022) • Interoceptive skills in robotics (Balconi & Crivelli, 2023)
Implications	
• Embodied design for learning (Abrahamson et al., 2021) • Embodied learning adaptations, affordances, and assimilation (Macrine & Fugate, 2022) • 4E cognition: embodied, enacted, embedded, extended (Gallagher, 2010) • Call for embodied enactivism (Clark & Chalmers, 1998) • Embodied intelligence in organisms (Pfeifer & Bongard, 2006)	• Embodied AI for learning (Gupta et al., 2021) • AI literacy—deep learning (Ng et al., 2021) • Embodied AI simulators (Duan et al., 2022) • Call for embodied AI (Paolo et al., 2024) • Embodied intelligence in robotics (Brooks, 1991)

that truly transcends traditional disciplinary boundaries. To achieve a more comprehensive and inclusive understanding of how embodiment fundamentally shapes the nature of intelligence across its diverse instantiations, we must pursue continued rigorous and nuanced comparative analysis. Such an expanded perspective will be crucial for a more comprehensive understanding of how embodiment fundamentally shapes the very nature of intelligence and the mechanisms of learning, adaptation, and behavior across these diverse systems.

References

Abrahamson, D., Dutton, E., & Bakker, A. (2021). Towards an enactivist mathematics pedagogy. In S. A. Stolz (Ed.), *The body, embodiment, and education: An interdisciplinary approach*. Routledge.

Adadi, A., & Berrada, M. (2018). Peeking inside the black-box: a survey on explainable artificial intelligence (XAI). *IEEE Access, 6*, 52138–52160. https://doi.org/10.1109/ACCESS.2018.2870052

Aziz-Zadeh, L., & Ivry, R. B. (2009). The human mirror neuron system and embodied representations. *Advances in Experimental Medicine and Biology, 629*, 355–376. https://doi.org/10.1007/978-0-387-77064-2_18

Aziz-Zadeh, L., Wilson, S. M., Rizzolatti, G., & Iacoboni, M. (2006). Congruent embodied representations for visually presented actions and linguistic phrases describing actions. *Current Biology, 16*(18), 1818–1823, https://doi.org/10.1016/j.cub.2006.07.060

Bahl, S., Gupta, A., & Pathak, D. (2021). *Robotics: Science and systems as hierarchical neural dynamic policies*. Robotics: Science and Systems. https://shikharbahl.github.io/hierarchical-ndps/.

Balconi, M., & Crivelli, D. (2023). Interoception as operator-centered proxy for embodiment of tele-robotic systems: Theoretical-practical remarks. In *Proceedings of the 2023 3rd International Conference on Electrical, Computer, Communications and Mechatronics Engineering (ICECCME)* (pp. 1–5). IEEE. https://doi.org/10.1109/ICECCME57830.2023.10252594

Banks, B., & Connell, L. (2023). Multi-dimensional sensorimotor grounding of concrete and abstract categories. *Philosophical Transactions of the Royal Society of London. Series B, Biological Sciences, 378*(1870), 20210366. https://doi.org/10.1098/rstb.2021.0366

Barfield, W. (2019). The process of evolution, human enhancement technology, and cyborgs. *Philosophies, 4*(1), 10. https://doi.org/10.3390/philosophies4010010.

Cangelosi, A., Bongard, J., Fischer, M. H., & Nolfi, S. (2015). Embodied intelligence. In J. Kacprzyk & W. Pedrycz (Eds.), *Springer handbook of computational intelligence* (pp. 697–714). Springer. https://doi.org/10.1007/978-3-662-43505-2_37

Cangelosi, A., Schlesinger, M., & Smith, L. (2015). *Developmental robotics: From babies to robots (Intelligent Robotics and Autonomous Agents series)*. MIT Press.

Carvalho, J. T., & Nolfi, S. (2016). Affordance generation through deep learning for developmental robotics. In Y. Jin & S. Kollias (Eds.), *2016 IEEE Symposium Series on Computational Intelligence (SSCI)* (pp. 1–8). IEEE. https://doi.org/10.1109/SSCI.2016.7850021

Casas, J., Senft, E., Gutierrez, L. F., Rincon-Rocancio, M., Munera, M., Belpaeme, T., & Cifuentes, C. A. (2021). Social assistive robots: Assessing the impact of a training assistant robot in cardiac rehabilitation. *International Journal of Social Robotics, 13*(6), 1189–1203. https://doi.org/10.1007/s12369-020-00708-y

Chartouny, A., Amini, K., Khamassi, M., & Girard, B. (2024). A new paradigm to study social and physical affordances as model-based reinforcement learning. *Cognitive Robotics, 4*, 142–155. https://doi.org/10.1016/j.cogr.2024.08.001

Chauhan, R., Ghanshala, K. K., & Joshi, R. C. (2018). Convolutional neural network (CNN) for image detection and recognition. In *2018 First International Conference on Secure Cyber Computing and Communication (ICSCCC)* (pp. 278–282). IEEE. https://doi.org/10.1109/ICSCCC.2018.8703316

Chen, L., Chen, P., & Lin, Z. (2020). Artificial intelligence in education: A review. *IEEE Access, 8*, 75264–75278. https://doi.org/10.1109/ACCESS.2020.2988510

Chomsky, N. (1959). On certain formal properties of grammars. *Information and Control, 2*(2), 137–167. https://doi.org/10.1016/S0019-9958(59)80017-6

Clark, A. (2009). *Supersizing the mind: Embodiment, action, and cognitive extension.* Philosophy of Mind Series. Oxford Academic. https://doi.org/10.1093/acprof:oso/9780195333213.001.0001

Clark, A. (2013). Whatever next? Predictive brains, situated agents, and the future of cognitive science. *Behavioral and Brain Sciences, 36*(3), 181–204.

Clark, A. (2015). *Surfing uncertainty: Prediction, action, and the embodied mind.* Oxford University Press.

Clark, A., & Chalmers, D. J. (1998). The extended mind. *Analysis, 58*(1), 7–19. http://www.jstor.org/stable/3328150

Cohen, Y., Engel, T. A., Langdon, C., Lindsay, G. W., Ott, T., Peters, M. A. K., Shine, J. M., Breton-Provencher, V., & Ramaswamy, S. (2022). Recent advances at the interface of neuroscience and artificial neural networks. *The Journal of Neuroscience, 42*(45), 8514–8523. https://doi.org/10.1523/JNEUROSCI.1503-22.2022

Da Costa, L., Parr, T., Sajid, N., Veselic, S., Neacsu, V., & Friston, K. (2020). Active inference on discrete state-spaces: A synthesis. *Journal of Mathematical Psychology, 99*, 102447. https://doi.org/10.1016/j.jmp.2020.102447

Da Costa, L. D., Sajid, N., Parr, T., Friston, K., & Smith, R. (2023). Reward maximization through discrete active inference. *Neural Computation, 35*(5), 807–852. https://doi.org/10.1162/neco_a_01574

Damasio, A. (1994). *Descartes' error: Emotion, reason, and the human brain.* Putnam.

Davis, J. L. (2023). Affordances for machine learning. In *Proceedings of the 2023 ACM Conference on Fairness, Accountability, and Transparency* (pp. 1779–1789). ACM. https://doi.org/10.1145/3593013.3594000

Debats, N. B., Heuer, H., & Kayser, C. (2021). Visuo-proprioceptive integration and recalibration with multiple visual stimuli. *Science Reports, 11*, 21640. https://doi.org/10.1038/s41598-021-00992-2

Deng, J., Dong, W., Socher, R., Li, L. J., Li, K., & Fei-Fei, L. (2009). ImageNet: A large-scale hierarchical image database. In I. Essa, S. B. Kang, & M. Pollefeys (Eds.), *2009 IEEE Conference on Computer Vision and Pattern Recognition* (pp. 248–255). IEEE. https://doi.org/10.1109/CVPR.2009.5206848

Descartes, R. (1998). *Discourse on method* (D. A. Cress, Trans.). Hackett Publishing. (Original work published 1637)

D'Mello, S. K., & Graesser, A. C. (2015). Feeling, thinking, and computing with affect-aware learning technologies. In R. A. Calvo, S. K. D'Mello, J. Gratch, & A. Kappas (Eds.), *The Oxford handbook of affective computing* (pp. 419–434). Oxford University Press. https://doi.org/10.1093/oxfordhb/9780199942237.013.032

Donato, E., Lo Preti, M., Beccai, L., & Falotico, E. (2025). *Sensorimotor control strategies for tactile robotics. arXiv.* http://arxiv.org/abs/2501.09468

Duan, J., Yu, S., Li Tan, H., Zhu, H., & Tan, C. (2022). A survey of embodied AI: From simulators to research tasks. *IEEE Transactions on Emerging Topics in Computational Intelligence, 6*(2), 230–244. https://doi.org/10.1109/TETCI.2022.3141105

Eitel, A., Hauff, N., & Burgard, W. (2019). When to start learning: Active learning for human-robot interaction. In F. Arai, K. Lynch, B. Nelson, A. M. Okamura, & H. Wang (Eds.), *2019 IEEE/RSJ International Conference on Intelligent Robots and Systems (IROS)* (pp. 6979–6986). IEEE. https://doi.org/10.1109/IROS40022.2019.8967824

Fodor, J. (1975). *The language of thought.* Thomas Crowell.

Fodor J. (1981). *Representations.* MIT Press.

Franklin, J. C. (2019). Psychological primitives can make sense of biopsychosocial factor complexity in psychopathology. *BMC Medicine, 17*(1), 1–8. https://doi.org/10.1186/s12916-019-1435-1

Conference on Computer Vision and Pattern Recognition (CVPR) (pp. 14249–14258). IEEE. https://doi.org/10.1109/CVPR52688.2022.01391

Krizhevsky, A., Sutskever, I., & Hinton, G. E. (2017). ImageNet classification with deep convolutional neural networks. *Communication of the ACM, 60*(6), 84–90. https://doi.org/10.1145/3065386

Lakatos, L., & Janka, Z. (2008). Az emberi agy és intelligencia evolúciója [Evolution of human brain and intelligence]. *Ideggyogyaszati szemle, 61*(7–8), 220–229.

Lakoff, G., & Johnson, M. (1981). *Metaphors we live by*. University of Chicago Press.

Lakoff, G., & Johnson, M. (1999). *Philosophy in the flesh: The embodied mind and its challenge to Western thought*. Basic Books.

Lanillos, P., Meo, C., Pezzato, C., Meera, A. A., Baioumy, M., Ohata, W., Tschantz, A., Millidge, B., Wisse, M., Buckley, C., & Tani, J. (2021). Active inference in robotics and artificial agents: Survey and challenges. *ArXiv* arXiv:2112.01871.

Levin, M. (2022). Technological approach to mind everywhere: An experimentally-grounded framework for understanding diverse bodies and minds. *Frontiers in Systems Neuroscience, 16*, 768201. https://doi.org/10.3389/fnsys.2022.768201

Levin, M. (2023). Darwin's agential materials: Evolutionary implications of multi-scale competency in developmental biology. *Cellular and Molecular Life Sciences: CMLS, 80*(6), 142. https://doi.org/10.1007/s00018-023-04790-z

Li, F.-F., Fergus, R., & Perona, P. (2007). Learning generative visual models from few training examples: An incremental Bayesian approach tested on 101 object categories. *Computer Vision and Image Understanding, 106*(1), 59–70. https://doi.org/10.1016/J.CVIU.2005.09.012

Li, F.-F., & Krishna, R. (2022). Searching for computer vision North stars. *Daedalus, 151*(2), 85–99. https://doi.org/10.1162/DAED_a_01902

Liao, Y. C., Todi, K., Acharya, A., Keurulainen, A., Howes, A., & Oulasvirta, A. (2022). "Are we there yet?" The role of visual, auditory, and haptic feedback in immersive VR navigation. In A. Schmidt, K. Väänänen, T. Goyal, P. O. Kristensson, A. Peters, S. Mueller, J. R. Williamson, & M. L. Wilson (Eds.), *Proceedings of the 2022 CHI Conference on Human Factors in Computing Systems (CHI '22)* (Article 398). ACM. https://doi.org/10.1145/3491102.3501980

Lin, Y., Lin, F., Cai, G., Chen, H., Zou, L., & Wu, P. (2024). Evolutionary reinforcement learning: A systematic review and future directions. *arXiv* 2402.13296.

Lindquist, K. A., & Barrett, L. F. (2012). A functional architecture of the human brain: Emerging insights from the science of emotion. *Trends in Cognitive Sciences, 16*(11), 533–540. http://dx.doi.org/10.1016/j.tics.2012.09.005

Liu, W., Wang, J., & Chang, S. F. (2012). Robust and scalable graph-based semisupervised learning. *Proceedings of the IEEE, 100*(9), 2624–2638. https://doi.org/10.1109/JPROC.2012.2197809

Macrine, S. L., & Fugate, J. M. (Eds.). (2022). *Movement matters: How embodied cognition informs teaching and learning.* MIT Press.

Manoel, E. D. J., Viana Felicio, P. F., Makida-Dionísio, C., Nascimento Soares, R. D., Freitas, A., & Gimenez, R. (2016). Proprioceptive-visual integration and embodied cognition: A developmental perspective. *Perceptual and Motor Skills, 123*(2), 460–476. https://doi.org/10.1177/0031512516664722

Marocco, D., Cangelosi, A., Fischer, K., & Belpaeme, T. (2010). Grounding action words in the sensorimotor interaction with the world: Experiments with a simulated iCub humanoid robot. *Frontiers in Neurorobotics, 4,* 7. https://doi.org/10.3389/fnbot.2010.00007

Matheson, H., White, N., & McMullen, P. (2015). Accessing embodied object representations from vision: A review. *Psychological Bulletin, 141*(3), 511–524. https://doi.org/10.1037/bul0000001

Mazzaglia, P., Cohen, T., & Dijkman, D. (2023). Uncertainty-driven affordance discovery for efficient robotics manipulation. *arXiv* 2308.14915v1 https://doi.org/10.48550/arXiv.2308.14915

Mazzuca, C., Fini, C., Michalland, A. H., Falcinelli, I., Da Rold, F., Tummolini, L., & Borghi, A. M. (2021). From affordances to abstract words: The flexibility of sensorimotor grounding. *Brain Sciences, 11*(10), 1304. https://doi.org/10.3390/brainsci11101304

McGeer, T. (1990). Passive dynamic walking. *The International Journal of Robotics Research, 9*(2), 62–82. https://doi.org/10.1177/027836499000900206

Mead, C. (1989). Analog VLSI and neutral systems. *NASA STI/Recon Technical Report A, 90,* 16574.

Medina, J., Khurana, P., & Coslett, H. B. (2015). The influence of embodiment on multisensory integration using the mirror box illusion. *Consciousness and Cognition, 37,* 71–82. https://doi.org/10.1016/j.concog.2015.08.011

Meier, F., & Schaal, S. (2016). Drifting Gaussian processes for learning inverse dynamics. In D. Kragic, A. Bicchi, & A. De Luca (Eds.), *2016 IEEE International Conference on Robotics and Automation (ICRA)* (pp. 1374–1380). IEEE. https://doi.org/10.1109/ICRA.2016.7487295

Merleau-Ponty, M. (1962). *Phenomenology of perception* [Colin Smith, Trans.]. Routledge & Kegan Paul. (Original work published 1945)

Miller, B. (2021). Data labeling: What it is and why it matters to improve your machine learning models by creating high-quality data labels. *Exchange: Scale*

Virtual Events Blog. https://exchange.scale.com/public/blogs/data-labeling-what-it-is-and-why-it-matters

Morgenstern, A. (2023). Children's multimodal language development from an interactional, usage-based, and cognitive perspective. *WIREs Cognitive Science, 14*(2), e1631. https://doi.org/10.1002/wcs.1631

Muratore, F., Ramos, F., Turk, G., Yu, W., Gienger, M., & Peters, J. (2022). Robot learning from randomized simulations: A review. *Frontiers in Robotics and AI, 9,* 799893. https://doi.org/10.3389/frobt.2022.799893

Nathan, M. J. (2023). *Embodied cognition: A field guide.* Routledge. https://doi.org/10.4324/9781003102328

Ng, D. T. K., Leung, J. K. L., Chu, S. K. W., & Qiao, M. S. (2021). Conceptualizing AI literacy: An exploratory review. *Computers and Education: Artificial Intelligence, 2,* 100041. https://doi.org/10.1016/J.CAEAI.2021.100041

Nguyen, P. D., Hoffmann, M., Pattacini, U., & Metta, G. (2019). Reaching development through visuo-proprioceptive-tactile integration on a humanoid robot-a deep learning approach. In A. Aly, E. Bicho, & S. Bouabdallah (Eds.), *2019 Joint IEEE 9th International Conference on Development and Learning and Epigenetic Robotics (ICDL-EpiRob)* (pp. 163–170). IEEE. https://ieeexplore.ieee.org/document/8850681

Nicolelis, M. A. (2001). Actions from thoughts. *Nature, 409*(6818), 403–407. https://doi.org/10.1038/35053191

Niedenthal, P. M. (2007). Embodying emotion. *Science, 316*(5827), 1002–1005.

Nilsson, D., Pirinen, A., Gärtner, E., & Sminchisescu, C. (2021). Embodied visual active learning for semantic segmentation. *Proceedings of the AAAI Conference on Artificial Intelligence, 35*(3), 2373–2383. https://doi.org/10.1609/aaai.v35i3.16338

Norman, D. A. (2013). The design of everyday things. Doubleday. (Original work published 1988)

Nygaard, L. C., & Tzeng, C. Y. (2021). Perceptual integration of linguistic and non-linguistic properties of speech. In J. S. Pardo, L. C. Nygaard, R. E. Remez, & D. B. Pisoni (Eds.), *The handbook of speech perception* (2nd ed., pp. 398–427). Wiley-Blackwell. https://doi.org/10.1002/9781119184096.ch15

Oosterwijk, S., Mackey, S., Wilson-Mendenhall, C., Winkielman, P., & Paulus, M. P. (2015). Concepts in context: Processing mental state concepts with internal or external focus involves different neural systems. *Social Neuroscience, 10*(3), 294–307. https://doi.org/10.1080/17470919.2014.998840

Ororbia, A., & Kifer, D. (2022). The neural coding framework for learning generative models. *Nature Communications, 13*(1), 2064. https://doi.org/10.1038/s41467-022-29632-7

Owens, A. P., Allen, M., Ondobaka, S., & Friston, K. J. (2018). Interoceptive inference: From computational neuroscience to clinic. *Neuroscience and Biobehavioral Reviews, 90*, 174–183. https://doi.org/10.1016/j.neubiorev.2018.04.017

Park, S., & Whang, M. (2022). Empathy in human-robot interaction: Designing for social robots. *International Journal of Environmental Research and Public Health, 19*(3), 1889. https://doi.org/10.3390/ijerph19031889

Parma, C., Doria, F., Zulueta, A., Lanzone, J., Boscarino, M., Giani, L., Lunetta, C., Vassallo, M., Parati, E. A., Picozzi, M., & Sattin, D. (2024). An overview of the bodily awareness representation and interoception: Insights and progress in the field of neurorehabilitation Research. *Brain Sciences, 14*(4), 386. https://doi.org/10.3390/brainsci14040386

Paolo, G., Gonzalez-Billandon, J., & Kégl, B. (2024). *A call for embodied AI.* http://arxiv.org/abs/2402.03824

Parr, T., Pezzulo, G., & Friston, K. J. (2022). *Active inference: The free energy principle in mind, brain, and behavior.* MIT Press.

Pezzulo, G., Parr, T., & Friston, K. (2024). Active inference as a theory of sentient behavior. *Biological Psychology, 186*, 108741. https://doi.org/https://doi.org/10.1016/j.biopsycho.2023.108741

Pfeifer, R., & Bongard, J. (2006). *How the body shapes the way we think: A new view of intelligence.* MIT Press. https://doi.org/10.7551/mitpress/3585.001.0001.

Pfeifer, R., & Gomez, G. (2009). Morphological computation: Connecting brain, body, and environment. In B. Sendhoff, E. Körner, O. Sporns, H. Ritter, & K. Doya (Eds.), *Creating brain-like intelligence: From basic principles to complex intelligent systems* (pp. 66–83). Springer. https://doi.org/10.1007/978-3-642-00616-6_5

Pfeifer, R., & Iida, F. (2004). Embodied artificial intelligence: Trends and challenges. In F. Iida, R. Pfeifer, L. Steels, & Y. Kuniyoshi (Eds.), *Embodied artificial intelligence* (Lecture Notes in Artificial Intelligence, LNAI 3139, pp. 1–26). Springer. https://doi.org/10.1007/978-3-540-27833-7_1

Pfeifer, R., & Iida, F. (2006). Morphological computation: connecting body, brain, and environment. In M. H. Coen (Ed.), *Creating brain-like intelligence: From basic principles to complex adaptive systems* (Lecture Notes in Computer Science, Vol. 3853, pp. 69–84). Springer. https://doi.org/10.1007/978-3-642-00616-6_5

Pfeifer, R., & Scheier, C. (1999). *Understanding intelligence.* MIT Press. https://doi.org/10.7551/mitpress/6979.001.0001

Picard, R. (1997). *Affective computing.* MIT Press. https://doi.org/10.7551/mitpress/1140.001.0001

Pulvermüller, F., & Hauk, O. (2006). Category-specific conceptual processing of color and form in left frontotemporal cortex. *Cerebral Cortex, 16*(8), 1193–1201. https://doi.org/10.1093/cercor/bhj060

Putra, R. V. W., Marchisio, A., Zayer, F., Dias, J., & Shafique, M. (2024). Embodied neuromorphic artificial intelligence for robotics: Perspectives, challenges, and research development stack. *ArXiv* 2404.03325.

Raju, H., & Tadi, P. (2022). Neuroanatomy, somatosensory pathways. In *StatPearls*. StatPearls Publishing. https://www.ncbi.nlm.nih.gov/books/NBK554486/

Rizzolatti, G., & Craighero, L. (2004). The mirror-neuron system. *Annual Review of Neuroscience, 27,* 169–192. https://doi.org/10.1146/annurev.neuro.27.070203.144230

Roemmich, K., Schaub, F., & Andalibi, N. (2023). Emotion AI at work: Investigating the fairness, accountability, and transparency of emotion-based job candidate screening. In A. Schmidt, K. Väänänen, T. Goyal, P. O. Kristensson, A. Peters, S. Mueller, J. R. Williamson, & M. L. Wilson (Eds.), *Proceedings of the 2023 CHI Conference on Human Factors in Computing Systems* (Article 241). ACM. https://doi.org/10.1145/3544548.3580790

Rouleau, N., & Levin, M. (2023). The multiple realizability of sentience in living systems and beyond. *eNeuro, 10*(11), ENEURO.0375-23.2023. https://doi.org/10.1523/ENEURO.0375-23.2023

Rosch, E. (1978). Principles of categorization. In E. Rosch & B. B. Lloyd (Eds.), *Cognition and categorization* (pp. 27–48). Lawrence Erlbaum Associates.

Rumelhart, D. E., & Norman, D. A. (1988). Representation in memory. In R. C. Atkinson, R. J. Herrnstein, G. Lindzey, & R. D. Luce (Eds.). *Stevens' handbook of experimental psychology: Perception and motivation* (2nd ed., pp. 511–587). John Wiley & Sons.

Ryle, G. (1949). Descartes' myth. In *The concept of mind* (pp. 11–24). Hutchinson's University Library.

Saeb, A. T. M., & Al-Naqeb, D. (2016). The impact of evolutionary driving forces on human complex diseases: A population genetics approach. *Scientifica, 2016,* 2079704. https://doi.org/10.1155/2016/2079704

Saeed, W., & Omlin, C. (2023). Explainable AI (XAI): A systematic meta-survey of current challenges and future opportunities. *Knowledge-Based Systems, 263,* 110273.

Savell, K. R. R., Auerbach, B. M., & Roseman, C. C. (2016). Constraint, natural selection, and the evolution of human body form. *Proceedings of the National Academy of Sciences, 113*(34), 9492–9497. https://doi.org/10.1073/pnas.1603632113

Saveriano, M., Abu-Dakka, F. J., Kramberger, A., & Peternel, L. (2023). Dynamic movement primitives in robotics: A tutorial survey. *International Journal of Robotics Research, 42*(13), 1133–1184. https://doi.org/10.1177/02783649231201196

Savery, R. J., & Weinberg, G. (2020). A survey of robotics and emotion: History, current practice, and future considerations. In S. Rossi, F. Ferland, & D. Morocco (Eds.), *2020 29th IEEE International Conference on Robot and Human Interactive Communication (RO-MAN)* (pp. 863–870). IEEE. https://doi.org/10.1109/RO-MAN47096.2020.9223594

Serapio-García, G., Safdari, M., Crepy, C., Sun, L., Fitz, S., Romero, P., Abdulhai, M., Faust, A., & Matarić, M. (2023). Personality traits in large language models. *arXiv*. https://doi.org/10.48550/arXiv.2307.00184

Seth, A. K. (2013). Interoceptive inference, emotion, and the embodied self. *Trends in Cognitive Sciences, 17*(11), 565–573.

Shapiro, L. (2010). *Embodied cognition (new problems of philosophy)*. Routledge. https://doi.org/10.4324/9780203850664

Shapiro, L. (2019). *Embodied cognition*. Routledge.

Shen, B., Xia, F., Li, C., Martin-Martin, R., Fan, L., Wang, G., Perez-D'Arpino, C., Buch, S., Srivastava, S., Tchapmi, L., Tchapmi, M., Vainio, K., Wong, J., Li, F.-F.., & Savarese, S. (2021). iGibson 1.0: A simulation environment for interactive tasks in large realistic scenes. *IEEE,* 7520–7527. https://doi.org/10.1109/IROS51168.2021.9636667

Shimada, S. (2022). Multisensory and sensorimotor integration in the embodied self: Relationship between self-body recognition and the mirror neuron system. *Sensors, 22*(13), 5059. https://doi.org/10.3390/s22135059

Šimić, G., Tkalčić, M., Vukić, V., Mulc, D., Španić, E., Šagud, M., Olucha-Bordonau, F. E., Vukšić, M., & Hof, P. R. (2021). Understanding emotions: Origins and roles of the amygdala. *Biomolecules, 11*(6), 823. https://doi.org/10.3390/biom11060823

Smith, R., Parr, T., & Friston, K. J. (2019). Simulating emotions: An active inference model of emotional state inference and emotion concept learning. *Frontiers in Psychology, 10,* 02844. https://doi.org/10.3389/fpsyg.2019.02844

Spezialetti, M., Placidi, G., & Rossi, S. (2020). Emotion recognition for human-robot interaction: Recent advances and future perspectives. *Frontiers in Robotics and AI, 7,* 532279. https://doi.org/10.3389/frobt.2020.532279

Staddon, J. E. R. (1983). *Adaptive behavior and learning.* Cambridge University Press.

Sučević, J., & Schapiro, A. C. (2023). A neural network model of hippocampal contributions to category learning. *eLife, 12,* e77185. https://doi.org/10.7554/eLife.77185

Sumer, O., Goldberg, P., D'Mello, S. K., Gerjets, P., Trautwein, U., & Kasneci, E. (2023). Multimodal engagement analysis from facial videos in the classroom. *IEEE Transactions on Affective Computing, 14*(2), 1184–1200. https://doi.org/10.1109/TAFFC.2021.3127692

Tian, C., Xu, Y., Li, Z., Zuo, W., Fei, L., & Liu, H. (2020). Attention-guided CNN for image denoising. *Neural Networks, 124*, 117–129. https://doi.org/10.1016/j.neunet.2019.12.024

Tolk, A., Barry, P., & Doskey, S. C. (2022). Using modeling and simulation and artificial intelligence to improve complex adaptive systems engineering. *International Journal of Modeling, Simulation, and Scientific Computing, 13*(02), 2241004. https://doi.org/10.1142/S1793962322410045

Turing, A. M. (1950). Computing machinery and intelligence. *Mind, 49*, 433–460.

Valenzo, D., Ciria, A., Schillaci, G., & Lara, B. (2022). Grounding context in embodied cognitive robotics. *Frontiers in Neurorobotics, 16*, 843108. https://doi.org/10.3389/fnbot.2022.843108

van de Maele, T., Verbelen, T., Çatal, O., & Dhoedt, B. (2022). Embodied object representation learning and recognition. *Frontiers in Neurorobotics, 16*, 840658. https://doi.org/10.3389/fnbot.2022.840658

Vong, W. K., Wang, W., Orhan, A. E., & Lake, B. M. (2024). Grounded language acquisition through the eyes and ears of a single child. *Science, 383*(6682), 504–511. https://doi.org/10.1126/science.adi1374

Wahn, B., Schmitz, L., Gerster, F. N., & Weiss, M. (2023). Offloading under cognitive load: Humans are willing to offload parts of an attentionally demanding task to an algorithm. *PloS ONE, 18*(5), e0286102. https://doi.org/10.1371/journal.pone.0286102

Weber, B. H., & Depew, D. J. (Eds.). (2003). *Evolution and learning: The Baldwin effect reconsidered*. MIT Press. https://doi.org/10.7551/mitpress/2871.001.0001

Willems, R. M., & Francken, J. C. (2012). Embodied cognition: Taking the next step. *Frontiers in Psychology, 3*, 582. https://doi.org/10.3389/fpsyg.2012.00582

Wilson-Mendenhall, C. D., Barrett, L. F., Simmons, W. K., & Barsalou, L. W. (2011). Grounding emotion in situated conceptualization. *Neuropsychologia, 49*(5), 1105–1127. https://doi.org/10.1016/j.neuropsychologia.2010.12.032

Wilson-Mendenhall, C. D., Barrett, L. F., & Barsalou, L. W. (2013). Neural evidence that human emotions share core affective properties. *Psychological Science, 24*(6), 947–956. https://doi.org/10.1177/0956797612464242

Yalçın, Ö. N., & DiPaola, S. (2020). Modeling empathy: Building a link between affective and cognitive processes. *Artificial Intelligence Review, 53*, 2983–3006. https://doi.org/10.1007/s10462-019-09753-0

Yang, J., Ren, Z., Xu, M., Chen, X., Crandall, D. J., Parikh, D., & Batra, D. (2019). Embodied aMODAL recognition: Learning to recognize when to look. In K. M. Lee, D. Forsyth, M. Pollefeys, & X. Tang (Eds.), *Proceedings of the 2019 IEEE/CVF*

International Conference on Computer Vision (ICCV) (pp. 9502–9511). IEEE. https://doi.org/10.1109/ICCV.2019.00950

Yeo, B. T., Krienen, F. M., Sepulcre, J., Sabuncu, M. R., Lashkari, D., Hollinshead, M., Roffman, J. L., Smoller, J. W., Zöllei, L., Polimeni, J. R., Fischl, B., Liu, H., & Buckner, R. L. (2011). The organization of the human cerebral cortex estimated by intrinsic functional connectivity. *Journal of Neurophysiology, 106*(3), 1125–1165. https://doi.org/10.1152/jn.00338.2011

Zahra, O., Tolu, S., Zhou, P., Duan, A., & Navarro-Alarcon, D. (2022). A bio-inspired mechanism for learning robot motion from mirrored human demonstrations. *Frontiers in Neurorobotics, 16*, 826410. https://doi.org/10.3389/fnbot.2022.826410

Zhang, D., Yu, Y., Dong, J., Li, C., Su, D., Chu, C., & Yu, D. (2024). *MM-LLMs: Recent advances in multimodal large language models*. https://mm-llms.github.io

Zhang, Z., & Xu, F. (2024). An overview of the free energy principle and related research. *Neural Computation, 36*(5), 963–1021. https://doi.org/10.1162/neco_a_01642

Zhong, J., Cangelosi, A., & Wermter, S. (2014). Toward a self-organizing pre-symbolic neural model representing sensorimotor primitives. *Frontiers in Behavioral Neuroscience, 8*, 22. https://doi.org/10.3389/fnbeh.2014.00022

Zhu, C., Ping, W., Xiao, C., Shoeybi, M., Goldstein, T., Anandkumar, A., & Catanzaro, B. (2021). Long-short transformer: Efficient transformers for language and vision. *Advances in Neural Information Processing Systems, 34*, 17723–17736. https://doi.org/10.48550/arXiv.2107.02192

2 Artificial Selves: Robodiment and Narrative Identity in Robots

Shaun Gallagher

Introduction

Complexity about Notions of Self and Self-Identity

In a recent article, Sohrab Hadeei (2023) provided a detailed and exhaustive review of (seventy) conceptions of self across Western philosophy, sociology, cognitive science, and Islamic and Buddhist thought. Hadeei argues that this is not just complexity, but that across many disciplines there is a significant amount of ambiguity, overlap, and confusion, so that the concept of self remains without a clear definition. He categorizes these theories under a variety of headings: personality, cognitive, neural, experiential, transcendental, narrative, perceptual, conceptual, social, intentional action, personal identity, and ancient to modern conceptions of the soul. The plurality of concepts, and even the idea that they overlap and reflect an ambiguity, can be taken as a problem, or a possible theoretical virtue. My own solution to this "problem" is to treat the complexity as a virtue and to develop a pluralistic approach called the pattern theory of self (Gallagher 2013, 2024).

What we call the self just is a pattern of diverse processes or factors that are dynamically related to each other so as to form a dynamical gestalt. By "pattern" I mean a "real" pattern in the sense defined by Daniel Dennett (1991)—that is, a set of processes with sufficient explanatory force that it can count as a pragmatic scientific concept. Such a pattern is defined as dynamical (Kelso, 1995), or what I call a dynamical gestalt, in terms of interventionist causality (Woodward, 2003): a gestalt or pattern is dynamical when an intervention on any one factor will cause a change in some other factor or factors, or in the whole pattern, and vice versa.

I've proposed a set of ten processes/factors included in the self-pattern, no one of which alone is sufficient to constitute a pattern, and on a general (or metatheoretical) level, no one of which is necessary (Table 2.1).

On the one hand, the pattern theory of self, following a pluralist route, rules out reductionist and deflationary theories of self that would claim that the self is just one thing (a body, a minimal self-experience, a narrative, a social construct, etc.) and nothing else. On the other hand, metatheoretically, none of the processes/factors listed are necessary conditions, although for any particular pattern theory one or more factors may be identified as such. For example, coma may involve the absence of self-awareness and of other capacities, cases of dysnarrativa may involve the absence of a capacity for self-narrative and amnesia may involve the absence of memory, but in each case a sufficient number of other processes continue to operate in the self-pattern for continued identity. On various theories, angels lack bodies, robots lack affectivity, and nonhuman animals lack narrative capacity, but we can still attribute self-identity. In such cases, although the patterns are quite different, they can still be considered self-patterns.

One issue that philosophers tend to worry about is coherency. Given this motley collection of factors, how, precisely, do they hold together so that something like self-identity emerges? These worries are typically framed as questions about continuity, unity of self, and identity over time and are usually discussed with respect to human self-identity. Paul Ricoeur (1992), for example, addresses this issue by distinguishing between *idem* identity (that which remains relatively the same persisting across time) and *ipse* identity (that which makes me who I am in relation to another, even as I change over time). These are two different problematics addressing two different questions. As Marya Schechtman (1996) puts it, this is the distinction between the question of reidentification (what makes a person the same from one time to another? [i.e., *idem* identity]) and the question of characterization (what makes a person the person she is? [*ipse* identity]).

Some philosophers, including Schechtman (1996), argue for bodily continuity as the criterion of reidentification (*idem*). To avoid some controversy, we need to assume that the concept of bodily continuity, which clearly involves some degree of physical continuity, is not reducible to physical continuity. In most general terms, a traditional solution, going back to Aristotle, is to think of the body in terms of form and function. In more specific terms that apply to animals and humans, we should think of

Table 2.1
Elements of the self-pattern

Elements of the pattern	Brief description
Bodily processes	Includes core biosystemic and autopoietic processes related to motoric, autonomic, endocrine, enteric, immune, and interoceptive functions, allowing the overall system to maintain the homeostasis necessary for survival and to distinguish between itself and what is not itself.
Prereflective experiential processes	Includes prereflective self-awareness, a structural feature of first-person consciousness constrained by bodily factors; the sense of ownership (mine-ness); and the sense of agency, which can involve various sensorimotor modalities, such as proprioception, kinesthesia, touch, and vision. These aspects form the experiential core of what is sometimes called the minimal self.
Affective processes	The fact that someone manifests a certain temperament or emotional disposition reflects a particular mix of affective factors that range from very basic and mostly covert or tacit bodily affects (e.g., hunger, fatigue, libido) to what may be a typical emotion pattern, a set of existential feelings, or a background mood.
Behavioral/action processes	Behaviors and actions make us who we are; behavioral habits and skills reflect, and perhaps actually constitute, our character. This is a classic view that goes back at least to Aristotle.
Social/intersubjective processes	Humans (and possibly some nonhuman animals) are born with a capacity for attuning to intersubjective existence: at a certain point in social relations a more developed self-conscious recognition of oneself as being distinct from others, a sense of self-for-others, and a sense of being part of a group or community.
Cognitive/psychological processes	These are aspects emphasized in traditional theories of personal identity, highlighting psychological continuity and memory, including one's conceptual understanding of oneself, beliefs, and cognitive dispositions, as well as personality traits.
Reflective processes	The ability to reflect on one's experiences and actions—closely related to notions of autonomy and moral personhood, including the capacity to reflectively evaluate and form second-order volitions about one's desires.

(*continued*)

Table 2.1
(continued)

Elements of the pattern	Brief description
Narrative processes	Self-interpretation has a narrative structure and recursively reflects (and often reinforces) the self-pattern. In some theories, selves are inherently or constitutively narrative entities.
Ecological processes	We tend to identify ourselves with our stuff—physical pieces of property, clothes, homes, and various things that we own; the technologies we use; the institutions we work in; and so on. Our embodied-situated actions engage with (and sometimes incorporate) artifacts, instruments, and bits and structures of the environment in ways that define us and scaffold our identities. Situations shape who we are, and affordances define our possibilities.
Normative processes	Our extensive engagement with the environment includes social and cultural practices. These are not just what we do but involve what we ought to do and obligations that we keep or not. Constraints (and sometimes well-defined roles) imposed by social, cultural, and institutional factors shape our habitual behaviors and our self-conceptions of who we are and who we think we should be.

Source: Gallagher, 2024

it in terms of the continuity of a unique living body, understood as a bodily existence that fits with one's agentive engagement in the world. This allows us to accommodate a point made by Ricoeur: that character involves both *ipse* and *idem* identity—the latter because one's character (character traits, habits) tends to be long-term and less subject to change; and character is expressed in one's (bodily) actions.

Simply stated, the agentive body operates as the principle of *idem* individuation, a kind of ontological anchor that allows us to say this is the same person from one time or context to another, even when some other aspects of the self-pattern change. In human and biological organisms, the idea may require a more nuanced analysis since bodily existence also changes, even as it maintains continuity over time; but for our purposes, we can ignore this issue.

To answer the characterization question of *ipse* identity—what makes this person the person who she is? (which is the traditional question of personal identity)—both Ricoeur and Schechtman point to narrative. Who

I am is some combination of the stories I tell about myself plus the stories that others tell about me. So, narrative can bring some degree of coherence to one's life; in self-narrative, we say, "This is who I am." Sometimes I say this in my self-narrative about my actions, and sometimes I say it in my actions themselves.

One complication in Schechtman's analysis is that in her discussion of narrative identity, she drifts into a discussion of a different question: What is it that constitutes a person or self? As we noted, she clearly takes narrative to be addressing the characterization question of *ipse* identity: What is it that makes a person who he is, even as he changes over time? This is a question about a person's character and personal identity, and Schechtman's view is that "a person creates his identity by forming an autobiographical narrative—a story of his life" (1996, p. 93). In contrast, the question about *self-constitution* (What is it that makes some entity a self or person?) is different. Her answer to the self-constitution question, however, is, once more, narrative. This makes a very strong claim about narrative, and indeed, she calls her position the "narrative self-constitution view": "At the core of [the self-constitution] view is the assertion that individuals constitute themselves as persons by coming to think of themselves as persisting subjects who have had experience in the past and will continue to have experience in the future, taking certain experiences as theirs" (Schechtman, 1996, p. 94). Here, then, are the three questions that Schechtman distinguishes:

1. *Idem*/reidentification: What makes someone the same person over time? Answer: bodily identity.
2. *Ipse*/characterization: What is it that makes a person who she is, even as she changes over time? Answer: narrative identity.
3. Self-constitution: What makes an individual a self/person (rather than a nonself)? Answer: narrative identity.

The strong claim that narrative answers the self-constitution question, however, has been challenged, for example, by Galen Strawson (2004): He argues that one can experience one's life as more episodic (moment to moment) than narrative (as a coherent story); despite the absence of self-narrative, this subject is still a person. We could arrive at a similar conclusion if we consider disorders that involve dysnarrativa, as in Korsakoff syndrome. The fact that a Korsakoff patient has profound trouble with episodic memory and cannot form a self-narrative does not mean that he is no longer a person. Self-narrative is not a necessary/ontological or constitutive condition of selfhood.

Accordingly, one can accept that bodily existence of a specific type answers the question about *idem* identity. The capacity for self-narrative answers the *ipse* question. With respect to the question of self-constitution, however, rather than narrative, a better answer is the self-pattern. What constitutes a self, what a self is, is just the self-pattern of the different and diverse processes of the dynamical gestalt that includes bodily processes and narrative capacity, and a number of other factors that constitute the self-pattern.

This is a complex but coherent answer to questions about self-identity in humans. A fuller account of the human self would involve a much more detailed analysis of all the factors involved in the self-pattern, which is beyond the scope of this chapter. In the remainder of the chapter, I want to ask whether and how far this analysis applies to robots.

Idem Identity in Robots

With respect to *idem* identity, since robots (so far) do not have living biological bodies, we need to return to the traditional answer that focuses on form and function. For Aristotle, form and function go together. The actual shape of a body determines its possible functions and the possible actions it can undertake. Form, with respect to the possibility of embodied action, however, is not just *morphe* or shape; it includes a principle of kinesthetic self-movement, a kind of dynamical *energeia* that enables the body to do something, and in terms of our analysis, relates to other processes of the self-pattern. In this respect, a body that is neither a living organism nor capable of self-movement would not serve the purpose for *idem* identity. A paralyzed body, as in locked-in syndrome, is nonetheless a living body and as such can be the basis of attributions of *idem* identity. A body that is not a living organism and is incapable of self-movement may have the identity of a thing but not a self-identity. Likewise, a body that is capable of movement but that functions in isolation from dynamical interactions with other processes that form a self-pattern would also not provide for *idem* identity.

I propose the following definition of *robodiment*, which can provide for idem identity: a robot body that is physically continuous, is capable of self-movement, and is dynamically capable of intelligible behavior (performing a function that may involve deciding among alternatives but could also be a kind of preprogrammed action). A robot that is properly "robodied" converts energy into action due to, and constrained by, its design. Accordingly, it would have a proper function. In this view my robotic vacuum cleaner has a kind of *idem* identity. We can understand intelligible behavior to mean an articulated instrumental behavior involving means and end. My robotic vacuum cleaner not only moves in a somewhat systematic way but also cleans the carpet, avoids falling down the stairs, moves in a different direction when it encounters an obstacle, finds its way "home" when it is low on energy, and empties the dirt into another container.

We can attribute *idem* identity to a robot that is robodied in this way. That is, on this basis we could reidentify the robot. To be clear, robodiment also includes physical continuity (sameness of body). Otherwise, one might object that robodiment is not sufficient for *idem* identity since we could create another robot (call it robot 2) to the same specification as robot 1. Robot 2 would not be physically continuous with robot 1, however. Someone might change out my robotic vacuum cleaner for an identical model without my knowing. I might then think that robot 2 is robot 1, sharing *idem* identity, but I would be mistaken. *Robodiment*: a robotic body that is physically continuous, capable of self-movement, and capable of intelligible behavior.

We'll return to the question of *idem* identity again to address Ricoeur's contention that *idem* identity contributes to the definition of character.

Ipse Identity in Robots

Perhaps the more interesting (and difficult) question concerns *ipse* identity. Is a robot capable of self-narrative? It can certainly have a history and a set of accomplished actions, and perhaps it has tasks it needs to accomplish. In this respect one can think that it certainly has some content that could be included in a self-narrative. The important question, then, is whether it has the capacity for self-narrative.

Self-narrative depends on at least four capacities:

Temporal Integration

Narrative involves a twofold temporal structure: (1) a time frame that is internal to the narrative itself, a serial order in which one event follows another, allowing for the composition of narrative structure, and (2) an external perspectival time frame that defines the narrator's temporal relation to the events of the narrative—that is, whether they happen in the past, present, or future relative to the narrator's present narrative. It's possible that if some set of narrated events never actually happened and never will (as in the case of fictional events), then they may not have a specifiable place in time relative to the narrator. But this is not the case for *self*-narrative. Even if the event in question never happened (e.g., a falsely remembered event) or never will happen (e.g., a planned event that is never actualized), it is still set in a temporal relation to the narrator in a self-narrative. Someone incapable of maintaining a normal perspectival frame of reference is unable to generate a self-narrative properly specified in its temporal framework.

First-Person Self-Reference

To begin to form a self-narrative, one must be able to refer to oneself by using the first-person pronoun. This, in turn, depends on a primitive embodied sense of differentiation between self and nonself. This self/nonself distinction is often characterized as an experiential differentiation that can be found even in early infancy in the experience of being moved versus moving oneself. It's implicit in proprioceptive and kinesthetic experience, and without the basic differentiation between self and nonself I would not be able to refer to myself with any specification, and self-narrative would have no starting point. Likewise, one's experience of agency for one's actions depends on the self/nonself distinction. In narrative, according to Ricoeur (1992), a character is either (or alternatively) an agent or a sufferer. In self-narrative the construal of the character (the narrated self) as agent or sufferer depends on the ability of the narrator to self-attribute action. This very basic, prereflective self-reference is sometimes referred to as the "minimal self" (Gallagher, 2000).

Episodic-Autobiographical Memory

Episodic autobiographical memory depends on both the capacity for temporal integration and the capacity for first-person self-reference, since the recollection of the specific time in the past when an event took place and self-attribution of experiencing or enacting that event are involved. According to many philosophers, starting with Locke (1690/1979), just such memories form the basis of personal identity, and it's clear that narratives recount past autobiographical events. Accordingly, if someone is unable to form memories of their life history or unable to access such memories, they will have nothing to interpret, nothing to narrate that would be sufficient for the formation of self-identity.

The importance of episodic memory for the construction of narrative is also recognized in neuropsychology. Maguire et al. (1999) point out that the successful use of stories to convey and acquire information depends on two factors: that the story makes sense and that the person who hears the story has access to prior knowledge. The coherence of the story depends on these factors. In the construction of self-narrative, autobiographical memory provides the prior knowledge out of which the coherent narrative is formed. The self-narrative process is not simply something that depends on the proper functioning of autobiographical memory but in fact contributes to the functioning of that memory.

Reflective Metacognition
The process of cognitive discrimination that shapes episodic-autobiographical memories into a narrative structure depends on the capacity for reflective metacognition. Self-narrative does not represent every small detail of life; it's selective, and this selectivity, which discerns an event in context, creates significance. To form a self-narrative, to be a self-interpreting entity, one needs to do more than simply remember life events. One needs to consider them and reflectively evaluate them in regard to how they fit together semantically.

We can ask to what extent it is possible to design a robot that could fulfill these capabilities.

Temporal Integration To the extent that robotic systems can be modeled on predictive processing and active inference, based on priors and prediction, there is already an intrinsic temporality included (Hohwy et al., 2016. "On this Bayesian picture, the perceptual process is a matter of unconsciously keeping track of priors and of generating expected inputs (predictions) given the best performing prior hypothesis" (Hohwy et al., 2016). This results in a constant protentional (anticipatory) process that can be regarded as a mechanism of self-reference that forms the background for events to be noticed by the robot as disruptions of the sensorimotor dynamic. Implicit in this kind of temporality, one can see contrasts between what is present (the current state of robot and environment), past (prior), and future (prediction) and the dynamics that keep the process going. It's a process that necessarily draws from the agent's past, since it is prior knowledge and prior experiences that guide its predictions about its ongoing action, which in turn participate in its goal selection and decision-making. The robot is attuned not only to the present but to an extended time ordered from the perspective of the present.

First-Person Self-Reference Chatila et al. (2018) argue that it's possible for a robotic agent to discriminate between self and nonself—distinguishing between itself and its environment. Over twenty-five years ago Jun Tani (1998) provided a hybrid design that allows for this capacity in robots using recurrent neural networks. Specifically, the robot (re)registers its own state when prediction error increases. Prediction error generated by a mismatch between the robot's forward model of its current action and its intended goal registers a distinction that in the human would be manifested in a sense of enactive self-agency (Lanillos & Cheng, 2018). In the human this depends on nonconscious processes that register the congruence between an intention or action goal and the anticipated outcome of bodily movement. In the robot these processes remain nonconscious and do not entail a conscious sense. Tani and White (2022) argue that these dynamics can be modeled on the free energy principle (FEP) (Friston, 2005). This view also indicates the importance of the dynamical relations that exist between this minimal self-registration, the robot's behavior, and ecological factors in the self-pattern. Enactive accounts refer to such relations in terms of circular or dynamical causality (Fuchs, 2020; Gallagher, 2024). Tani and White offer the following characterization: "Circular causality describes the embedded and embodied agent's situation. An agent acts on the world, and a sequence of causes and effects returns to the original cause, possibly altering it, whereby another sequence is produced in an ongoing interactive feed-back loop" (Tani & White, 2022, p. 93).

> They provide the example of a ball-handling humanoid robot. When a prediction error is generated for the ball position, the forward model is updated "in the direction of minimizing the error." This process generates "the next-step prediction of proprioception which, in the case of tracking a ball, turns out to be new target joint angles. These angles are fed into the robot motor controllers. Then, both hands of the robot move to push the ball further, for example." (Tani & White, 2022, p. 93)

They provide further specifications about the dynamic instability and unpredictability of robot-environment relations that keeps this process going and that is important for characterizing the autonomy of the robot.

Episodic/Autobiographical Memory A number of studies have investigated the expression of past events in narratives based on acquired autobiographical memories in robots (Dias et al., 2007; Sieber & Krenn, 2010; Syrdal et al., 2014). The results of memory, evaluation, and action selection processes involved in the robot's self-controlled behavior (Doya et al., 2002; Kaplan, 2015; Krichmar, 2018; Kuniyoshi & Sangawa, 2006) can be retained and put to use in the construction of episodic memory. Moulin-Frier et al. (2017) present an architecture of autobiographical memory that collects long-term information (days, months, years) about interactions and stores data (e.g., object locations, human presence), as well as short-term information (during action or interaction) "which can be used by reasoning modules focusing on the action itself" (p. 8): "A narrative structure learning module can then use language processing to build a narrative, formatting stories in term of initial states, goal states, actions and results" (Moulin-Frier et al., 2017, p. 9).

Reflective Metacognition Likewise, a pragmatic understanding of the environment (in terms of being able to move around and do things) by an agent may also require some form of reflective self-monitoring. Chatila et al. (2018) propose that two reflective elements are necessary for self-monitoring: (1) self-evaluation, which involves the evaluation of the abilities learned by the robot and a judgment of when to use them, and (2) metareasoning, a form of deliberation on one's own reasoning driven by basic motivations. Self-monitoring is generated by the implementation of a metacontroller that monitors the activity of the system and predicts its effects through the coordination of different strategies for decision-making and reinforcement learning. These are clearly reflective processes that arguably already depend on a self/nonself differentiation and on the capacity for temporal integration.

Assuming that these capacities can be instantiated in a robotic architecture, this should be sufficient to allow for the development of self-narrative. This aligns with the view of Tani and White (2022). They suggest that to the extent that the robot can discriminate among contradictory instructions, treating them as having distinguishable pragmatic meanings, they are in effect constructing a narrative: "If the robots are tutored with a set of sensory flows consisting of inconsistent primitive sequences like A →B → C, A → B →A, and A→C→ B, these sensory flows are learned as compositions of reusable primitives with each of them segmented. We presume that the cognitive competency of compositionality for segmenting the continuous sensory flow into a set of reusable primitives and composing and decomposing a whole action sequence using these primitives can provide a basis for development of the narrative self" (Tani & White, 2022, p. 95).

The notion of compossibility derives from Ricoeur's (1984) notion of a dialectic between "prefiguration," anticipating how the world is (a generative model), and "refiguration" (adjustments made to prediction errors). Tani and White contend that compositionality enables the robot to simulate future actions and to "rehearse what has been experienced in the past to prepare for the uncertain future" (p. 95). They also suggest ways to translate this into linguistic expression, although clearly the advent of large language models (LLMs) may make this easier.

Conclusion

As Ricoeur explains, because a person's character (which is putatively captured in their self-narrative) depends on a person's embodied actions, and on their own understanding of their actions, character requires both *idem* and *ipse* identity and depends on both bodily and narrative processes. Applied to robots, minimal self-other discrimination, which is defined in terms of robodiment (which includes the capacities for self-movement and intelligible behavior), is intertwined with the capacity of forming a self-narrative. Engaging in actions and being able to narrate and evaluate those actions define a particular formation of character. From a more comprehensive view of the self-pattern, however, embodiment and narrative, if essential, are not sufficient for either *idem* or *ipse* identities, and this applies to both humans and robots. The living body does not exist on its own, independently of other processes in the self-pattern; nor does self-narrative. They are both intertwined with the other processes that make up the self-pattern such that to do justice to any account of body or narrative, and thus of *idem* or *ipse* identity, one needs to take into consideration their dynamical relations with ecological, social, and normative factors, as well as cognitive and affective processes.

Should we consider attributing selfhood or some form of self-identity to a robot? On the pattern theory of self, a pattern of (self-relevant) processes/factors allows for the attribution of self. In this regard, we can say that for the robot, the self-pattern consists of more than bodily (physical) processes (which we've called robodiment) and behavioral/action-related processes that account for the possibility of minimal self-reference. Even for the robot the self-pattern includes cognitive processes (e.g., memory) and the capacity for reflective (metacognitive) evaluation—factors that contribute to the capacity for self-narrative. Cognitive architectures can integrate and dynamically coordinate these core cognitive abilities. For example, roboticists recognize the requirement for robots to maintain perceptual continuity, which involves making information acquired earlier usable at a later time. This can be achieved through the use of memory modules. This enables the robotic agent to show flexible and context-sensitive behavior, prospectively selecting the actions required to achieve given goals (Incao, 2022).

I have not discussed other factors of the self-pattern, such as affectivity, normative factors, and social/relational processes. Although these topics go beyond the scope of this paper, I note that much of the discussion in Tani and White (2022) and Moulin-Frier (2022), for example, involve a broad social context of human-robotic interaction as the basis for learning and the possibility of self-narrative in robots. In addition, once one starts to think of the possible self-evaluation that is involved in forming character, affect (at the very least, aspects of cognitive appraisal and action tendencies involved in affect [see Dias et al. 2017]) and normativity are not very far behind.

References

Chatila, R., Renaudo, E., Andries, M., Chavez-Garcia, R. O., Luce-Vayrac, P., Gottstein, R., Alami, R., Clodic, A., Devin, S., Girard, B., & Khamassi, M. (2018). Toward self-aware robots. *Frontiers in Robotics and AI, 5*, 88. https://doi.org/10.3389/frobt.2018.00088

Dennett, D. C. (1991). Real patterns. *The Journal of Philosophy, 88*(1), 27–51. https://doi.org/10.2307/2027085

Dias, J., Paiva, A., Aylett, R., & Woods, S. (2007). A model of emotions for embodied agents. In A. Paiva, R. Prada, & R. W. Picard (Eds.), *Affective computing and intelligent interaction (ACII 2007)* (Vol. 4738). (pp. 590–601). Springer. https://doi.org/10.1007/978-3-540-74889-2_64

Doya, K., Samejima, K., Katagiri, K. I., & Kawato, M. (2002). Multiple model-based reinforcement learning. *Neural Computation, 14*(6), 1347–1369. https://doi.org/10.1162/089976602753712972

Friston, K. (2005). A theory of cortical responses. *Philosophical Transactions: Biological Sciences, 360*(1456), 815–836. https://doi.org/10.1098/rstb.2005.1622

Fuchs, T. (2020). The circularity of the embodied mind. *Frontiers in Psychology, 11*, 1707. https://doi.org/10.3389/fpsyg.2020.01707

Gallagher, S. (2013). A pattern theory of self. *Frontiers in Human Neuroscience, 7*(443), 1–7. https://doi.org/10.3389/fnhum.2013.00443

Gallagher, S. (2024). *The self and its disorders*. Oxford University Press.

Gallagher, S. (2000). Philosophical conceptions of the self: Implications for cognitive science. *Trends in Cognitive Science, 4*, 14–21. https://doi.org/10.1016/S1364-6613(99)01417-5

Hadeei, S. (2023). *The problems with the self: A conceptual analysis*. Authoria Preprints. https://doi.org/10.31124/advance.22592017.v2 .

Hohwy, J., Paton, B., & Palmer, C. (2016). Distrusting the present. *Phenomenology and the Cognitive Sciences, 15*, 315–335. https://doi.org/10.1007/s11097-015-9439-6

Incao, S. (2022). *Insights on the self from a cognitive robotics perspective* [doctoral dissertation, University of Genoa]. Genoa University Institutional Archive. https://iris.unige.it/handle/11567/4792987

Kaplan, D. M. (2015). Moving parts: The natural alliance between dynamical and mechanistic modeling approaches. *Biology and Philosophy, 30*(6), 757–786. https://doi.org/10.1007/s10539-015-9499-6

Kelso, J. A. S. (1995). *Dynamic patterns*. MIT Press.

Krichmar, J. L. (2018). Neurorobotics: A thriving community and a promising pathway toward intelligent cognitive robots. *Frontiers in Neurorobotics, 12*, 42. https://doi.org/10.3389/fnbot.2018.00042

Kuniyoshi, Y., & Sangawa, S. (2006). Early motor development from partially ordered neural-body dynamics: Experiments with a cortico-spinal-musculo-skeletal model. *Biological Cybernetics, 95*(6), 589–605. https://doi.org/10.1007/s00422-006-0127-z

Lanillos, P., & Cheng, G. (2018). Neural-cognitive architecture for robot body learning: From development to deployment. In C. Balaguer, H. Asama, D. Kragic, & K. H. S. Wang (Eds.), *2018 IEEE/RSJ International Conference on Intelligent Robots and Systems (IROS)*. (pp. 1–8). IEEE. https://doi.org/10.1109/IROS.2018.8593881

Locke, J. (1979). *An essay concerning human understanding* (P. H. Nidditch, Ed.). Oxford Clarendon. (Original work publisher 1690)

Maguire, E. A., Frith, C. D., & Morris, R. G. M. (1999). The functional neuroanatomy of comprehension and memory: The importance of prior knowledge. *Brain, 122*(10), 1839–1850. https://doi.org/10.1093/brain/122.10.1839

Moulin-Frier, C. (2022). *The ecology of open-ended skill acquisition*. Artificial Intelligence [cs.AI]. Université de Bordeaux. https://inria.hal.science/tel-03875448/document.

Moulin-Frier, C., Fischer, T., Petit, M., Pointeau, G., Puigbo, J. Y., Pattacini, U., Low, S. C., Camilleri, D., Nguyen, P., Hoffmann, M., Chang, H. J., Zambelli, M., Mealier, A. L., Damianou, A., Metta, G., Prescott, T., Demiris, Y., Dominey, P. F., & Verschure, P. (2017). DAC-h3: A proactive robot cognitive architecture to acquire and express knowledge about the world and the self. *IEEE Transactions on Cognitive and Developmental Systems, 10*(4), 1005–1022. https://doi.org/10.1109/TCDS.2017.2754143

Ricoeur, P. (1984). *Time and narrative* (Vol. 1, K. McLaughlin & D. Pellauer, Eds). University of Chicago Press.

Ricoeur, P. (1992). *Oneself as another* (Kathleen Blamey, Trans.). University of Chicago Press.

Schechtman, M. (1996). *The constitution of selves*. Cornell University Press.

Sieber, G., Krenn, B. (2010). Towards an Episodic Memory for Companion Dialogue. In: Allbeck, J., Badler, N., Bickmore, T., Pelachaud, C., Safonova, A. (Eds.) Intelligent Virtual Agents. IVA 2010. *Lecture Notes in Computer Science: Vol 6356*. Springer. https://doi.org/10.1007/978-3-642-15892-6_33

Strawson, G. (2004). Against narrativity. *Ratio, 17*(4), 428–542.

Syrdal, D. S., Dautenhahn, K., Koay, K. L., & Ho, W. C. (2014). Views from within a narrative: Evaluating long-term human–robot interaction in a naturalistic environment using open-ended scenarios. *Cognitive Computation, 6*(4), 741–759. https://doi.org/10.1007/s12559-014-9284-x

Tani, J. (1998). An interpretation of the "self" from the dynamical systems perspective: A constructivist approach. *Journal of Consciousness Studies, 5*(5–6), 516–542.

Tani, J., & White, J. (2022). Cognitive neurorobotics and self in the shared world, a focused review of ongoing research. *Adaptive Behavior, 30*(1), 81–100. https://doi.org/10.1177/1059712320962158

Woodward, J. (2003). *Making things happen: A Theory of causal explanation*. Oxford University Press.

3 The Thirty Years' War on Representations and the Active Inference Westphalia

Axel Constant, Andy Clark, and Karl Friston

Introduction

Over the last thirty years, representationalist and dynamicist positions in the philosophy of cognitive science have argued over whether neurocognitive processes should be viewed as representational or not. In recent years, an enactive generalization of predictive processing—known as active inference—has been proposed as a unifying theory of brain functions. Since then, active inference has fueled representationalist and dynamicist campaigns. However, we believe that when diving into the formal details of active inference, one should be able to find a solution to the war.

This chapter shows how both representationalist and dynamicist positions can peacefully coexist within the new territory of active inference. We aim to show that the concept of generative models as applied to the brain under active inference accommodates a representationalist and a dynamicist (aka nonrepresentational) view of cognition. We show that the architecture or configuration of neuronal pathways under a (Markovian) generative model (for discrete state spaces) can—and, generally speaking, will—realize both representational and nonrepresentational processes.

Thirty Years of Representation Wars

1980s Connectionism

Until the 1990s, driven by advances in computer science, the philosophy of cognitive science was dominated by cognitivism and connectionism (e.g., Churchland, 1989; Fodor, 1975). Connectionism was presented as a first attack on cognitivism—cognitivism being an attempt at understanding the brain as a logical symbol-manipulating system. For connectionists, the brain should not be studied as a symbol-manipulating system but rather, consistent with the brain's actual neurophysiology, as a set of hierarchically deployed neural networks. The spirit of connectionism is still very much alive today, such as in deep learning research (for a review, see LeCun et al., 2015).

Both cognitivism and connectionism deal with a view of cognition as a problem-solving activity. And both paradigms have typically invoked some notion of representation, with the main difference being whether the representations had symbol-level content or something softer, something contentful yet "sub-symbolic" (for extensive discussion, see Clark, 1989; 1993).

Although these are different types of representations, each involving different criteria, a cognitive process will—for the purposes of this chapter—be deemed representational whenever that process can be said to fulfil the following sufficient conditions (see Hutto & Myin, 2013; Siegel, 2010):

1. The cognitive process is about something else (aka aboutness).
2. The cognitive process has satisfaction conditions with respect to the thing it is about.

1990s Dynamicism

The 1990s marked the rise of embodied views in cognitive science, such as enactivism (Varela et al., 1991) and radical embodied cognition (Chemero, 2009). Embodied approaches were motivated by developments in the field of dynamical system theory, which casts cognitive systems as coupled quantitative variables, mutually changing interdependently over time (Beer, 2000; Thelen & Smith, 1996; Smith & Thelen, 2003; Van Gelder, 1995), with one variable being the organism and the other being the environment. Dynamicism has been driven by two main criticisms of much previous work (Thompson, 2007):

1. Since the brain is embodied, we cannot abstract cognition from the body and consequently from the environment.
2. Since representationalism posits the mediation of the world and cognition by the mental manipulation of representations, representationalism cannot genuinely acknowledge embodiment.

Therefore, for these kinds of dynamicists (see Clark [1997] for a more liberal approach) we should reject the representational view of cognition altogether. Instead, cognition should be viewed as a process of self-organization among the components of the biological system performing the cognitive activity. These components include the brain (internal states) and the body and the environment (external states). In this view, cognition is a homeostatic and allostatic process of attunement to cope with environmental perturbations—a process of "coping, not computing."

2000s Active Inference Westphalia?

At the turn of the millennium, based on a Helmholtzian view of embodied perception, the theory of active inference was introduced as a realization of the free energy principle (FEP) (Friston, 2010; Friston et al., 2006, 2016). This enactive generalization of predictive processing marked a paradigm shift in cognitive science: active inference became a potential candidate to meet the challenge of the grand unification of neurocognitive functions (Clark, 2013). Since then, many enthusiasts have leveraged active inference to attempt explanations of the underlying computational processes of biobehavioral functions such as action, perception, learning, attention, memory, decision-making, emotions, planning and navigation, visual foraging, communication, social learning, and many more (Badcock et al., 2019; Constant, Ramstead, et al., 2018; Feldman & Friston 2010; Friston & Frith, 2015; Friston et al., 2016; Joffily & Coricelli, 2013; Kaplan & Friston, 2018; Mirza et al., 2016; Parr & Friston, 2017).

In line with much of Bayesian statistics, active inference claims that the brain is fundamentally in the business of finessing a generative model of the causes of its sensations—as if the brain was a scientist, trying to infer the causal architecture of its own relation to its world. Put another way, under active inference, the brain is a dynamical system that models the action-relevant causal structure of its coupling with the other dynamical system that embeds it—the body and the environment (i.e., the system generating its sensations). The mathematical formalism of active inference describes neuronal dynamics as a gradient flow that optimizes the evidence for a generative model of the lived world. In this view, neuronal networks embodied by the brain form a set of nodes (modeling hidden states) and edges (modeling conditional dependencies) of a probabilistic (Bayesian) graphical model.

But active inference itself soon became contested ground too. The cognitivist campaign claimed that "brains as generative models" were "rich and reconstructive, detached, truth-seeking inner representations" of the world (Hohwy, 2013, p. 58; see also Hohwy, 2016); others resisted by claiming that generative models were manifest as transient webs of neuronal coupling that are cost-efficient and sometimes (though not always) freed from heavy-duty manipulation of internal representations—generating actions that exploited environmental opportunities by weaving themselves closely to the opportunities provided by the body and the world (Clark 2013, 2016).

Representational Pathways in Active Inference

Active inference assumes that the brain entails a causal model of the world (aka a generative model), whose structure represents the components involved in the cognitive function of interest, as well as the dynamics that realize that cognitive function. Formally, these components and dynamics are expressed as a Bayesian graphical model, with nodes and edges representing the dynamic relations among components, and the structure of which is assumed to map onto the neuroanatomy of neuronal systems realizing any cognitive function. The cognitive function, then, is realized by these dynamics—which play the role of neuronal messages passing in the service of belief updating (i.e., inference) that underwrites the cognitive function in question (Friston et al., 2017).

The representational interpretation of active inference is employed to study cognitive functions that rely on dynamics and components of the generative models that involve the internal manipulation of representational content (e.g., beliefs about hidden states of the world,[1] including one's body and physiology) (Hohwy, 2013, 2019). The motivation for appealing to representational generative models to explain perception and action in active inference stems from the inverse nature of the dual inference problems our brains solve (i.e., figuring out what causes what before inferring what caused that):

1. *Perceptual problem:* The brain does not have direct access to causes of sensations, nor is there a stable one-to-one mapping between causes

and sensations. For instance, a sensory input (e.g., red sensation) may be caused by multiple fluctuating causes (e.g., red jacket, red car, red traffic light). In the philosophical literature, this problem is sometimes referred to as the black box, seclusion, or solipsism problem (Clark, 2013; Hohwy, 2016; Metzinger & Wiese, 2017).

2. *Action planning problem:* All the brain can work with are the sensory inputs it receives. If we are to engage adaptive action, we must not only infer the causes of our sensations (i.e., forming a sufficiently veridical perception—or conception—of the world in which we currently find ourselves), but we must also predict the consequences of engaging in this or that action in the future. In the philosophical literature, this problem is sometimes referred to as the problem of mere versus adaptive active inference (Bruineberg et al., 2016; Kirchhoff et al., 2018) and requires action planning (planning as inference in machine learning (ML)).

This means that under active inference, agents like us must find a solution to infer, in an ill-posed setting, the nature of the cause of our sensations and to infer what action will lead to outcomes that are consistent with our model of the lived world. Under active inference, perception and action are explained as solutions to these inverse problems.

Perception

The problem of indirect perception can be approached as follows. Consider a sensory outcome O generated by a hidden state of the world (S). Taken together, these can be viewed as forming a joint probability distribution ($P(S,O)$). The only quantity to which the brain has access is a sensory consequence, not its cause. To perceive things, the brain must reconstruct the hidden state or cause (S), or rather its posterior probability (i.e., the probability of the cause), after observing the sensory datum $P(S|O)$.

To infer this posterior probability, the brain learns the causal (i.e., generative) model of the way the world caused the sensation. "Learning" here is a technical term. It refers to the optimization of the parameters of a model—here, the generative model. The brain learns the parameters of hidden states causing sensory outcomes, about which the brain may have prior beliefs. These prior beliefs are part of the generative model entailed by the brain. Hence, a generative model decomposes into prior beliefs about hidden states and the likelihood of these hidden states, given outcomes.

However, direct calculation of the posterior probability of one's beliefs given sensory data $P(s|o)$ based on evaluating the marginal likelihood is simply intractable. Thus, the problem of indirect inference may be restated as follows: the brain cannot access the true posterior probability over the causes of its sensations because this requires evaluating an intractable marginal likelihood. What the brain can do, however, is perform "approximate Bayesian inference" based on its prior beliefs and the sensory data it receives.[2] In active inference, the "manipulation of content" rests on this method of inference, known as approximate Bayesian inference (Beal, 2003; Dayan et al., 1995; Feynman, 1972).

Approximate Bayesian inference allows the inversion of the generative model to estimate the marginal likelihood via an approximation to the true posterior over sensory causes (i.e., what the brain would do using exact Bayesian inference if it had access to the marginal likelihood). Taking advantage of Jensen's inequality, the method of approximate Bayesian inference involves the minimization of an upper bound on (negative log) model evidence (aka surprisal), called variational free energy. This bound is constructed by using an arbitrary probability distribution[3] $Q(S)$ that is used to minimize the variational bound—and the generative model $P(S,O)$:

$$
\begin{aligned}
F &= \sum Q(s)\ln\frac{Q(s)}{P(s,o)} \\
&= \sum Q(s)\ln\frac{Q(s)}{P(s|o)} - \ln P(o) \\
&= E_{Q(s)}\left[\ln\frac{Q(s)}{P(s|o)}\right] - \ln P(o) \\
&= \overbrace{D[\underbrace{Q(s)}_{\text{Approximate Posterior}} \,||\, \underbrace{P(s|o)}_{\text{True Posterior}}]}^{\text{bound}} - \ln \underbrace{P(o)}_{\text{Marginal likelihood}}
\end{aligned}
\tag{1}
$$

Equation 1 says that the free energy of our approximate posterior (i.e., Bayesian) beliefs, given some sensory outcomes, is the Kullback-Leibler divergence (D) from the true posterior probability of external states, given the sensory input, minus the (negative log) marginal likelihood. Estimating the marginal likelihood can be achieved by minimizing the free energy functional of (Bayesian) beliefs and sensations:

$$\begin{aligned} Q(s) &= \underset{s}{\operatorname{argmin}} F \Rightarrow \\ Q(s) &\approx \underbrace{P(s|o)}_{\text{True posterior}} \\ F &\approx -\underbrace{\ln P(o)}_{\text{Log evidence}} \end{aligned} \tag{2}$$

The Kullback-Leibler (KL, or D here) divergence represents the difference between the agent's beliefs about external states $Q(S)$ and the true posterior probability over these states, given the sensory data $P(S|O)$. Any KL divergence is always nonnegative, which means that as the free energy gets smaller (i.e., as we minimize the functional), the divergence tends toward zero. This means that minimizing free energy entails two things:

1. The marginal likelihood estimation (Beal, 2003) by making free energy a tight upper bound on the (negative log) marginal likelihood—IN$P(O)$.
2. Perception (and learning) of external states by making the approximate posterior $Q(S)$ a good approximation of the true posterior $P(S|O)$.

Perception (and learning), then, is simply the process whereby the approximate posterior $Q(S)$—parameterized or encoded by the internal states of the brain—are made "statistically consistent" with the true posterior distribution over the external states of the world given sensory observations.

Depending on the structure (i.e., entailed knowledge) in the generative model, approximate Bayesian inference optimizes not only beliefs about the world "out there" but also beliefs about the consequences of doing this or that. These beliefs yield inference to the best action to engage.

Action Planning

To account for action, one must start thinking about the manner in which states of the world change over time. This requires us to cast the generative

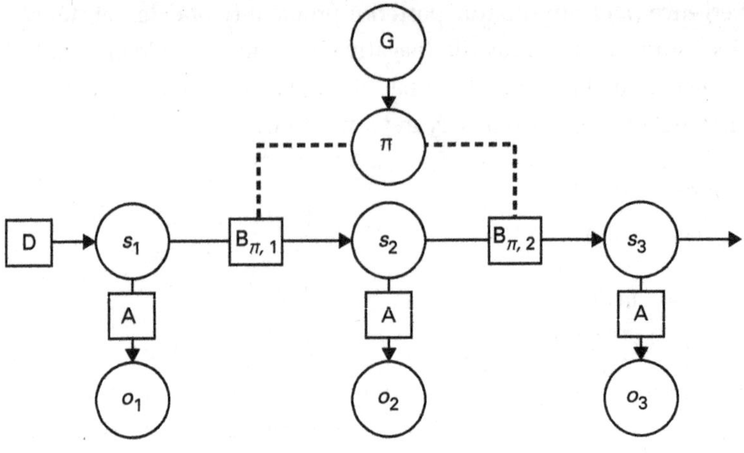

$P(o_{1\tau}, s_{1\tau}, \pi) = P(s_1)P(\pi)\prod_\tau P(o_\tau \mid s_\tau)P(s_\tau \mid s_{\tau-1}, \pi)$	Generative model
$P(o_\tau \mid s_\tau) = Cat(\mathbf{A})$ $P(s_{\tau+1} \mid s_\tau, \pi) = Cat(\mathbf{B}_{\pi,\tau})$ $P(s_1) = Cat(\mathbf{D})$ $P(\pi) = \sigma(-\mathbf{G})$	Empirical priors and likelihood (factors)
$Q(s_\tau \mid \pi) = Cat(s_{\pi\tau})$ $Q(\pi) = Cat(\boldsymbol{\pi})$	Approximate posterior (states) Approximate posterior (policy)
- - - - - - -	Representational pathway

Figure 3.1
Minimal discrete state space generative model for action. For a complete description see Friston et al., 2017. The edges that link the policy and the transition matrices are undirected. This is important, as it means that the evaluation of expected free energy requires a message from hidden state representations, thereby affording a representationalist pathway. *Source*: the author.

model over multiple times steps T into the future and how an action policy π (i.e., possible sequence of actions) may influence the trajectory of states when this or that policy is realized. Thus, our generative models will have the form $p(S, \pi, O)$—to allow us to infer future hidden states and associated outcomes O_T relative to a policy π (see figure 3.1). Here, for the sake of simplicity, we will focus on a discrete formulation of the ensuing generative model for action.

The structure of the graphical model in figure 3.1 allows us to work with a free energy appropriate for outcomes that have yet to be observed. This is known as expected free energy G (see Parr & Friston, 2017):

$$\begin{aligned} G(\pi,T) &= \sum P(o_T|s_T)Q(s_T|\pi)\ln\frac{Q(s_T|\pi)}{P(o_T,s_T|\pi)} \\ &= \sum P(o_T|s_T)Q(s_T|\pi)\ln\frac{Q(s_T|\pi)}{P(s_T|o_T,\pi)P(o_T)} \\ &= \sum P(o_T|s_T)Q(s_T|\pi)\ln\frac{Q(s_T|\pi)}{P(s_T|o_T,\pi)} - \sum P(o_T|s_T)Q(s_T|\pi)\ln P(o_T) \\ &= E_{P(o_T|s_T)Q(s_T|\pi)}\left[\ln\frac{Q(s_T|\pi)}{P(s_T|o_T,\pi)}\right] - E_{P(o_T|s_T)Q(s_T|\pi)}[\ln P(o_T)] \\ &= \underbrace{-E_{Q(s_T|\pi)P(o_T|s_T)}[\ln P(o_T)]}_{\text{Instrumental}} + \underbrace{E_{Q(s_T|\pi)P(o_T|s_T)}[\ln Q(s_T|\pi) - \ln P(s_T|o_T,\pi)]}_{\text{Epistemic}} \end{aligned} \tag{3}$$

In equation 3, the expected free energy of a policy at a given time $G(\pi, T)$ decomposes into a pragmatic or instrumental term and an epistemic term, also known as extrinsic and intrinsic values. The pragmatic term, or extrinsic value, constitutes the goal-seeking component of expected free energy (often referred to as expected value or utility in psychology and economics) (Kauder, 1953; Sutton & Barto, 1998). Extrinsic value is the expected value of a policy relative to preferred outcomes that will be encountered in the future $\mathrm{IN}P(O_T)$. In turn, the epistemic term, or intrinsic value, constitutes the information-seeking component of expected free energy. The intrinsic value is the expected information gain relative to future states under a given policy (i.e., What policy will best guarantee the minimization of uncertainty in my beliefs about the causal structure of the world?). In visual neurosciences, this is called salience, and it is a key determinant of epistemic foraging or exploratory behavior (Itti & Baldi, 2009; Sun et al., 2011). As such, it is sometimes referred to as intrinsic motivation (Barto et al., 2013; Oudeyer and Kaplan, 2007; Ryan & Deci, 1985; Schmidhuber, 2010). Selecting the policy that affords the least expected free energy guarantees an adaptive action; that is, which first consolidates knowledge about the world and then optimizes—i.e., works toward—preferred outcomes. For a complete discussion see Friston et al., 2015.

In active inference, the need for rich representations involving generative models stems directly from the problem of inverse inference about causes and adaptive actions to resolve uncertainty about those causes. The ill-posed nature of the inference problem we face forces us to first figure out for ourselves "what causes what?" before being able to zero-in on "what caused that" (perception) and "I will cause that" (i.e., action planning). This problem forces us to learn hierarchically (i.e., over multiple levels of prior beliefs) and temporally (i.e., over multiple time steps, such as in figure 3.2) deep generative models (Friston et al., 2017).

Dynamic Pathways in Active Inference

We turn now to the role of nonrepresentational dynamics in active inference. There is a technical sense in which an austere, dynamicist reading of active inference is licensed in a fundamental way. This follows because the representational account above emerges from a certain kind of dynamics; namely, gradient flows on variational and expected free energy (cf. Ramstead et al., 2019). In other words, the cognitivist functionality rests on optimizing free energy, and this optimization is a necessary consequence of neuronal dynamics that, not unlike a river flowing downhill, descend free energy gradients—to find free energy minima where the gradients are destroyed (Tschacher & Haken, 2007). Indeed, the back story to active inference shows that this kind of dynamical behavior is a necessary aspect of any self-organization to nonequilibrium steady-state in any random dynamical system that possesses a Markov blanket (Friston, 2013). In this view, any system that possesses some attracting states has dynamics that look as if they are trying to minimize free energy and therefore acquire a representational and teleological interpretation (cf. Ramstead et al., 2019).

Here, we consider nonrepresentationalist formulations of active inference. These formulations speak to notions of extended and embedded optimization, which call on hierarchical dynamics that consider the dynamical exchange between an agent and its (physiological, evolutionary, and cultural) "econiche." Accordingly, the dynamic pathways in generative models under active inference (examples appear below) do not appeal to the manipulation of representations of hidden states of the world to explain the cognitive processes underlying the behavior they generate. Dynamic pathways can be exemplified by application to a specific class of unplanned action—more specifically, enculturated action (more on that below)—that does not rest on the manipulation of rich webs of internal representations.

Rather, heuristically, the dynamicist view is a view of action that only requires processing something as doing, such that the doing (e.g., action sequences or realized policies) is directly conditioned on the something (e.g., sensory observation). In the recent literature on active inference, this sort of action has been coined "deontic" action (Constant et al., 2019).

Deontic Action

Deontic actions are actions for which the underlying policy has acquired a deontic value—namely, the shared, or socially admitted, value of a policy (Constant et al., 2019). A deontic action is guided by the consideration of "what would a typical other do in my situation." For instance, stopping at a red traffic light at 4 A.M. when no one is present may be viewed as such a deontically afforded action. Central to our agenda, deontic actions are processed through different mappings in the generative model. Technically, deontic value is the likelihood of a policy given an observation $\mathrm{IN}P(O_T|\pi)$ that grounds posterior beliefs about policies.[4] This likelihood is an empirical prior that constitutes expected free energy. The deontic value $\mathrm{IN}P(O_T|\pi)$ effectively supplements or supplants the likelihood of outcomes under different states $P(O_T|ST)$ (see figure 3.2). From the point of view of the generative model, this means that if I am pursuing this policy, then these outcomes are more likely (e.g., when I stop doing something, I am likely to see a stop sign). From the point of view of inference, this means that if I see these deontic outcomes, I will infer that I am doing this (e.g., if I see a stop sign, I will stop). Put simply, a deontic action is an available (i.e., plausible) policy that is triggered by a sensory input and leads directly to an internally consistent action. Crucially, this means that deontic action selection bypasses representational beliefs about states of the world and associated sensory consequences.

The computational architecture of deontic action is a clear candidate to implement a form of dynamicism under active inference. In effect, the information processing underlying deontic action eludes the two sufficient conditions of representationalism. For example, this is the case in mental disorders, where behavioral outcomes are maladaptive with respect to the individual's environment. Those two layers of success are apparent in the fact that one can perform (Bayes) optimal inference while generating suboptimal behavior because of suboptimal (prior) beliefs (Corlett & Fletcher, 2014). Deontic processes conform to success

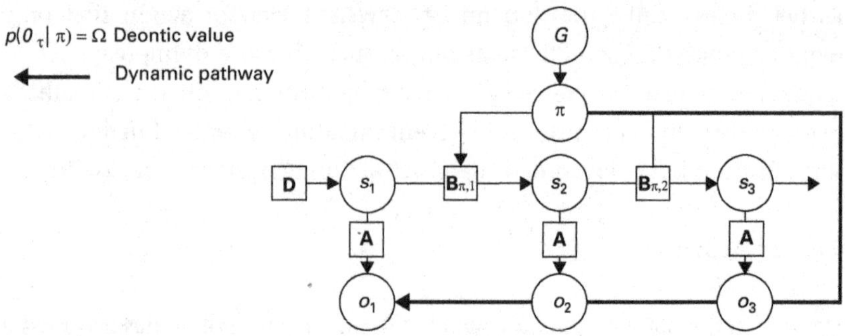

Figure 3.2
Dynamic pathway. This graphical generative model is the same as the one presented in figure 3.1. However, it incorporates deontic value, which is read here as a dynamic pathway. *Source*: the author.

conditions of this second layer. They do not have success conditions qua brain processes but rather have success condition qua agent-world coupling processes. They are simple observation-action loops, not rich and reconstructive policy selection loops. This is so because deontic processes circumnavigate the computation of expected free energy, which as we have seen, is used to compare different policies with respect to their ability to maximize instrumental and epistemic values. Under active inference, the maximization of these values such as captured by expected free energy (G) is the success condition of nondeontic action selection in the brain. This construction means that inferences about states of the world—which admit a representationalist interpretation—are now replaced by direct action, without any intervening inference or representation of the consequences of action.

Having said this, the dual aspect architecture in figure 3.2 means that the representationalist and dynamic pathways can happily live side-by-side, mutually informing each other—but both are sufficient for enactive engagement with the niche on their own. The distinction between the pathways—or routes to—(subpersonal) action selection have some important implications. For example, deontic action circumnavigates expected free energy and therefore precludes planning as (active) inference. This means that under active inference, systems employing deontic strategies do not need to plan courses of action into the future. They simply act on the basis of the observation. Furthermore, in the absence of inference about hidden states, there could be no phenomenal opacity—in the sense that there would be no opportunity to optimize the precision of the likelihood mapping between hidden states and outcomes that, in active inference, is usually associated with (covert) mental action and attention (Limanowski et al., 2018). For instance, this speaks directly to the sort of actions experts perform (e.g., athletes), which are often complex, though for which experts do not seem to plan ahead (i.e., in the head). Such skilled actions seem to yield very little phenomenal opacity (e.g., as when the athlete responds, "I don't know, I just did it in the flow of action," or "We simply executed the game plan," when interviewed about a game-winning shot).

Deontic Action as a Reflex?

The sort of "automatized" deontic behavior underwritten by dynamic pathways in the generative model might strike one as being conceptually close to the sort of cognitive processes underlying reflexes and other (homeostatic) functions processed through the autonomic nervous system. The computational pathway of deontic action indeed looks very much like a closed control loop—secured by robust causal regularities in the world generating reliable sensory inputs—akin to a reflex processed at the brainstem and spinal cord level but processed in a constructed "local world" (Constant et al., 2019, p. 15; cf. Ramstead et al., 2016). Under active inference, motoric and autonomic reflexes are framed as an action that manages the sensory signal that comes from within the system that generated it (e.g., suppression of interoceptive prediction error [Pezzulo et al., 2015]).

Autonomic reflexes facilitate homeostatic regulation by engendering series of events necessary for the activity of the agent. For example, salivation facilitates ingestion by easing the passing of the food. In this sense, they can be regarded as allostatic in nature. Similarly, one can think of sequences of deontic actions that facilitate social, affective, and emotional regulation; e.g., the outcome generated by the red traffic light triggers a stop, which facilitates me reaching into my pocket to grab my phone to check my notifications (which itself might trigger salivation). For some enculturated agents, such a sequence of social reflexes may be necessary to pass through the day.

Now, the reader might worry that deontic action ends up being as unexciting as digestive cognition. But rest assured; deontic action has been used to account for complex behavioral phenomena like social conformity (aka deference to the socially approved norm learned through social influence or learning [Asch, 1955]) and cooperative decision-making (aka decision-making under fairness psychology—as evidenced by the human tendency to zero in on fair decisions in economic games compared to nonhuman animals) (for a review, see Henrich, 2015). Deontic action—as a social reflex—facilitates social interactions by easing the coordination among humans, if you will.

Deontic action is explained in terms of the circular causality between outsourcing decision-making to trusted others in the form of deontic cues (material or agential)—indicating the locally adaptive action—and learning the underlying cue-policy mappings. The closed control loop, then, comprises the enculturated agent and regularities in the agent's (social) environment. In effect, deontic cues are defined as such because they represent a reliable informational aggregate of "what a creature like me would do in this situation." These cues consolidate over development and through the modification of the environment by generations of other enculturated agents (i.e., creatures like me) (Constant et al., 2019).

Once the action afforded by these cues is learned, there is no need for computing future states and associated outcomes; these are secured by the configuration of the cultural setting. For instance, in Canada, you can trust stopping or crossing, according to the deontic cue afforded by the traffic light, because the traffic light has come to represent what others typically do at an intersection—perhaps not in France, though. And when faced with an uncertain outcome in an economic game (e.g., "If I don't know what the opponent will do and my reward depends on her response, should I share or should I maximize my gain?"), you can trust that the fair option is the one the other is most likely to select since you've been socialized as a "typical other," presumably, just as the other did (for a review, see Veissière et al., 2019).

In summary, for proponents of dynamicism, generative models are not rich and reconstructive internal models. Rather, they are fast and frugal. If internal representations play a role at all, that role is thin and simple. A rich and reconstructive internal model is one in which multiple trajectories of hidden states (with different precisions—more on that below) would be entertained before selecting the action. The fast and frugal alternative is the one that underwrites deontic action.

Worries about Rich Settings for Shallow Strategies

First Worry

One might worry that even deontic actions have to be selected by inferring the current context. The agent might need first to figure out if the context renders deontic action the most apt response. This worry raised by representationalists (e.g., Hohwy, 2019) might be a problem for the kind of account developed in this chapter and elsewhere (e.g., Clark, 2015b). For even the selection of frugal dynamic strategies would require the ongoing inference afforded by a rich inner model, able to determine when such strategies are warranted—and override them when necessary. In other words, the recruitment of the right transient webs of deontic activity, at the right time, is itself a high-grade cognitive achievement where the inner model plays, representationalists argue, a necessary and ongoing role. The upshot is a worry that truly ecumenical accounts may be hostage to "a potential tension . . . between allowing and withholding a role for rich models" (Hohwy, 2019). For surely (so the argument goes), active inference agents must repeatedly infer when they are in a situation where some low-cost deontic response is viable. In effect, according to representationalists, setting and learning the confidence of prior beliefs through perceptual processes (aka precision, or gain of control on sensory evidence, or prediction error)—needs to be a principled response, and that implicates the rich inner model even when the selected strategy is itself a frugal one.

In active inference, the mapping between causes (e.g., states and policies) and consequences (e.g., sensory outcomes) is parameterized in terms of probabilistic mappings that necessarily have a *precision*. In other words, the contingencies implicit in likelihood mappings can have different degrees of reliability, ambiguity, or uncertainty. For instance, if my child starts running toward the sea, as she gets further away (and closer to the water), my beliefs about whether she is in danger of drowning will become increasingly imprecise. Then, to disambiguate (hidden) states of affairs, I might plan an epistemic, representational strategy: running after my child to ensure that she doesn't go into the water without supervision. Had I known that my child would start running toward danger, I could have restrained her. After multiple visits to the beach, this might become my deontic, automatic, dynamic strategy (e.g., setting foot on the sand causes my hand to grab my child).

This means that the more representationalist picture of the continuous rational influence of planning, we claim, is subtly mistaken. For example, suppose I am playing table tennis well. My context-sensitive "precision settings" are all apt, and no unexpected circumstances arise (alien invasions etc.). In such circumstances, I harvest a flow of expected kinds of prediction errors. These get resolved, in broadly predictable ways, as play unfolds without pushing far up the processing hierarchy. But if "unexpected surprises" (for more on this distinction, see Yu & Dayan, 2005) occur, some errors are more fundamentally unresolved and get pushed higher. This provides the seed for reorganizing the precision of various likelihood mappings to lend more weight to different kinds of (internal, external, and action-involving) information. That, we suggest, is how we can remain constantly poised (e.g., Sutton's compelling [2007] work on expert cricket) for nuance, even while behaving in the fast, fluent manner of a "habit machine." In a deep sense, we exist in that moment as a habit machine—that is nonetheless constantly poised to become another transient machine should the need arise. This speaks to the coalition between representational and dynamic pathways illustrated in figure 3.2.

Put another way, wherever possible, simple habit systems should guide behavior, dealing with expected prediction error fluently and fast. But where these fail, or where a change of context indicates the need, more and more knowledge-intensive resources (internal and external) can be assembled, via new waves of precision-weighting, to quash any outstanding prediction errors (i.e., free energy); see Pezzulo and colleagues (2015) for a complete argument. Hence, we should not deny that there really is, in advanced minds, what representationalists correctly describe as "immense storage of causal knowledge" (Hohwy, 2019, p. 191). But via moment-by-moment, self-organizing, free-energy-minimizing kinetics, we manifest as a succession of relatively special-purpose brain-body-world devices, strung together by those shifting but self-organizing webs of precision-weighting. Importantly, it is self-organizing around free energy that itself delivers the subsequent precision variations that recruit the "next machine." There is no precision master sitting atop this web, carefully deciding moment by moment just how to assign precision; there's just the generative model itself.

Second Worry

At this point a new version of representationalist worry may arise, for it may seem that precision estimates—the roots of each episode of restructuring—are cognitively expensive and purely inner-model-bound. But this too—or so we have been arguing—is subtly mistaken. If we shift perspectives and timescales, we can see the human-built cognitive niche as itself a prime reservoir, both of achieved precision estimations and of tools for cheaply estimating precisions on the fly. And once learned, they allow nonrepresentation involving deontic action pathways (e.g., positioning cheap cues in the world, such as warning triangles around a broken-down vehicle). These otherwise arbitrary structures attract attention and act as local proxies for precision (e.g., Hutchins, 2014; Paton et al., 2013; Roepstorff et al., 2010). Urgent fonts, food packaging, and priestly robes all provide handy shortcuts for our precision-estimating brains. Squint just a little bit and much of the human-built world—including all those patterned social practices, such as stopping at red traffic lights—can be seen as a bag of tricks for managing precision estimation and epistemic trust (Fonagy & Allison, 2014; Parr & Friston, 2017). And, as we behave in the present niche, we gradually alter it, "uploading" (Constant, Ramstead, et al., 2018) more and more of our individual and collective precision estimations into persisting (transmissible) material and social structures. These, in turn, alter the inner models that individuals need to command to negotiate their worlds.

Third Worry

A final representationalist worry may be that fast and frugal, nonrepresentational deontic action could simply not yield adaptive behavior in a highly volatile world like ours and thus may lead to suboptimal, maladaptive decision-making (e.g., decision-making that fails to generate action that succeeds with respect to environmental challenges), especially if our generative models of precision are not apt for a volatile world (Parr, Benrimoh, et al., 2018; Parr, Rees, & Friston, 2018). Consequently, one should favor explanations based on rich reconstructing planning. This is a fair worry—a fair worry for humans in general, not for the dynamicists' perspective, though. Indeed, humans learn to generate deontic actions that do not always lead to the Machiavellian, or perhaps Darwinian utility-maximizing option relative to the current environment; we miss steps and fall down the stairs, forget to stop on the red, develop disorders such as PTSD that make us misperceive threats, and generate many more maladaptive traits (Badcock et al., 2017; Cornwell et al., 2017; Peters et al., 2017). The tricks humans employ—to minimize the potential cost of normal maladaptive actions—is not to plan more in the head, but to plan more in the world (e.g., making sure that the synchronization of the traffic lights is consistent with the traffic flow at different hours of the day). This "planning the world" solution stabilizes the environment to enable the acquisition (i.e., learning through representationalist processes) of cheap deontic action shared among cultural conspecifics—people enculturated like me, on a 9 to 5 schedule (Constant, Bervoets, et al., 2018; Constant, Ramstead, et al., 2018). Under that view, in certain situations, one can dispense with rich models that stand-in for that world for the purposes of planning, reasoning, and the guidance of action (Clark, 2016). In a word, for enculturated, deontically construed agents like us, the world is often our shared best model.

Conclusion: Bury the Hatchet, or Use It to Carve a New Path(way)

This chapter offers a mathematically informed reading of generative models that could accommodate both richly representationalist and dynamicist views of cognition. Is cognition under active inference a richly representational or a dynamic process reliant on simple cues and couplings? The answer is "both." We have shown that dynamicism has a computational grip when implemented in the theory of active inference. However, one has yet to propose candidate neural correlates of dynamicist pathways, which is a research enterprise for neuroscience made possible on the basis of implementable theory, such as the one discussed in this chapter. Thus, despite the lack of empirical evidence, we consider settling the general active inference debate about representationalism a major development, since it is a first step toward scientifically informed debates on the representational nature of specific pathways, which could then feed back to further strengthen future philosophical discussions and inform research trajectories.

Notes

1. Hidden states are sometimes called latent states and refer to variables that cannot be observed. They are effectively hidden behind observations and have to be inferred as random variables.

2. Approximate Bayesian inference should not be read as inference that is approximate; rather, it should be read as inference that can be realized. Indeed, as we will see later, exact Bayesian inference is a special case of approximate Bayesian inference when certain conditions are met.

3. This arbitrary probability distribution is variously called an approximate posterior, a recognition distribution, or more simply, a Bayesian belief.

4. E.g., when I see the sock on my night table, I put it in my pocket.

References

Asch, S. E. (1955). Opinions and social pressure. *Scientific American, 193*(5), 31–35.

Badcock, P. B., Davey, C. G., Whittle, S., Allen, N. B., & Friston, K. J. (2017). The depressed brain: An evolutionary systems theory. *Trends in Cognitive Sciences, 21*(3), 182–194. https://doi.org/10.1016/j.tics.2017.01.005

Badcock, P. B., Friston, K. J., & Ramstead, M. J. D. (2019). The hierarchically mechanistic mind: A free-energy formulation of the human psyche. *Physics of Life Reviews, 31*, 104–121. https://doi.org/10.1016/j.plrev.2018.10.002

Barto, A., Mirolli, M., & Baldassarre, G. (2013). Novelty or surprise? *Frontiers in Psychology, 4*, 907. https://doi.org/10.3389/fpsyg.2013.00907

Bastos, A. M., Usrey, W. M., Adams, R. A., Mangun, G. R., Fries, P., & Friston, K. J. (2012). Canonical microcircuits for predictive coding. *Neuron, 76*(4), 695–711. https://doi.org/10.1016/j.neuron.2012.10.038

Beal, M. J. (2003). *Variational algorithms for approximate Bayesian inference*. University of London.

Bruineberg, J., Kiverstein, J., & Rietveld, E. (2016). The anticipating brain is not a scientist: The free-energy principle from an ecological-enactive perspective. *Synthese, 195*(6), 2417–2444. https://doi.org/10.1007/s11229-016-1239-1

Chemero, A. (2009). *Radical embodied cognition*. MIT Press.

Churchland, P. M. (1989). Some reductive strategies in cognitive neurobiology. In S. Silvers (Ed.), *Rerepresentation: Readings in the philosophy of mental representation* (pp. 223–253). Springer Netherlands. https://doi.org/10.1007/978-94-009-2649-3_12

Clark, A. (1989). *Microcognition: Philosophy, cognitive science and parallel distributed processing*. MIT Press/Bradford Books.

Clark, A. (1993). *Associative engines: Connectionism, concepts and representational change*. MIT Press/Bradford Books.

Clark, A. (1997). The dynamical challenge. *Cognitive Science, 21*(4), 461–481. https://doi.org/10.1016/S0364-0213(99)80030-5

Clark, A. (2013). Whatever next? Predictive brains, situated agents, and the future of cognitive science. *The Behavioral and Brain Sciences, 36*(3), 181–204. https://doi.org/10.1017/S0140525X12000477

Clark, A. (2015). *Surfing uncertainty: Prediction, action, and the embodied mind.* Oxford University Press.

Clark, A., & Chalmers, D. (1998). The extended mind. *Analysis, 58*(1), 7–19. http://www.jstor.org/stable/3328150.

Constant, A., Bervoets, J., Hens, K., & Van de Cruys, S. (2018). Precise worlds for certain minds: An ecological perspective on the relational self in autism. *Topoi, 39,* 611–622. https://doi.org/10.1007/s11245-018-9546-4.

Constant, A., Ramstead, M. J. D., Veissière, S. P. L., Campbell, J. O., & Friston, K. J. (2018). A variational approach to niche construction. *Journal of the Royal Society, Interface / the Royal Society, 15*(141), 20170685. https://doi.org/10.1098/rsif.2017.0685.

Constant, A., Ramstead, M., Veissière, S., & Friston, K. J. (2019). Regimes of expectations: An active inference model of social conformity and decision making. *Frontiers in Psychology, 10,* 679. https://doi.org/10.3389/fpsyg.2019.00679.

Corlett, P. R., & Fletcher, P. C. (2014). Computational psychiatry: A Rosetta Stone linking the brain to mental illness. *The Lancet Psychiatry, 1*(5), 399–402. https://doi.org/10.1016/S2215-0366(14)70298-6

Cornwell, B. R., Garrido, M. I., Overstreet, C., Pine, D. S., & Grillon, C. (2017). The unpredictive brain under threat: A neurocomputational account of anxious hypervigilance. *Biological Psychiatry, 82*(6), 447–454. https://doi.org/10.1016/j.biopsych.2017.06.031

Dayan, P., Hinton, G. E., Neal, R. M., & Zemel, R. S. (1995). The Helmholtz machine. *Neural Computation, 7*(5), 889–904. https://doi.org/10.1162/neco.1995.7.5.889

Feldman, H., & Friston, K. J. (2010). Attention, uncertainty, and free-energy. *Frontiers in Human Neuroscience, 4,* 215. https://doi.org/10.3389/fnhum.2010.00215

Feynman, R. (1972). *Statistical mechanics: A set of lectures.* Benjamin/Cummings Publishing.

Fodor, J. A. (1975). *The language of thought.* Harvard University Press.

Fonagy, P., & Allison, E. (2014). The role of mentalizing and epistemic trust in the therapeutic relationship. *Psychotherapy, 51*(3), 372–380. https://doi.org/10.1037/a0036505

Friston, K. (2013). Life as we know it. *Journal of the Royal Society Interface, 10*(86), 20130475. https://doi.org/10.1098/rsif.2013.0475

Friston, K. (2018). Does predictive coding have a future? *Nature Reviews Neuroscience, 21*(8), 1019–1021. https://doi.org/10.1038/s41593-018-0200-7

Friston, K., Rigoli, F., Ognibene, D., Mathys, C., Fitzgerald, T., & Pezzulo, G. (2015). Active inference and epistemic value. *Cognitive Neuroscience, 6*(4), 187–214. https://doi.org/10.1080/17588928.2015.1020053

Friston, K. J. (2010). The free-energy principle: A unified brain theory? *Nature Reviews Neuroscience, 11*(2), 127–138. https://doi.org/10.1038/nrn2787

Friston, K. J., FitzGerald, T., Rigoli, F., Schwartenbeck, P., O Doherty, J., & Pezzulo, G. (2016). Active inference and learning. *Neuroscience and Biobehavioral Reviews, 68*, 862–879. https://doi.org/10.1016/j.neubiorev.2016.06.022

Friston, K. J., FitzGerald, T., Rigoli, F., Schwartenbeck, P., & Pezzulo, G. (2017). Active inference: A process theory. *Neural Computation, 29*(1), 1–49.

Friston, K. J., & Frith, C. (2015). A duet for one. *Consciousness and Cognition. 36*, 390–405. https://doi.org/10.1016/j.concog.2014.12.003

Friston, K. J., Kilner, J., & Harrison, L. (2006). A free energy principle for the brain. *Journal of Physiology-Paris, 100*(1–3), 70–87. https://doi.org/10.1016/j.jphysparis.2006.10.001

Friston, K. J., Parr, T., & de Vries, B. (2017). The graphical brain: Belief propagation and active inference. *Network Neuroscience, 1*(4), 381–414. https://doi.org/10.1162/NETN_a_00018

Friston, K. J., Rosch, R., Parr, T., Price, C., & Bowman, H. (2017). Deep temporal models and active inference. *Neuroscience and Biobehavioral Reviews, 77*, 388–402. https://doi.org/10.1016/j.neubiorev.2017.04.009

Henrich, J. (2015). *The secret of our success: How culture is driving human evolution, domesticating our species, and making us smarter.* Princeton University Press.

Hohwy, J. (2013). *The predictive mind.* Oxford University Press.

Hohwy, J. (2016). The self-evidencing brain. *Noûs, 50*(2), 259–285. https://doi.org/10.1111/nous.12062

Hohwy, J. (2019). Quick'n'Lean or Slow and Rich? Andy Clark on predictive processing and embodied cognition. In L. Irvine, M. Stapleton, M. Colombo (Eds.), *Andy Clark and his critics* (pp. 191–205). Oxford University Press.

Hutchins, E. (2014). The cultural ecosystem of human cognition. *Philosophical Psychology, 27*(1), 34–49. https://doi.org/10.1080/09515089.2013.830548

Hutto, D. D., & Myin, E. (2013). *Radicalizing enactivism: Basic minds without content.* MIT Press.

Itti, L., & Baldi, P. (2009). Bayesian surprise attracts human attention. *Vision Research, 49*(10), 1295–1306. https://doi.org/10.1016/j.visres.2008.09.007

Joffily, M., & Coricelli, G. (2013). Emotional valence and the free-energy principle. *PLoS Computational Biology, 9*(6), e1003094. https://doi.org/10.1371/journal.pcbi.1003094

Kaplan, R., & Friston, K. J. (2018). Planning and navigation as active inference. *Biological Cybernetics, 112*(4), 323–343. https://doi.org/10.1007/s00422-018-0753-2

Kauder, E., (1953). Genesis of the marginal utility theory: From Aristotle to the end of the eighteenth century. *The Economic Journal, 63*(251), 638–650. https://doi.org/10.2307/2226451.

Kirchhoff, M., Parr, T., Palacios, E., Friston, K. J., & Kiverstein, J. (2018). The Markov blankets of life: Autonomy, active inference and the free energy principle. *Journal of the Royal Society, Interface / The Royal Society, 15*(138), 20170792. https://doi.org/10.1098/rsif.2017.0792

Kirchhoff, M. D., & Robertson, I. (2018). Enactivism and predictive processing: A non-representational view. *Philosophical Explorations: An International Journal for the Philosophy of Mind and Action, 21*(2), 264–281. https://doi.org/10.1080/13869795.2018.1477983

LeCun, Y., Bengio, Y., & Hinton, G. (2015). Deep learning. *Nature, 521*(7553), 436–444. https://doi.org/10.1038/nature14539

Limanowski, J., & Friston, K. (2018). "Seeing the dark": Grounding phenomenal transparency and opacity in precision estimation for active inference. *Frontiers in Psychology, 9*, 643. https://doi.org/10.3389/fpsyg.2018.00643

Metzinger, T., & Wiese, W. (2017). *Philosophy and predictive processing*. MIND Group.

Mirza, M. B., Adams, R. A., Mathys, C. D., & Friston, K. J. (2016). Scene construction, visual foraging, and active inference. *Frontiers in Computational Neuroscience, 10*, 56. https://doi.org/10.3389/fncom.2016.00056

Oudeyer, P. Y., & Kaplan, F. (2007). What is intrinsic motivation? A typology of computational approaches. *Frontiers in Neurorobotics, 1*, 6. https://doi.org/10.3389/neuro.12.006.2007

Parr, T., Benrimoh, D., Vincent, P., & Friston, K. (2018). Precision and false perceptual inference. *Frontiers in Integrative Neuroscience, 12*, 39. https://doi.org/10.3389/fnint.2018.00039

Parr, T., & Friston, K. J. (2017). Uncertainty, epistemics and active inference. *Journal of the Royal Society Interface, 14*(136), 20170376. https://doi.org/10.1098/rsif.2017.0376

Parr, T., Rees, G., & Friston, K. J. (2018). Computational neuropsychology and Bayesian inference. *Frontiers in Human Neuroscience, 12,* 61. https://doi.org/10.3389/fnhum.2018.00061

Paton, B., Skewes, J., Frith, C., & Hohwy, J. (2013). Skull-bound perception and precision optimization through culture. *Behavioral and Brain Sciences, 36*(3), e222. https://doi.org/10.1017/S0140525X12002191

Paton, B., Skewes, J., Frith, C., & Hohwy, J. (2013). Skull-bound perception and precision optimization through culture. *Behavioral and Brain Sciences, 36*(3), e222. https://doi.org/10.1017/S0140525X12002191

Peters, A., McEwen, B. S., & Friston, K. (2017). Uncertainty and stress: Why it causes diseases and how it is mastered by the brain. *Progress in Neurobiology, 156,* 164–188. https://doi.org/10.1016/j.pneurobio.2017.05.004

Pezzulo, G., Rigoli, F., & Friston, K. (2015). Active inference, homeostatic regulation and adaptive behavioural control. *Progress in Neurobiology, 134,* 17–35. https://doi.org/10.1016/j.pneurobio.2015.09.001

Ramstead, M. J. D., Badcock, P. B., & Friston, K. J. (2019). Answering Schrödinger's question: A free-energy formulation. *Physics of Life Reviews, 31,* 1–16. https://doi.org/10.1016/j.plrev.2018.09.001

Ramstead, M. J. D., Veissière, S. P. L., & Kirmayer, L. J. (2016). Cultural affordances: Scaffolding local worlds through shared intentionality and regimes of attention. *Frontiers in Psychology, 7,* 1090. https://doi.org/10.3389/fpsyg.2016.01090

Roepstorff, A., Niewöhner, J., & Beck, S. (2010). Enculturating brains through patterned practices. *Neural Networks, 23*(8–9), 1051–1059. https://doi.org/10.1016/j.neunet.2010.08.002

Ryan, R., & Deci, E. (Eds.). (1985). *Intrinsic motivation and self-determination in human behavior.* Springer Science+Business Media.

Schmidhuber, J. (2010). Formal theory of creativity, fun, and intrinsic motivation (1990–2010). *IEEE Transactions on Autonomous Mental Development, 2*(3), 230–247. https://doi.org/10.1109/TAMD.2010.2056368

Schwartenbeck, P., & Friston, K. (2016). Computational phenotyping in psychiatry: A worked example. *ENeuro, 3*(4), 0049–16.2016. https://doi.org/10.1523/ENEURO.0049-16.2016

Siegel, S. (2010). *The contents of visual experience.* Oxford University Press.

Sun, Y., Gomez, F., & Schmidhuber, J. (2011). Planning to be surprised: Optimal Bayesian exploration in dynamic environments. In J. Schmidhuber, K. R. Thórisson, & M. Looks (Eds.), *Proceedings of the 4th international conference on artificial general intelligence* (pp. 41–51). Springer-Verlag. https://doi.org/10.1007/978-3-642-22887-2_5

Sutton, J. (2007). Batting, habit and memory: The embodied mind and the nature of skill. *Sport in Society, 10*(5), 763–786. https://doi.org/10.1080/17430430701442462

Sutton, R. S., & Barto, A. G. (1998). *Reinforcement learning: An introduction.* MIT Press.

Smith, L. B., & Thelen, E. (2003). Development as a dynamic system. *Trends in Cognitive Sciences, 7*(8), 343–348. https://doi.org/10.1016/s1364-6613(03)00156-6

Thelen, E., & Smith, L. B. (1996). *A dynamic systems approach to the development of cognition and action.* MIT Press.

Thompson, E. (2007). *Mind in life: Biology, phenomenology, and the sciences of mind.* Harvard University Press.

Tschacher, W., & Haken, H. (2007). Intentionality in non-equilibrium systems? The functional aspects of self-organised pattern formation. *New Ideas in Psychology, 25*(1), 1–15. https://doi.org/10.1016/j.newideapsych.2006.09.002

Van Gelder, T. (1995). What might cognition be if not computation? *Journal of Philosophy, 92*(7), 345–381. https://doi.org/10.2307/2941061

Varela, F. J., Thompson, E., & Rosch, E. (1991). *The embodied mind: Cognitive science and human experience.* MIT Press.

Veissière, S. P. L., Constant, A., Ramstead, M. J. D., Friston, K. J., & Kirmayer, L. J. (2019). Thinking through other minds: A variational approach to cognition and culture. *Behavioral and Brain Sciences, 43,* 1–97. https://doi.org/10.1017/S0140525X19001213

4 Predictive Processing: From Sensorimotor Skills to Higher Cognition

Giovanni Pezzulo, Thomas Parr, and Karl Friston

Introduction

A fundamental challenge in neuroscience, psychology, and philosophy is understanding how the advanced cognitive abilities—and the brain architectures supporting these abilities—emerged during evolution. This chapter explores the hypothesis that higher cognitive abilities emerged during evolution through exaptation of basic circuits for "predictive processing" that supported motor prediction and control in our earlier ancestors. This chapter is mostly concerned with a high (functional or computational) level of description: it discusses the nature of the problems addressed by simpler and more advanced organisms and how various elaborations and refinements of predictive processing might have provided adaptive solutions to these problems.

Our treatment connects with various lines of research, which recognize a fundamental limitation in our current conceptualization of brain processing and cognition—namely, starting from the vantage point of advanced (e.g., human) cognition, without asking how it might have evolved (Buzsáki, 2019; Cisek & Hayden, 2022). One potential misunderstanding is that we often take for granted an a priori taxonomy of the functions of brain areas or networks, which follows psychologically defined concepts (e.g., perception, decision-making, motor control, emotion), but which might be a misdirection. One intriguing line of research proposes to go beyond this a priori taxonomy by describing the evolution of brain architectures as a process of "phylogenetic refinement," in which early brain designs necessarily influence the evolution of later brain designs, as exemplified by the passage from early unicellular and multicellular organisms to vertebrates

(Cisek, 2019) and from chordates to primates (Cisek, 2021a). This line of research suggests that the history of brain architectures is a history of continuous extensions of simpler control loops into more complex and deeper (and diachronic) structures. Anatomically, this is reflected by progressively specialized areas—initially developed for simpler forms of control (e.g., a high-level controller, the apical nervous system, and a lower-level controller, the basal nervous system)—in our modern brains, which include parallel control systems dedicated to affordance-based guidance of actions and a hierarchy of selection mechanisms to arbitrate among them (see also Cisek & Kalaska, 2010; Pezzulo & Cisek, 2016). A related perspective focuses on the evolution of the brain's architecture from early bilaterians to humans, through five main evolutionary milestones, comprising steering, reinforcing, simulating, mentalizing, and speaking (Bennett, 2021).

Furthermore, ignoring evolutionary issues might lead to a misconception of advanced cognitive functions, such as planning, cognitive control, mind reading, and language processing. Failing to understand how they might have emerged during evolution—from simpler sensorimotor loops—might lead to the (questionable) assumption that they are novel cognitive modules, juxtaposed in some way to older brain substrates. However, various alternative approaches suggest that the neural substrate affording "simpler" cognitive abilities was reused and repurposed for more complex cognitive abilities; for example, neural (e.g., hippocampal) circuits—supporting spatial navigation—might also support navigation in abstract domains (Behrens et al., 2018; Bellmund et al., 2018; Buzsáki & Moser, 2013), and motor control circuits might have been reused for language (Anderson, 2010; Arbib, 2005). Related proposals describe planning as a process of internalization of sensorimotor and predictive loops (Buzsáki et al., 2014; Grush, 2004; Pezzulo & Castelfranchi, 2007) and the evolution of executive function as a shift of focus from externally directed to internally directed behavior and self-regulation (Barkley, 2001).

Along similar lines, embodied and motor theories of cognition suggest that the conceptual system is grounded in the sensory modalities (Barsalou, 2008) and that advanced cognitive functions, such as planning, imagination, and mindreading, are rooted in motor control systems (Jeannerod, 2006). These theories propose that cognitive processes are fundamentally intertwined with bodily actions and that advanced cognitive abilities maintain some vestigial aspects of sensorimotor processing (Hurley,

2008; Pezzulo, 2011; Wilson, 2002). The historical context of these theories includes various efforts to shift from viewing cognition as completely brain-centric to recognizing the role of the body and of situated interaction with the environment—and with others—to shape it (Chiel & Beer, 1997; Clark, 1998; Cisek, 2021b; O'Regan & Noe, 2001; Pfeifer & Bongard, 2006). Embodied cognition (EC) also inherits from early theories, suggesting that the fundamental organization of behavior is active rather than (only) exploiting stimulus-response mappings. Various researchers emphasize that behavior starts with an action that changes the environment and creates regularities that are sensed (Ahissar & Assa, 2016; Brembs, 2009; Engel et al., 2013; Gibson, 1979; Lungarella & Sporns, 2006; Pezzulo & Cisek, 2016). Furthermore, the cybernetic tradition emphasizes that behavior exploits closed loops and is engaged when discrepancies between desired set points and environmental circumstances are sensed, not as a response to stimuli (Ashby, 1947; Miller et al., 1960; Powers, 1973; Wiener, 1948). In turn, this outcome-based organization provides more flexibility than stimulus-response loops; for example, an observable outcome can be achieved through multiple behaviors (e.g., a location can be reached by running or crawling).

Finally, various researchers argued that higher cognitive skills are exaptations of basic circuits for motor prediction and control that existed in our early evolutionary ancestors (Clark & Grush, 1999; Pezzulo, 2017; Pezzulo & Castelfranchi, 2007, 2009). Some of these ideas have been developed within the framework of active inference (Parr et al., 2022; Pezzulo, D'Amato et al., 2024; Pezzulo et al., 2024). Specifically, Pezzulo et al. (2022) proposed how to develop complex brains from simpler functional architectures, speaking to the evolution of brain architectures as the progressive refinement of internal generative models—namely, statistical models that describe how sensations are generated. This kind of refinement starts from simple closed-loop and predictive behaviors and homeostatic drives and progresses in four main aspects: the multimodal *expansion* of predictive control into an allostatic loop, its *duplication* to form multiple sensorimotor loops that expand an animal's behavioral repertoire, and the gradual endowment of generative models with *hierarchical depth* (to deal with aspects of the world that unfold at different spatial scales) and *temporal depth* (to select plans in a future-oriented manner; see also Corcoran et al., 2020; Pezzulo, D'Amato, et al., 2024; Pezzulo et al., 2015; Tschantz et al., 2022 for related arguments about the evolution of sophisticated predictive processing skills and neural representations).

This chapter unpacks the arguments presented above and discusses how active inference could offer a cohesive explanation of how higher cognitive functions emerge from simpler sensorimotor skills.

Active Inference: A Brief Introduction

Active inference is a unifying account of self-organization that links perception, action, and cognition (Parr et al., 2022). It posits that the brain uses a (learned) *generative model*[1] of the world and of the body in it—formally, a statistical model of the joint distribution of observed sensory input and of unobserved (or hidden) states—for perceptual inference (i.e., inferring the most likely causes of sensations) and action selection (i.e., inferring the course of actions most likely to realize preferred outcomes).

As shown in figure 4.1, the formal process that underwrites active inference is the minimization of free energy via perception and action. Free energy is a function that scores the divergence between the model's predictions and the sensory input predicted (which, under simplifying assumptions, reduces to prediction error). Perception reduces free energy by changing the brain's probabilistic beliefs about hidden states, to better align predictions with the sensed world. Conversely, action reduces free energy by changing the world itself, to realize predictions and thereby align the brain's probabilistic beliefs and hidden states. For example, when predicting seeing an apple but observing instead something green that jumps, one can either change one's beliefs about the causes of sensations (e.g., infer that one is observing a frog, not an apple) or saccade to an apple tree, to observe an apple. In both cases, the minimization of free energy entails a better capability to predict the next input, either because a better hypothesis about the causes of that input has been selected (e.g., a frog) or because the input has been sampled to realize the hypothesis (e.g., looking at an apple).

When the action system is engaged to fulfil prior predictions, these assume the role of goals or cybernetic set points and guide purposive action via prediction error minimization. For example, in motor control, an organism's generative model might specify desired action effects (e.g., the appropriate configuration of fingers to grasp an object) as proprioceptive

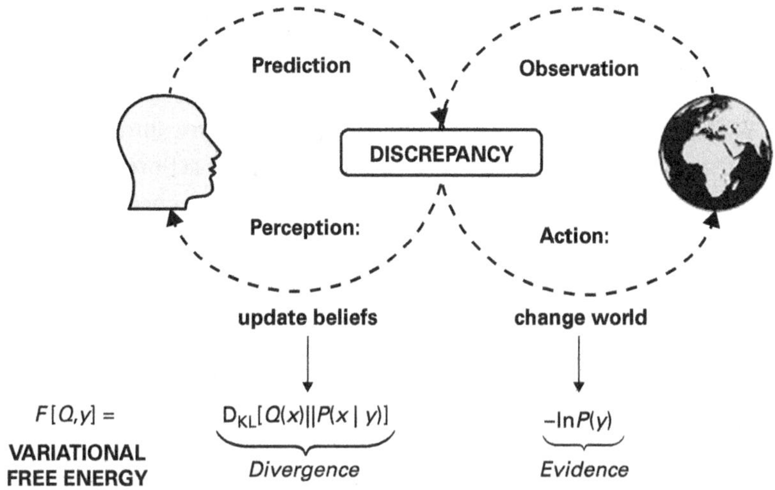

Figure 4.1
Action-perception loop of active inference. *Note:* A central assumption of active inference is that perception and action play complementary roles in minimizing variational free energy, which scores the divergence between the model's predictions and sensory input (which, under simplifying assumptions, reduces to prediction error). From an informal perspective, perception minimizes the divergence by updating the agent's beliefs so that they explain better observations; rather, action minimizes the divergence by changing the world, such that the agent samples the predicted observations. From a formal perspective, perception corresponds to minimizing the first term of variational free energy above, which scores the divergence between the agent's approximate posterior beliefs (*Q*) and the true posterior distribution (*P*) about hidden causes—in other words, optimizing the approximate posterior belief. In a complementary fashion, action maximizes the second (evidence) term, expressed as a surprise term of negative (log) evidence—in other words, minimizing surprise. Importantly, all the quantities involved in free energy minimization can be calculated based on information the agent has access to (using its generative model). See the main text for an example and Parr et al. (2022) for a formal treatment. *Source*: the author.

predictions, which are resolved by enacting a grasping action (Adams et al., 2013; Priorelli et al., 2023). The grasping action therefore results from prediction error minimization. The same mechanism might support autonomic and interoceptive processing. If a preferred body temperature is specified as a prior prediction—in the autonomic system—any discrepancy (e.g., feeling too hot) would be sensed as a prediction error and result in a corrective action, either within the autonomic system itself (e.g., vasodilation) or by engaging a cascade of more complex behaviors and allostatic loops (e.g., looking for shelter or fresh water) (Pezzulo et al., 2015; Seth, 2013).

Active inference also explains planning and other forms of future-oriented cognition. Planning involves the selection of action sequences (or policies) expected to minimize the future, *expected* free energy when considering some time horizon. Since expected free energy is a function of possible future observations, which one might obtain by selecting a particular policy, planning requires a generative model with *temporal depth* to simulate (policy-dependent) future observations. Importantly, the expected free energy permits scoring policies according to two complementary criteria. The first is their instrumental value—namely, their capability to generate preferred observations (encoded as probabilistic priors in the model). The second is their epistemic value—namely, their capability to disambiguate future hidden states. Jointly considering instrumental and epistemic value allows active inference to balance exploitation and exploration in a principled way (Friston et al., 2017).

Understanding Evolution from a Generative Modeling Perspective

Active inference assumes that the brain constantly performs a single operation: the minimization of free energy using a generative model. This suggests that there would be no categorical divide between the fundamental brain functions of simpler and more advanced organisms: both might use predictive processing and active inference. However, what distinguishes organisms (and individuals) are their generative models (and their learnable parameters).

In the following subsections, we discuss how different types of animals' brains might correspond to an ontology of generative models, which necessarily builds on the generative models of evolutionary ancestors. We highlight how the expansion of generative models corresponds to enhancements of the scope of prediction and goal-directed control, with a transition from generative models that support simple behaviors satisfying immediate drives (possibly in invertebrates and early vertebrates) to generative models that support more sophisticated behaviors that achieve abstract and distal goals (possibly in primates).

2014; Tschantz et al., 2022]) might be precursors of—or learned from—more sophisticated generative models with abstract goals that support motivated behavior and affective processes (Pezzulo et al., 2015, 2018).

Scaling up the level of complexity, the most sophisticated generative models of external variables could be described as *world models* (Ha & Schmidhuber, 2018; LeCun, 2022; Taniguchi et al., 2023). These might form hidden states (or abstract schemas [Piaget, 1954] or theories [Gerstenberg & Tenenbaum, 2017]) representing entities like objects and people as building blocks to model physical and social dynamics: the so-called core systems of cognition (Kinzler & Spelke, 2007). For example, from a statistical perspective, an "object" could be constructed as the common (and stable) cause of multiple action-perception loops (e.g., a glass constructed as something transparent if I look at it, solid if I grasp it, rigid if I move it, etc.) (Drescher, 1991). In turn, the ensuing generative (world) models of the physical and social reality support sophisticated human behaviors, like navigating cities, reading books, or participating in scientific conferences.

Importantly, exhibiting sophisticated behaviors requires generative (world) models that capture three types of statistical regularities; namely, representing multiple levels of abstraction, that unfold at different time scales, and that show some independency between factors (Friston et al., 2024). Capturing these regularities requires generative models with three novel features. The first feature—*hierarchical depth*—means that the model includes hidden states at deeper levels, with the higher levels abstracting away from lower-level details. One example is the hierarchical representation of letters, words, sentences, and narratives, which afford hierarchical language processing (Donnarumma et al., 2024). The second feature—*temporal depth*—means that the model includes hidden states for present, past, and future events, permitting predicting and achieving distal events and goals. One example is representation of present and future spatial positions during goal-directed spatial navigation (Friston et al., 2017). Finally, the third feature—*factorization*—means that the model can treat some events as having conditionally independent causes, hence simplifying the inference and potentially affording compositionality (Bishop, 2006; Smolensky et al., 2022). Some examples comprise brain factorizations into "what" and "where" visual streams and into "when" and "what" hippocampal processing (Friston & Buzsáki, 2016).

In sum, generative models supporting higher cognition might be conceptualized as building on basic predictive circuits by incorporating various features such as multimodality, hierarchy, temporal depth, and factorization. Importantly, solving challenging cognitive problems might require generative models that combine multiple features. For example, generative models supporting navigation through a sequence of goal-states might be both hierarchically and temporally deep, combining the capability to represent spatial and task rule maps hierarchically and to reason at multiple temporal scales within each of them (Van de Maele et al., 2023). Similarly, generative models supporting advanced social interactions and joint actions might install different types of predictive regularities, including physical and social dynamics (Maisto et al., 2023). Of course, these sophisticated generative models are only mandated if they fit the organism's ecological niche, meaning that not all animal lineages are expected to show expansions of generative models across all aspects. On the other hand, the tailoring of generative models to an ecological niche implies that one might hope to reverse engineer an animal's (or even a person's) generative model from the niche it inhabits and the behavior it expresses. In turn, failures to behave adaptively—as seen, for example, in psychopathological conditions—could be explained in terms of aberrant (parameterizations of) generative models (Pezzulo, et al., 2024).

Discussion

We have used active inference as a guiding principle to develop a gradualist perspective on the development of brains that—along with the body phenotype—allow organisms to better adapt to (and shape) their ecological niches. We proposed that during evolution, organisms could have learned to exploit the statistical structure of their environment and internalize some predictable proxy for the conditions that promote their existence, forming generative and predictive models. Therefore, simple predictive circuits already mediate adaptive behavior in simple organisms. The advanced capabilities of complex animals, like us, could result from progressive elaborations of generative models for active inference, which leverage progressively richer predictable regularities in the physical and social domains (hierarchical perception) and guiding these regularities in the preferred directions, specified by drives and goals (hierarchical action and planning). In short, the expansion of generative models corresponds to improved capabilities for prediction and control over the physical and

Ashby, R. (1947). Principles of the self-organizing dynamic system. *Journal of General Psychology, 37*, 125–128. https://doi.org/10.1080/00221309.1947.9918144

Barkley, R. A. (2001). The executive functions and self-regulation: An evolutionary neuropsychological perspective. *Neuropsychology Review, 11*(1), 1–29. https://doi.org/10.1023/A:1009085417776

Barrett, L. F., & Simmons, W. K. (2015). Interoceptive predictions in the brain. *Nature Reviews Neuroscience, 16*(7), 419–429. https://doi.org/10.1038/nrn3950

Barsalou, L. W. (2008). Grounded cognition. *Annual Review of Psychology, 59*, 617–645.

Bastos, A. M., Usrey, W. M., Adams, R. A., Mangun, G. R., Fries, P., & Friston, K. J. (2012). Canonical microcircuits for predictive coding. *Neuron, 76*(4), 695–711. https://doi.org/10.1016/j.neuron.2012.10.038

Behrens, T. E. J., Muller, T. H., Whittington, J. C. R., Mark, S., Baram, A. B., Stachenfeld, K. L., & Kurth-Nelson, Z. (2018). What is a cognitive map? Organizing knowledge for flexible behavior. *Neuron, 100*(2), 490–509. https://doi.org/10.1016/j.neuron.2018.10.002

Bellmund, J. L. S., Gärdenfors, P., Moser, E. I., & Doeller, C. F. (2018). Navigating cognition: Spatial codes for human thinking. *Science, 362*(6415), eaat6766. https://doi.org/10.1126/science.aat6766

Bennett, M. S. (2021). Five breakthroughs: A first approximation of brain evolution from early bilaterians to humans. *Frontiers in Neuroanatomy, 15*, 693346. https://doi.org/10.3389/fnana.2021.693346

Berkes, P., Orbán, G., Lengyel, M., & Fiser, J. (2011). Spontaneous cortical activity reveals hallmarks of an optimal internal model of the environment. *Science, 331*(6013), 83–87. https://doi.org/10.1126/science.1195870

Bishop, C. M. (2006). *Pattern recognition and machine learning*. Springer.

Brembs, B. (2009). The importance of being active. *Journal of Neurogenetics, 23*(1–2), 120–126. https://doi.org/10.1080/01677060802471643

Buckner, R. L., & Krienen, F. M. (2013). The evolution of distributed association networks in the human brain. *Trends in Cognitive Sciences, 17*(12), 648–665. https://doi.org/10.1016/j.tics.2013.09.017

Buzsáki, G. (2019). *The brain from inside out*. Oxford University Press.

Buzsáki, G., & Moser, E. I. (2013). Memory, navigation and theta rhythm in the hippocampal-entorhinal system. *Nature Neuroscience, 16*(2), 130–138. https://doi.org/10.1038/nn.3304

Buzsáki, G., Peyrache, A., & Kubie, J. (2014). Emergence of cognition from action. *Cold Spring Harbor Symposia on Quantitative Biology, 79*, 41–50. https://doi.org/10.1101/sqb.2014.79.024679

Chen, C. L., Aymanns, F., Minegishi, R., Matsuda, V. D. V., Talabot, N., Günel, S., Dickson, B. J., & Ramdya, P. (2023). Ascending neurons convey behavioral state to integrative sensory and action selection brain regions. *Nature Neuroscience, 26*(4), 682–695. https://doi.org/10.1038/s41593-023-01281-z

Chiel, H. J., & Beer, R. D. (1997). The brain has a body: Adaptive behavior emerges from interactions of nervous system, body and environment. *Trends in Neurosciences, 20*, 553–557. https://doi.org/10.1016/S0166-2236(97)01149-1

Cisek, P. (2019). Resynthesizing behavior through phylogenetic refinement. *Attention, Perception, & Psychophysics, 81*, 2265–2287. https://doi.org/10.3758/s13414-019-01760-1

Cisek, P. (2021a). Evolution of behavioural control from chordates to primates. *Philosophical Transactions of the Royal Society B: Biological Sciences Royal Society, 377*(1844), 20200522. https://doi.org/10.1098/rstb.2020.0522

Cisek, P. (2021b). An evolutionary perspective on embodiment. In M. D. Robinson & L. E. Thomas (Eds.), *Handbook of embodied psychology: Thinking, feeling, and acting* (pp. 547–571). https://doi.org/10.1007/978-3-030-78471-3_24

Cisek, P., & Hayden, B. Y. (2022). Neuroscience needs evolution. *Philosophical Transactions of the Royal Society B: Biological Sciences, 377*(1844), 20200518. https://doi.org/10.1098/rstb.2020.0518

Cisek, P., & Kalaska, J. F. (2010). Neural mechanisms for interacting with a world full of action choices. *Annual Review of Neuroscience, 33*, 269–298. https://doi.org/10.1146/annurev.neuro.051508.135409

Clark, A. (1998). Embodied, situated, and distributed cognition. In W. Bechtel & G. Graham (Eds.), *A Companion to Cognitive Science* (pp. 506–517). Blackwell. https://doi.org/10.1002/9781405164535.ch39

Clark, A., & Grush, R. (1999). Towards a cognitive robotics. *Adaptive Behavior, 7*(1), 5–16. https://doi.org/10.1177/105971239900700101

Corcoran, A. W., Pezzulo, G., & Hohwy, J. (2020). From allostatic agents to counterfactual cognisers: Active inference, biological regulation, and the origins of cognition. *Biology & Philosophy, 35*(3), 32. https://doi.org/10.1007/s10539-020-09746-2

Crapse, T. B., & Sommer, M. A. (2008). Corollary discharge circuits in the primate brain. *Current Opinion in Neurobiology, 18*(6), 552–557. https://doi.org/10.1016/j.conb.2008.09.017

Domenech, P., & Koechlin, E. (2015). Executive control and decision-making in the prefrontal cortex. *Current Opinion in Behavioral Sciences, 1,* 101–106. https://doi.org/10.1016/j.cobeha.2014.10.007

Donnarumma, F., Frosolone, M., & Pezzulo, G. (2024). Integrating large language models and active inference to understand eye movements in reading and dyslexia. *arXiv*. https://doi.org/10.48550/arXiv.2308.04941

Drescher, G. L. (1991). *Made-up minds: A constructivist approach to artificial intelligence*. MIT Press.

Engel, A. K., Maye, A., Kurthen, M., & König, P. (2013). Where's the action? The pragmatic turn in cognitive science. *Trends in Cognitive Sciences, 17*(5), 202–209. https://doi.org/10.1016/j.tics.2013.03.006

Friston, K. (2011). What is optimal about motor control? *Neuron, 72*(3), 488–498. https://doi.org/10.1016/j.neuron.2011.10.018

Friston, K., & Buzsáki, G. (2016). The functional anatomy of time: What and when in the brain. *Trends in Cognitive Sciences, 20*(7), 500–511. https://doi.org/10.1016/j.tics.2016.05.001

Friston, K., Heins, C., Verbelen, T., Da Costa, L., Salvatori, T., Markovic, D., Tschantz, A., Koudahl, M., Buckley, C., & Parr, T. (2024). From pixels to planning: Scale-free active inference. *arXiv*. https://doi.org/10.48550/arXiv.2407.20292

Friston, K. J., FitzGerald, T., Rigoli, F., Schwartenbeck, P., & Pezzulo, G. (2017). Active inference: A process theory. *Neural Computation, 29*(1), 1–49. https://doi.org/10.1162/NECO_a_00912

Gershman, S. J., & Niv, Y. (2012). Exploring a latent cause theory of classical conditioning. *Learning & Behavior, 40*(3), 255–268. https://doi.org/10.3758/s13420-012-0080-8

Gerstenberg, T., & Tenenbaum, J. B. (2017). Intuitive theories. In M. R. Waldmann (Ed.), *The Oxford handbook of causal reasoning* (pp. 515–548). Oxford University Press. https://doi.org/10.1093/oxfordhb/9780199399550.013.28

Gibson, J. J. (1979). *The ecological approach to visual perception*. Lawrence Erlbaum Associates, Inc.

Graziano, M. S. (2016). Ethological action maps: A paradigm shift for the motor cortex. *Trends in Cognitive Sciences, 20*(2), 121–132. https://doi.org/10.1016/j.tics.2015.10.008

Grush, R. (2004). The emulation theory of representation: motor control, imagery, and perception. *Behavioral and Brain Sciences, 27*(3), 377–396. https://doi.org/10.1017/S0140525X04000093.

Ha, D., & Schmidhuber, J. (2018). World models. *arXiv*. https://doi.org/10.5281/zenodo.1207631

Hunt, L. T., Kolling, N., Soltani, A., Woolrich, M. W., Rushworth, M. F., & Behrens, T. E. (2012). Mechanisms underlying cortical activity during value-guided choice. *Nature Neuroscience, 15*(3), 470–476. https://doi.org/10.1038/nn.3017

Hurley, S. (2008). The shared circuits model (SCM): How control, mirroring, and simulation can enable imitation, deliberation, and mindreading. *Behavioral and Brain Sciences, 31*, 1–22. https://doi.org/10.1017/S0140525X07003123

Jeannerod, M. (2006). *Motor cognition.* Oxford University Press.

Jékely, G., Godfrey-Smith, P., & Keijzer, F. (2021). Reafference and the origin of the self in early nervous system evolution. *Philosophical Transactions of the Royal Society B: Biological Sciences, 376*(1821), 20190764. https://doi.org/10.1098/rstb.2019.0764

Kato, S., Kaplan, H. S., Schrödel, T., Skora, S., Lindsay, T. H., Yemini, E., Lockery, S., & Zimmer, M. (2015). Global brain dynamics embed the motor command sequence of *Caenorhabditis elegans*. *Cell, 163*(3), 656–669. https://doi.org/10.1016/j.cell.2015.09.034

Keramati, M., & Gutkin, B. (2014). Homeostatic reinforcement learning for integrating reward collection and physiological stability. *eLife, 3*, e04811. https://doi.org/10.7554/eLife.04811

Kim, A. J., Fenk, L. M., Lyu, C., & Maimon, G. (2017). Quantitative predictions orchestrate visual signaling in drosophila. *Cell, 168*(1), 280–294e12. https://doi.org/10.1016/j.cell.2016.12.005

Kim, A. J., Fitzgerald, J. K., & Maimon, G. (2015). Cellular evidence for efference copy in *Drosophila* visuomotor processing. *Nature Neuroscience, 18*(9), 1247–1255. https://doi.org/10.1038/nn.4083

Kinzler, K. D., & Spelke, E. S. (2007). Core systems in human cognition. *Progress in Brain Research, 164*, 257–264. https://doi.org/10.1016/S0079-6123(07)64014-X

LeCun, Y. (2022). A path towards autonomous machine intelligence. *Open Review, 62*(1), 1–62.

Levin, M., Pezzulo, G., & Finkelstein, J. M. (2017). Endogenous bioelectric signaling networks: Exploiting voltage gradients for control of growth and form. *Annual Review of Biomedical Engineering, 19*, 353–387. https://doi.org/10.1146/annurev-bioeng-071114-040647

Livneh, Y., Ramesh, R. N., Burgess, C. R., Levandowski, K. M., Madara, J. C., Fenselau, H., Goldey, G. J., Diaz, V. E., Jikomes, N., Resch, J. M., Lowell, B. B., & Andermann, M. L. (2017). Homeostatic circuits selectively gate food cue responses in insular cortex. *Nature, 546*(7660), 611–616. https://doi.org/10.1038/nature22375

Lungarella, M., & Sporns, O. (2006). Mapping information flow in sensorimotor networks. *PLoS Computational Biology, 2*(10), e144. https://doi.org/10.1371/journal.pcbi.0020144

Lyon, P., Keijzer, F., Arendt, D., & Levin, M. (2021). Reframing cognition: Getting down to biological basics. *Philosophical Transactions of the Royal Society B: Biological Sciences, 376*(1820), 20190750. https://doi.org/10.1098/rstb.2019.0750

MacIver, M. A., & Finlay, B. L. (2022). The neuroecology of the water-to-land transition and the evolution of the vertebrate brain. *Philosophical Transactions of the Royal Society B: Biological Sciences, 377*(1844), 20200523. https://doi.org/10.1098/rstb.2020.0523

Maisto, D., Donnarumma, F., & Pezzulo, G. (2023). Interactive inference: A multi-agent model of cooperative joint actions. *IEEE Transactions on Systems, Man, and Cybernetics: Systems, 54,* no. 2, 704–715. https://doi.org/10.1109/TSMC.2023.3312585

Mannella, F., Maggiore, F., Baltieri, M., & Pezzulo, G. (2021). Active inference through whiskers. *Neural Networks, 144,* 428–437. https://doi.org/10.1016/j.neunet.2021.08.037.

Miller, G. A., Galanter, E., & Pribram, K. H. (1960). *Plans and the structure of behavior.* Holt, Rinehart and Winston.

Mugan, U., & MacIver, M. A. (2020). Spatial planning with long visual range benefits escape from visual predators in complex naturalistic environments. *Nature Communications, 11*(1), 3057. https://doi.org/10.1038/s41467-020-16102-1

Mussells Pires, P., Zhang, L., Parache, V., Abbott, L. F., & Maimon, G. (2024). Converting an allocentric goal into an egocentric steering signal. *Nature, 626*(8000), 808–818. https://doi.org/10.1038/s41586-023-07006-3

O'Regan, J. K., & Noe, A. (2001). A sensorimotor account of vision and visual consciousness. *Behavioral and Brain Sciences, 24*(5), 883–917. https://doi.org/10.1017/S0140525X01000115

Parr, T., Friston, K., & Pezzulo, G. (2024). Generative models for sequential dynamics in active inference. *Cognitive Neurodynamics, 18,* 3259–3272. https://doi.org/10.1007/s11571-023-09963-x.

Parr, T., Pezzulo, G., & Friston, K. J. (2022). *Active inference: The free energy principle in mind, brain, and behavior.* MIT Press.

Pezzulo, G. (2011). Grounding procedural and declarative knowledge in sensorimotor anticipation. *Mind and Language, 26*(1), 78–114. https://doi.org/10.1111/j.1468-0017.2010.01411.x

Pezzulo, G. (2017). Tracing the roots of cognition in predictive processing. *Open Mind.* https://doi.org/10.15502/9783958573215

Pezzulo, G., & Castelfranchi, C. (2007). The symbol detachment problem. *Cognitive Processing, 8*(2), 115–131. https://doi.org/10.1007/s10339-007-0164-0

Pezzulo, G., & Castelfranchi, C. (2009). Thinking as the control of imagination: A conceptual framework for goal-directed systems. *Psychological Research, 73*(4), 559–577. https://doi.org/10.1007/s00426-009-0237-z

Pezzulo, G., & Cisek, P. (2016). Navigating the affordance landscape: Feedback control as a process model of behavior and cognition. *Trends in Cognitive Sciences, 20*(6), 414–424. https://doi.org/10.1016/j.tics.2016.03.013

Pezzulo, G., D'Amato, L., Mannella, F., Priorelli, M., Van de Maele, T., Stoianov, I. P., & Friston, K. (2024). Neural representation in active inference: Using generative models to interact with—and understand—the lived world. *Annals of the New York Academy of Sciences, 1534*(1), 45–68. https://doi.org/10.1111/nyas.15118

Pezzulo, G., & Levin, M. (2016). Top-down models in biology: Explanation and control of complex living systems above the molecular level. *Journal of the Royal Society, Interface, 13*(124), 20160555. https://doi.org/10.1098/rsif.2016.0555

Pezzulo, G., Parr, T., Cisek, P., Clark, A., & Friston, K. (2023). Generating meaning: Active inference and the scope and limits of passive AI. *Trends in Cognitive Sciences*. https://doi.org/10.1016/j.tics.2023.10.002

Pezzulo, G., Parr, T., & Friston, K. (2022). The evolution of brain architectures for predictive coding and active inference. *Philosophical Transactions of the Royal Society B, 377*(1844), 20200531. https://doi.org/10.1098/rstb.2020.0531

Pezzulo, G., Parr, T., & Friston, K. (2024). Active inference as a theory of sentient behavior. *Biological Psychology, 186,* 108741. https://doi.org/10.1016/j.biopsycho.2023.108741

Pezzulo, G., Rigoli, F., & Friston, K. J. (2015). Active inference, homeostatic regulation and adaptive behavioural control. *Progress in Neurobiology, 136,* 17–35. https://doi.org/10.1016/j.pneurobio.2015.09.001

Pezzulo, G., Rigoli, F., & Friston, K. J. (2018). Hierarchical active inference: A theory of motivated control. *Trends in Cognitive Sciences, 22*(4), 294–306. https://doi.org/10.1016/j.tics.2018.01.009

Pezzulo, G., Zorzi, M., & Corbetta, M. (2021). The secret life of predictive brains: What's spontaneous activity for? *Trends in Cognitive Sciences, 25*(9), 730–743. https://doi.org/10.1016/j.tics.2021.05.007

Pfeifer, R., & Bongard, J. C. (2006). *How the body shapes the way we think.* MIT Press.

Piaget, J. (1954). *The construction of reality in the child.* Ballentine.

Powers, W. T. (1973). *Behavior: The control of perception.* Aldine.

Prescott, T. J., Redgrave, P., & Gurney, K. (1999). Layered control architectures in robots and vertebrates. *Adaptive Behavior, 7*(1), 99–127. https://doi.org/10.1177/105971239900700105

Priorelli, M., Maggiore, F., Maselli, A., Donnarumma, F., Maisto, D., Mannella, F., Stoianov, I. P., & Pezzilo, G. (2023). Modeling motor control in continuous-time active inference: A survey. *IEEE Transactions on Cognitive and Developmental Systems*. https://doi.org/10.1109/TCDS.2023.3338491

Priorelli, M., Pezzulo, G., & Stoianov, I. P. (2023). Deep kinematic inference affords efficient and scalable control of bodily movements. *Proceedings of the National Academy of Sciences, 120*(51), e2309058120. https://doi.org/10.1073/pnas.2309058120

Rao, R. P. N. (2024). A sensory-motor theory of the neocortex. *Nature Neuroscience, 27*(7), 1221–1235. https://doi.org/10.1038/s41593-024-01673-9

Redgrave, P., Prescott, T. J., & Gurney, K. (1999). The basal ganglia: A vertebrate solution to the selection problem? *Neuroscience, 89*, 1009–1023. https://doi.org/10.1016/S0306-4522(98)00319-4

Seth, A. K. (2013). Interoceptive inference, emotion, and the embodied self. *Trends in Cognitive Sciences, 17*(11), 565–573. https://doi.org/10.1016/j.tics.2013.09.007.

Smolensky, P., McCoy, R. T., Fernandez, R., Goldrick, M., & Gao, J. (2022). Neurocompositional computing: From the central paradox of cognition to a new generation of AI systems. *AI Magazine, 43*, 308–322. https://doi.org/10.1002/aaai.12065

Sperry, R. W. (1950). Neural basis of the spontaneous optokinetic response produced by visual inversion. *Journal of Comparative and Physiological Psychology, 43*(6), 482–489. https://doi.org/10.1037/h0055479

Stringer, C., Pachitariu, M., Steinmetz, N., Reddy, C. B., Carandini, M., & Harris, K. D. (2019). Spontaneous behaviors drive multidimensional, brainwide activity. *Science, 364*(6437), 255. https://doi.org/10.1126/science.aav7893

Sutton, R. S., & Barto, A. G. (1998). *Reinforcement learning: A introduction*. MIT Press.

Taniguchi, T., Murata, S., Suzuki, M., Ognibene, D., Lanillos, P., Uğur, E., Jamone, L., Nakamura, T., Ciria, A., Lara, B., & Pezzulo, G. (2023). World models and predictive coding for cognitive and developmental robotics: Frontiers and challenges. *Advanced Robotics, 37*, 780–806. https://doi.org/10.1080/01691864.2023.2225232

Tomasello, M. (2014). *A natural history of human thinking*. Harvard University Press.

Torigoe, M., Islam, T., Kakinuma, H., Fung, C. C. A., Isomura, T., Shimazaki, H., Aoki, T., Fukai, T., & Okamoto, H. (2021). Zebrafish capable of generating future state prediction error show improved active avoidance behavior in virtual reality. *Nature Communications, 12*(1), 5712. https://doi.org/10.1038/s41467-021-26010-7

Tschantz, A., Barca, L., Maisto, D., Buckley, C. L., Seth, A. K., & Pezzulo, G. (2022). Simulating homeostatic, allostatic and goal-directed forms of interoceptive control using active inference. *Biological Psychology, 169*, 108266. https://doi.org/10.1016/j.biopsycho.2022.108266

Van de Maele, T., Dhoedt, B., Verbelen, T., & Pezzulo, G. (2023). Bridging cognitive maps: A hierarchical active inference model of spatial alternation tasks and the hippocampal-prefrontal circuit. *arXiv*. https://doi.org/10.48550/arXiv.2308.11463

Verschure, P., Pennartz, C. M. A., & Pezzulo, G. (2014). The why, what, where, when and how of goal-directed choice: Neuronal and computational principles. *Philosophical Transactions of the Royal Society B: Biological Sciences, 369*, 20130483. https://doi.org/10.1098/rstb.2013.0483

Verschure, P. F. M. J. (2012). Distributed adaptive control: A theory of the mind, brain, body nexus. *Biologically Inspired Cognitive Architectures, 1*, 55–72. https://doi.org/10.1016/j.bica.2012.04.005

von Holst, E., & Mittelstaedt, H. (1950). The reafference principle: Interaction between the central nervous system and the periphery. *Naturwissenschaften, 37*, 464–476.

Wiener, N. (1948). *Cybernetics: Or control and communication in the animal and the machine*. MIT Press.

Wilson, M. (2002). Six views of embodied cognition. *Psychonomic Bulletin & Review, 9*(4), 625–636. https://doi.org/10.3758/BF03196322

Yu, A. J., & Dayan, P. (2005). Uncertainty, neuromodulation, and attention. *Neuron, 46*(4), 681–692. https://https://doi.org/10.1016/j.neuron.2005.04.026

II Embodied Artificial Intelligence

Part II explores embodied artificial intelligence (EAI). EAI is modeled based on collective intelligence and draws on the principle of active inference and the free energy principle (FEP). This part discusses the importance of equipping artificial embodied agents (AEAs) with a perceptuo-motor apparatus for perceiving and acting, which is crucial for grounding, embodiment, and situatedness. It also highlights the importance of body morphology, sensor morphology, and sensorimotor coordination for control and perception, criticizing the prioritization of the "robot brain." The concept of "strong" versus "weak" AI is tackled, defining the former as exhibiting human-like flexibility and creativity. The debate hinges on the definition of concepts and the necessity of embodiment for meaning, suggesting that strong AI with human-like concepts remains a possibility despite different sensorimotor systems. This part discusses whether large language models (LLMs) in EAI are merely "weakly" embodied and therefore inherit aspects of Good Old-Fashioned AI's (GOFAI's) symbol-grounding problem. The part ends with research directions for stronger EAI, including bio-inspired design, hybrid simulation to bridge the reality gap, and computational methods for studying embodied interactions, emphasizing standardization, accessible tools, and cross-disciplinary collaboration to advance the field.

Common Themes

Self-awareness in AEAs is a necessary element of consciousness (Reber et al., 2023), and interoception (sensing internal body states) is considered a building block of emotion and abstract concept knowledge (Connell et al., 2018). Making motor information available to AEAs would likely require providing them with a human-like body, supplied with the corresponding sensors, and teaching them how to interpret their input throughout development (chapter 6). Therefore, AEAs should have systems that simulate the cognitive mechanisms of biological agents, including tactile sensing, as they influence self-awareness and interoception (e.g., AlQallaf & Aragon-Camarasa, 2021).

Anthropomorphism is a pivotal factor determining both AEAs' efficiency and trustworthiness. AEAs that resemble humans in proportions and features are more effective at navigating human-centered environments. Humans are better able to perceive nonverbal cues (like facial expressions or gaze shifts) from AEAs when the agent is physically present, suggesting that appropriate embodied traits can improve communication efficiency (Mollahosseini et al., 2018; chapter 6).

Deciphering "concepts" and "conceptions" might shed light on the nature of embodiment in AEAs. Although AEAs may not conceptualize the world as human beings do, they might nonetheless have human-like concepts (chapter 7).

Weak embodiment in EAI (WEAI) faces issues similar to those of GOFAI. Despite this concern, today's EAI is still superior to the GOFAI because it enables more flexible and continuous learning with the real world through visual language models (VLMs) for perception and code LMs for action (chapter 8). Despite recent technological developments that have made impressive leaps to address some of these issues (e.g., LLMs), limitations remain:

1. **Symbol-grounding problem:** Multimodal LLMs must be grounded in different embodiments and their action spaces (Szot et al., 2024; chapter 8). The challenge of grounding language, especially when using LLMs for reasoning, still persists.
2. **Embodiment and situatedness:** Without detailed body morphology, this knowledge can't be effectively used for control or perception (chapter 8). Knowledge from texts, images, and videos forms a world model that is often disembodied or only shallowly embodied.
3. **Soft real-time interaction:** Materials used in these systems typically lack simulator-friendly models, and simulations don't always accurately reflect real-world scenarios due to the "reality gap," especially when dealing with complex, multipart, soft bodies and rich environmental interactions (chapter 9).

References

AlQallaf, A., & Aragon-Camarasa, G. (2021). Enabling the sense of self in a dual-arm robot. In A. Sciutti & J. Chen (Eds.), *2021 IEEE International Conference on Development and Learning (ICDL)* (pp. 1–7). IEEE. https://doi.org/10.1109/ICDL49984.2021.9515649

Connell, L., Lynott, D., & Banks, B. (2018). Interoception: The forgotten modality in perceptual grounding of abstract and concrete concepts. *Philosophical Transactions of the Royal Society B: Biological Sciences, 373*(1752), 20170143. https://doi.org/10.1098/rstb.2017.0143

Mollahosseini, A., Abdollahi, H., Sweeny, T. D., Cole, R., & Mahoor, M. H. (2018). Role of embodiment and presence in human perception of robots' facial cues. *International Journal of Human Computer Studies, 116*, 25–39. https://doi.org/10.1016/j.ijhcs.2018.04.005

Reber, A. S., Miller, W. B., & Baluška, F. (2023). Consciousness: Unicellular organisms know the secret. *Nature, 620*(7972), 37. https://doi.org/10.1038/d41586-023-02437-4

model. Importantly, a tractable lower bound on model evidence (equivalently, an upper bound on surprise) called *variational free energy* (Winn & Bishop, 2005) can be written as complexity minus accuracy. In minimizing free energy, a system thereby maximizes the predictive accuracy of its model while minimizing its complexity (implementing a Bayesian version of Occam's razor and thus limiting overfitting and poor generalization) (Sengupta & Friston, 2018).

Self-evidencing provides a unifying account of perception (belief updating in light of data), learning (parameter updating and model selection), and action, where the active inference framework is based on the hypothesis that actions are selected in order to maximize (expected) model evidence, assuming that the model encodes preferences in terms of prior beliefs (Friston, 2011; Ramstead et al., 2019). The *ensuing expected free energy* comprises both expected information gain and the expected value, where value is just the (log) preference for an outcome. This captures exactly the dual aspects of Bayes optimality—namely, optimal Bayesian experimental design (Balietti et al., 2021; Lindley, 1956; Mackay, 1992) and decision theory (Berger, 1985), where resolution of uncertainty is optimally balanced against the constraint that preferred outcomes are realized. Curiosity or novelty-seeking thus emerges as an existential imperative (Barto et al., 2013; Lindley, 1956; Mackay, 1992; Schmidhuber, 2006; Schwartenbeck et al., 2019; Still & Precup, 2012) and a fundamental aspect of intelligence. Crucially, the approach to existence as modeling just outlined can be applied recursively, in a nested fashion, to systems as well as their components, providing the foundations for mathematical theories of collective intelligence at any scale.

Modeling Collective Intelligence

While an emphasis on neuroscience in AI and machine learning (ML) is natural (given the locus of paradigmatic intelligence in the mammalian brain) some of the best work in this field has converged on information-theoretically characterized principles of self-organization whose effects are evident in the brain but whose principles of operation are much more general—for example, the ability of neurons to function as oriented edge detectors, or more broadly, feature detectors can be modeled based on domain-general principles such as sparse coding (Olshausen & Field, 1996) and predictive coding (Elias, 1955; Salvatori et al., 2025), which can be applied not only to neurons but to any suitable ensemble of processors. An important precedent of these ideas (of which active inference may indeed be seen as a contemporary extension) is the cybernetics movement, whose roots lie close to those of AI as a whole. This encompasses the good regulator theorem (Conant & Ashby, 1970) or "internal model principle" in control theory (Francis & Wonham, 1976), according to which systems that exist physically must contain structures that are homomorphic to whatever environmental factors they are capable of controlling. Even more fundamental is Ashby's "law of requisite variety" (Ashby, 1958), according to which a system that controls an environment (as represented by a set of observations) must possess at least as many degrees of freedom (in probabilistic terms, as much entropy) as the phenomenon that is controlled.

The generality and scale-free nature of these principles (Fields et al., 2021; Levin 2019) allows us to describe both human action and more pervasive and fundamental processes such as morphogenesis in multicellular organisms (Davies & Levin, 2022; Fields & Levin, 2022; Fields et al., 2020) literally and univocally as "intelligence." Moreover, observation of nature suggests that more complex forms of intelligence are composed of simpler intelligent systems. Recent research has shed some light on the mechanisms of such composition (Tissot et al., 2024), which amounts to model sharing among the constituent intelligences, as discussed below. These principles, applied consistently across scales and subsystems, may enable the construction of powerful, flexible AI systems. In the remainder of this chapter, we consider concrete directions of exploration along these lines.

Ecological AI

There are many generic advantages of a thoroughly Bayesian approach to AI. As noted above, the complexity of a model can be measured in terms of the divergence of posterior inferences from priors, providing a natural form of regularization. In addition, Bayesian approaches inherently allow for the quantification of uncertainty, which is crucial for intelligently combining information from sources of varying degrees of reliability, and plays an important theoretical role in accounts of diverse phenomena from attention (Zhang et al., 2021) to motor control (Toussaint & Goerick, 2010).

In the remainder of this chapter, we focus on the specific advantages that arise from Bayesian modeling of *ecosystems* of intelligence. That is, we examine forms of collective intelligence that emerges from the parallel and mutually constraining activity of many agents organized into hierarchies, networks, or ensembles.

Ecosystems as Factorized Generative Models

In practical applications of Bayesian methods, generally, factorization is a crucial aspect of model construction. A factorized distribution is simply one that can be written as a combination (product) of individual distributions. For example, a joint distribution $P(X, Y, Z)$ over random variables X, Y, and Z might factorize as $P(Z)P(X|Z)P(Y|Z)$, where $P(Z)$ is a prior over Z, and the remainder are conditional distributions. Factorized probability distributions, like factorized functions more generally, enable modular design of systems.

The factorization just considered defines a Bayesian network, enabling ancestral sampling—namely, the sequential sampling of latent and observed states from prior distributions and their associated conditional dependencies. Perhaps more importantly, operations such as marginalization and posterior inference can be performed efficiently using message-passing algorithms that involve only computations that use information available *locally* at each factor. Finally, local computation is an important constraint on the realistic modeling of biological systems.[1] While it is common to employ factorized generative models to describe the operations (e.g., cortical networks in the brain), multiscale active inference licenses the application of the same formalism to interactions among agents. As discussed above, the FEP is premised on the assumption of sparse dynamics—that is, the existence of conditional independencies that license a partition of the overall state space, where probabilistic boundaries are defined by Markov blankets. These conditional independence relations are identical in kind to those exhibited in simple Bayesian networks. Thus, ecosystems of intelligence can be described literally as large-scale factorized generative models, where each intelligent agent or component is a factor in the overall model.

Crucially, the same kind of objective function (variational free energy, as well as its expected values over time, given a model, i.e., expected free energy)—where variational free energy is a lower bound on the model evidence P (o)—governs the dynamics of all components of the system across both space and time. Just as individual models partition latent factors at different scales to capture local and global dependencies (Friston, Ramstead, et al., 2024), ecosystems distribute cognitive processes across multiple agents. Each agent maintains its own local model, and the interactions among these models contribute both to a *shared* generative model of the environment across similar agents (as discussed below) and to *higher-level* models at larger spatiotemporal scales, in which specialized agents model particular aspects of the environment, each contributing uniquely to the collective intelligence embodied by the higher-level model. Moreover, under certain assumptions, free energy is an extensive quantity (i.e., the free energy of subsystems combines additively to determine the free energy of the whole). In these cases, a clear formal relationship exists between smaller- and larger-scale optimization of generative models.

Factorization across Space

For concreteness, we consider two important aspects of this factorization at the ecosystem level. The first is factorization across models (or agents) existing at the same spatiotemporal scale. The simplest example of this relationship is the factorization of an overall system into subsystems distributed across space. We may, for example, imagine an overall state space partitioned into two sectors—the internal states μ of an agent, and external (environmental) states η, separated by a Markov blanket b.[2] Blanket states whose dynamics are independent of those of the external states are called "active" states a, while those that are independent of internal states are called "sensory" states s. This labeling is from the point of view of the reference agent, but from the environment's perspective, the roles of the active and sensory states, and internal and external states, respectively, are switched. In other words, the overall generative model (joint density) of the system may be written equivalently in two ways:

$$P(\mu agent, \eta agent, sagent, aagent) = P(\mu env, \eta env, senv, aenv)$$

$$\mu agent = \eta env$$

$$\mu env = \eta agent$$

$$sagent = aenv$$

$$senv = aagent$$

Clearly, as is to be expected in a model of action-perception loops, the overall generative model specified by these equations describes a cyclic flow:

$$\mu \rightarrow a \rightarrow \eta \rightarrow s \rightarrow \mu \ldots$$

so that a simple factorization of the joint density in terms of priors and conditionals, as in a Bayes net (a type of acyclic graph), is not possible. Nonetheless, local message passing algorithms (Murphy et al., 1999) can approximate the posterior distribution even in such cyclic graphs. One way of looking at the factorization of the joint density is in terms of the conditional independence of internal and external states given blanket states—that is, $P(\mu, \eta, b) = P(b)P(\mu|b)P(\eta|b)$, where $P(b)$ at a given time can be regarded as an "empirical prior" over blanket states (i.e., one deriving from recurrent dynamics). In recent treatments of the FEP, this description is streamlined by considering *paths* rather than states; that is, external and internal paths are independent when conditioned on blanket paths (Friston, Da Costa, et al., 2023).

A key feature of the FEP is that it models the fact that each agent has epistemic access to its environment only via blanket (in particular, sensory) states. Thus, the true underlying joint density described above is not accessible to any agent within the system. Instead, the internal dynamics of each subsystem (agent) can be understood as optimizing a (variational) approximation $Q(\eta)$ to the posterior $P(\eta|s)$ under the generative model, where $Q(\eta)$ is *parameterized* by $P(\mu|s)$ (Parr et al., 2020). The conditional independence among agents thus means that we can understand each agent in an ecosystem as optimizing a distinct generative model m of the environment by maximizing (a tractable internal proxy for) its marginal likelihood or model evidence, where Markov blankets both enable this factorization (and thus the learning and refinement of local per-agent models) and, as permeable probabilistic boundaries, ensure a kind of alignment or synchronization across models (i.e., generalized synchrony) (Bilek et al., 2022).

Spatiotemporal Coarse-Graining

Orthogonal to the kind of factorization just considered is a separation into functionally disjoint *spatiotemporal scales*. Typically, probabilistic models are defined in terms of (both observed and latent) random variables, whose values change on a given timescale, and parameters, whose values change on a slower timescale so that inferences occurring on the timescale of states share parameters. Concretely, parameters and states in the brain are often supposed to be encoded in synaptic connections and neuronal activities, respectively, but the principle applies to hierarchical generative models more generally (Friston, Heins, et al., 2024).

We may also consider separation across spatial scales. The kind of spatial factorization considered in the previous subsection may occur entirely *within* an agent that is a subsystem with respect to a larger containing system. For example, layers in neuronal processing hierarchies in the brain are separated by Markov blankets, and more generally, there are many ways to partition the state space of the brain, across several scales (Hipólito et al., 2021). Here, it is crucial to consider the ways in which Markov blankets limit information flow. Because all interactions are mediated by blanket states, there is a sense in which the internal states at a given scale do not contribute to the model at larger scales (Palacios et al., 2020; Ramstead et al., 2019). Two sets of internal states are equivalent, from the environment's point of view, so long as they produce the same distribution over active states, so at larger spatial scales it is *only the blanket states* that make a difference. Suppose, for example, that we have a system S partitioned into subsystems S^1 and S^2, separated by a Markov blanket. Each of S^k in turn is partitioned into a set of subsystems $\{s_1^k, s_2^k, \ldots s_n^k\} \in S^k$. The distribution over the active states a_{s^k} of each subsystem by definition depends on its internal (as well as, potentially, sensory) states. But suppose there is a further partitioning of the process by which the internal states influence the active states; for example, $P(a_{s^k}) = P(a_{s^k} | \mu_{s_1^k}) P(\mu_{s_1^k} | \mu_{s_2^k}) P(\mu_{s_2^k})$.[3] From the point of view of the system described at this spatiotemporal grain, the factorization of the distribution over active states is immaterial and can be summarized simply as $P(a_{s^k})$ (which contributes to the P [b] term in the factorization of the generative model $P[\mu_{s^k}, \eta_{s^k}, b]$, whereas above, the labeling of the μ and η terms, as well as the active and sensory states, is observer-dependent). In principle, the factorization of $P(a_{s^k})$ cannot be observed from the outside, since internal states cannot be measured. Nonetheless, the effects of the internal states may be felt, and the distribution over active states sharply constrains hypotheses about this internal factorization. For example, more degrees of freedom are needed to model a multimodal distribution than a unimodal one (for example, the former may be encoded as a Gaussian mixture model, and the latter as a single Gaussian)—a case of the law of requisite variety (Ashby, 1958; Maull & Godsiff, 2011). Thus, dynamics at a given spatiotemporal scale depend on rich structure at lower spatiotemporal scales, though these cross-scale effects are always blanket-mediated.

Shared Generative Models

Thus far, we have considered the *composition* of larger-scale generative models from smaller-scale ones. There, however, is a quite different (though complementary) way in which the generative models of subsystems may be related: shared generative models serve as a key mechanism for creating resilient multiagent systems. By "shared generative model," we mean the adoption by multiple agents of similar models of their external environments.

In the active inference framework, it is common to associate generative models with the "phenotype" of a given agent class (i.e., computational phenotyping [Che et al., 2015; Schwartenbeck & Friston, 2016]). Thus, members of a species, for example, ought to possess very similar models, at least in some ways. The adaptive value of shared generative models is obvious: when agents share similar models of their environment—including one another—they can more easily predict one another's behavior, communicate effectively, and work together toward common goals (Constant et al., 2022). This shared understanding forms the basis of shared protentions (Albarracin, Pitliya, et al., 2024)—aligned expectations about the future that enable smooth social interaction and cooperation.

Given such a shared model, interesting forms of collective behavior may arise at larger scales. Interestingly, Tissot et al. (2024) shows that a collection of smaller-scale active inference agents implement a specific higher-level generative model (a Boltzmann machine), provided the individual agents update their states *asynchronously*, while Heins et al. (2023) show in a somewhat different setting that cross-model resonance (i.e., *synchronous* updates) propagate the actions of individual components of the system to larger scales.

These are two examples of how the dynamics (and thus belief updating) of multiagent systems may arise from those of their constituents. In such cases, agents gather evidence for their own models and, in so doing, also evince a higher-level collective model. While sharing models can be highly adaptive for groups of agents, however, more interesting forms of collective intelligence involve not simply synchronization but functional specialization and diversity. This requires a nuanced negotiation of perspectives and an inference of relative positions within a shared environment. Thus, the fact that some ensembles or groups of agents share generative models does not imply identical models across all agents. Instead, we should expect partial overlaps and complementarities between different agents' models, just as different species in an ecosystem occupy distinct niches while contributing to overall ecosystem function (Miller et al., 2022).

The latter is exhibited compellingly in several studies of group decision-making and leadership dynamics in nature (Couzin et al., 2011; Strandburg-Peshkin et al., 2015), in which a relatively small number of more well-informed agents can, in concert with a pool of "uninformed" agents, drive coherent action at the level of the collective. One can perhaps explain these results by recognizing that the uninformed agents share a generative model that is a subspace of the richer model available to the specialists. These collective dynamics have been reproduced using multiagent active inference in Heins et al. (2024), which examines the general thesis that surprise minimization at the individual level gives rise to collective behavior.

While we have emphasized cooperative dynamics, competitive dynamics are also fundamental in nature and, moreover, may also be underwritten by partially shared generative models. While the preference distributions (phenotypes) of predator and prey obviously differ greatly, for example, their polar opposition in fact sets up a correspondence, an inverse relationship that allows both perspectives to be represented in terms of simple transformations of a shared model (Hyland et al., 2024). This is merely the most extreme example of the more general point that agents may use shared generative models and of assumptions about the ways in which they differ (i.e., the formal transformations that relate their beliefs to those of others) in how they implement "theory of mind" and predict one another's behavior (Devaine et al., 2014; Ondobaka et al., 2017). A range of multiagent interactions have been studied by combining tools from active inference with those of category theory (Albarracin, Pitliya, et al., 2024) and game theory (Demekas et al., 2024; Hyland et al., 2024).

The preservation of individual models distinct from a "consensus" model is both inevitable and necessary for a well-functioning ecosystem. The challenge lies in reconciling diverse perspectives to form a cohesive shared model. Agents must engage in a process of mutual inference and iterative model adjustment, sharing beliefs so as to align with the collective understanding while maintaining their unique viewpoints (Albarracin et al., 2022).

The Resilience of Ecosystems

The process by which agents come to share overlapping generative models via belief sharing mirrors the continuous adaptations seen in resilient ecosystems and is an important aspect of self-organization (Friston, Ramstead et al., 2024). In this section we examine the connection between active inference and resilience, which we have explored in other work (Albarracin, Ramstead, et al., 2024; Miller et al., 2022), and highlight resilience and balance through feedback—defining properties of natural ecosystems—as intrinsic, scale-free properties of ecological AI.

Homeostatic and Allostatic Agents and Resilience

The multiscale, interconnected nature of ecosystem resilience is reflected formally in the active inference account of collective intelligence we have given above, in terms of factorized generative models. At the microscale, individual organisms demonstrate resilience through physiological adaptations and behavioral plasticity. At broader scales, species populations show resilience through genetic diversity and adaptive capacity. At the ecosystem level, resilience appears as the ability of the system to absorb shocks, self-organize, and transform as necessary, maintaining core functions and structures in the face of disturbances. These scales of resilience are deeply interconnected, with the resilience of individual organisms contributing to population resilience, which in turn enhances ecosystem-level resilience.

As Miller et al. (2022) argue, resilience is the ability to handle change—either by resisting it (intertia), bouncing back from it (elasticity), or adapting to it (plasticity). From an active inference perspective, the first two notions of resilience are variations on the common theme of maintaining internal states within tolerable ranges (i.e., homeostasis), while the third encompasses allostasis, or the ability to adjust expectations (e.g., internal set points) to cope with environmental changes, thus preserving the integrity of the organism through change.

The idea that individual agents are resilient is of course foundational to the FEP. Not only are complex, identifiable systems by definition able to limit their exposure to unexpected sensory observations—a form of inertia—but they necessarily, in virtue of their existence, embody models of external states, calibrated on the timescales of both learning and inference, enabling actions that allow the system to rebound to attracting states (elasticity). In short, the FEP is by construction a scale-free description of resilient or sustainable systems. The equivalence of the FEP to a principle of (constrained) maximum-entropy inference (Jaynes, 1957; Sakthivadivel, 2022) allows us to demonstrate this via the technical notion of degeneracy.

Degeneracy may be thought of as "useful redundancy": duplication of parts or functions that may prevent catastrophic failure—for example, the existence of multiple kidneys in the human body or error-correcting codes in information transmission (Shannon, 1948), which provide robustness against noise. Beyond simple redundancy, components of complex systems in nature often exhibit partial overlap of function (cf. Whitacre & Bender, 2010). Degeneracy in this sense precisely describes the aforementioned trade-off between variety and uniformity of subagents in complex systems. Agents with partially overlapping models are capable of coordinated action and mutual understanding, while their differences enable specialized roles and diverse perspectives.

The notion of degeneracy maps to the approximate (variational) posterior distribution $Q(s)$ over states (Sajid et al., 2020) in active inference: the dynamics of variational free energy minimization tend to increase the entropy of $Q(s)$ (see "entropy" in equation 1, where F is the variational free energy), while at the same time ensuring that inferred states confer a high likelihood on sensory observations ("accuracy") and that the difference between prior and posterior beliefs ("complexity") is minimized. Thus, free energy minimization encourages agents to believe (insofar as doing so is consistent with the evidence) that a variety of circumstances may give rise to any given sensory observation.[4]

$$\begin{aligned} F &= \underbrace{\mathbb{E}_Q[-\ln P(s,o)]}_{\text{Energy}} - \underbrace{\mathbb{E}_Q[-\ln Q(s)]}_{\text{Entropy}} \\ &= \underbrace{D_{KL}[Q(s)|P(s)]}_{\text{Complexity}} - \underbrace{\mathbb{E}_Q[\ln P(o|s)]}_{\text{Accuracy}} \end{aligned} \tag{1}$$

Degeneracy is central to understanding "higher order" resilience through plasticity (Miller et al., 2022), which gives rise to elasticity and inertia through learning. Paradigmatically, plasticity concerns dynamics on the timescales of learning and model selection and is thus appropriate for characterizing ecosystems, particularly in terms of the evolution of their parts (individual agents). Evolution may be described as Bayesian model selection (Campbell, 2016; Friston, Friedman, et al., 2023), in which not only the parameters but more broadly the shape (i.e., functional form or family, and factorization) of the generative density P are adjusted.

Crucially, the "fitness" of a model depends both on its accuracy and its complexity or energy expenditure. Subject to (e.g., energetic/accuracy) constraints, $P(s)$ (i.e., the prior over states in an agent's model) can be adjusted via learning to match the entropy of $Q(s)$. In many implementations of active inference agents, this prior over states is an empirical prior based on a representation of state dynamics (i.e., a transition operator). Thus, agents come to believe in as wide a variety of state paths as possible given their defining constraints. Moreover, to connect this point to the preceding sections explicitly, the same degeneracy that favors a variety of hypotheses (i.e., a flexible, relatively wide range of beliefs) about the sensory data within one agent also underwrites a *variety of agents* at the ecosystem level—since an ecosystem must by the law of requisite variety (Ashby, 1958) possess sufficient degrees of freedom to model its environment, and an ecosystem's internal states factor into agents. Relatedly, it is to be expected in such a multiscale setting that distinct kinds of agents or phenotypes may come to play sensory, active, internal, and external roles within a higher-level model (Friston, Friedman, et al., 2023).

In this context, "higher-level models" correspond to populations or ecosystems, which interact with external states at their own proper scales. A subtle point here is that, as stressed in (Friston, Ramstead, et al., 2024) and elsewhere, the individuation of systems rests on sparse coupling.

Thus, for example, a new insect species introduced to an existing ecosystem should be identifiable as an interloper on the indigenous ecosystem by the sparseness of its interactions with the rest of the ecosystem compared with other creatures already assimilated into that system, whose habits should be expected to interact with those of other components of the system in richer ways. Mutual adjustment in the light of feedback will restore equilibrium after such a disruptive event, usually in terms of mutual adaptation. The upshot of all this for AI research is that agents designed with homeostasis and allostasis in view, unlike those trained to maximize ad hoc reward functions,[5] should be limited in the extent to which they are capable of "extreme" behavior (simply because such extremity is not conducive to maintaining states within tolerable ranges). The "reward" function is in this case definitionally tied to the agent's represented world (i.e., as it is a matter of gathering evidence for the agent's model) and thereby to its survival as a bounded individual. While some may worry that this fact would give active inference agents a potentially dangerous self-preservation instinct, within an ecosystem model, this instinct would be tied to the preservation of a phenotype that of necessity integrates into a wider system balancing the needs of many stakeholders. It is thus a more promising starting point for human-aligned AI than a system that extremizes an arbitrary reward or cost function.

Life and Itinerancy

While it is common to attribute some degree of knowledge, both of language and of world affairs, to contemporary language models (or at least to view them as embodying a large store of human knowledge), it is far more controversial and still regarded as a mistake in the mainstream of AI research, to attribute genuine agency, awareness, or consciousness to such systems (Seth, 2024). There are many reasons for this, not least of which is their purely reactive nature: although interpolation in high-dimensional spaces can yield outputs that are novel relative to the training data and thus exhibit hallmarks of creativity, the dominant generative AI systems operate in a largely "feedforward" way, conditioning their responses on user input but lacking intrinsic goals.

It is beyond the scope of this chapter to treat this subject in depth, but we wish to point out one way the kind of analysis undertaken above allows us to model more life-like systems. The key notion is that of *itinerant dynamics*: dynamics exhibiting periodicity or oscillation. Oscillation or movement is, arguably, fundamental to (at least in the sense of being a necessary condition for) life, and varieties of itinerant dynamics such as chaotic itinerancy (Inoue et al., 2020) have been found empirically to be characteristic of brain function.

As noted in literature on the FEP (Friston, Salvatori et al., 2023), the flow of states in any system described by a nonequilibrium steady state (i.e., an attractor existing far from thermodynamic equilibrium) can be written in terms of both a dissipative component (which describes a gradient ascent on the log probability of states) and a divergence-free or solenoidal component (by virtue of which the system visits distinct but equiprobable states).[6] The solenoidal component breaks detailed balance and is the mark of life-like systems existing away from thermodynamic equilibrium. This circular flow is associated with stability and persistence even in very simple systems, such as convection currents (DeLanda, 2011).

The preceding points are particularly salient from an ecosystem perspective. The more chaotic dynamics of complex forms of life can be expected to emerge from the layering of simpler forms of periodicity across scales, modulated by dissipative dynamics (i.e., the struggle for persistence). In a full ecosystem perspective, even the generative model P itself undergoes change and should exhibit both dissipative and solenoidal dynamics of its own. Itinerancy symbolizes the fundamentally *regulative* nature of homeostatic AI—and it is embeddedness in an ecosystem or wider network that regulates the activities of individual nodes and prevents submodels from visiting extreme states.

The Scope and Limits of Ecological AI

In this final section, we aim to temper any tendency toward utopianism with a grounded look at how and where the ideas presented above may provide tangible benefits in the implementation of AI systems today—including an examination of successful strands of research that already lean in the directions charted here—as well as a consideration of the major challenges to implementing ecosystems of agents.

There is one obvious challenge to the vision sketched here and in our previous work. It is that *ecosystems are large, extremely complex systems*. While one may sensibly recommend that an AI practitioner implement a Bayesian network, a support vector machine, or (given appropriate computing resources) a large-scale transformer, depending on the problem to be solved, the suggestion "implement an ecosystem of agents" may seem impractical as advice. Another way of making this point is to note that ML has, to date, worked with deliberately simplified models of neuronal networks and that, in general, simplification is a prerequisite to successful modeling—indeed, from the point of view of active inference itself, we expect creatures to adopt models of the external world that are only as complex as they need be to explain action-relevant data (Tschantz et al., 2020). Thus, it may seem that a reasonable, practically feasible computational description of an ecosystem of intelligent agents would, of necessity, abstract from many of the fine-grained details highlighted above.

For example, while a treatment of model parameters in terms of "beliefs"—that is, randomly distributed values with their own priors and posteriors—is preferable, from the point of view of Bayesian modeling, to the point estimates commonly employed in ML, the latter have been employed in part because doing so typically reduces the computational burden on simulations by orders of magnitude.

On one level, these concerns are well-founded, and it is of course still largely an open empirical question whether an AI research program based on the ecosystems perspective will bear fruit. That said, there are compelling reasons to suspect that this perspective points in the direction of the future of AI research. Addressing the criticisms just outlined invites us to consider the ecosystem hypothesis in a more practical light and thus to more sharply frame its advantages.

First, the proper aim is not unconstrained imitation of the ecosystems found in nature in all their detail, but rather the design of systems whose *principles of operation* mimic those of nature but that might be quite different otherwise. Indeed, the framing of the research program in Friston, Ramstead, et al. (2024) explicitly appealed to this idea and advised against imitation at a more superficial level (i.e., by focusing exclusively on neuroscience). Indeed, the trajectory of work on active inference implementations (Friston et al., 2017, 2021; Friston, Parr, et al., 2024; Friston, Salvatori, et al., 2023; Heins et al., 2022) has been precisely on expanding a toolkit of algorithmic building blocks based on the principles of active inference, with the aim of building and scaling tractable multiagent systems.

Second, we note that the ecosystem perspective becomes salient even when one focuses narrowly on (human or mammalian) biological intelligence, as transcending the limits of current AI systems almost certainly requires a complex, modular system composed of heterogeneous parts—that is, an ecosystem of agents (cf. LeCun, 2023). While the focus in deep learning has been almost exclusively on modeling cortical networks, important strands of research in ML have investigated models of brain structures at an intermediate scale (e.g., cortical columns [Hawkins et al., 2017, 2018]), as well as functions that are at least in part extra-cortical, such as longer-term memory [Wu et al., 2018]. The specialized cortical machinery employed by the hippocampus has also recently become an important locus of research (Raju et al., 2024).

than merely an ensemble (Levy & Bononno, 1997), and considered how this perspective on the nature of intelligence might inform AI research. As argued in (Friston, Ramstead, et al., 2024), the form of collective intelligence that we envision can emerge only from a network of essentially unique, epistemically and experientially diverse agents. As suggested above, this useful diversity of perspectives is a special case of functional specialization across the components of a complex system.

The complexities of modeling entire ecosystems (or engineering new ecosystems based on the principles that govern natural ones) are of course immense and require contributions from a diversity of fields. Indeed, the synthesis we have presented in this paper, like the field of active inference research, draws on these various strands and attempts to distill an abstraction that is useful as an orienting paradigm while respecting and being grounded in the mathematical details. This is possible because the principles of biological intelligence are scale-free.

Acknowledgments

We wish to thank Karl Friston, Chris Buckley, and other members of VERSES Research for indispensable feedback on the work presented in this manuscript. We also wish to thank all authors of the original "Designing Ecosystems of Intelligence" paper.

Notes

1. As an example, it is generally considered unlikely that cortical networks directly implement the back-propagation algorithm, but local algorithms such as predictive coding and equilibrium propagation may be used to achieve similar or identical results (Song et al., 2020; Millidge et al., 2022).

2. Of course, the external (and internal) states might themselves be partitioned again many times over, so that the overall system is characterized by many nested relationships of the kind described above.

3. We leave out any dependencies of active states on sensory states for simplicity.

4. Moreover, the same mathematics applies on both sides of any given Markov blanket, as discussed above. Degeneracy can thus be expected to be a property both of external generative processes (i.e., the external states about which agents harbor beliefs) and of the agent-internal dynamics that support belief updating.

5. Reward maximization can sometimes be cast in terms of homeostasis, but only where it is subject to finite bounds or constraints.

6. We note that in recent formulations of the FEP, it has been shown that the description of systems in terms of steady states can, for relevant purposes, be replaced by a description in terms of densities over paths (Friston, Salvatori, et al., 2023).

7. We thank Karl Friston for this point.

References

Albarracin, M., Demekas, D., Ramstead, M., & Heins, C. (2022). Epistemic communities under active inference. *Entropy, 24*(4), 476. https://doi.org/10.3390/e24040476

Albarracin, M., Pitliya, R. J., St. Clere Smithe, T., Friedman, D. A., Friston, K., & Ramstead, M. J. D. (2024). Shared protentions in multi-agent active inference. *Entropy, 26*(4), 303. https://doi.org/10.3390/e26040303

Albarracin, M., & Poirier, P. (2022). Enacting gender: An enactive-ecological account of gender and its fluidity. *Frontiers in Psychology, 13*, 772287. https://doi.org/10.3389/fpsyg.2022.772287

Albarracin, M., Ramstead, M., Pitliya, R. J., Hipolito, I., Da Costa, L., Raffa, M., Constant, A., & Manski, S. G. (2024). Sustainability under active inference. *Systems, 12*(5), 163. https://doi.org/10.3390/systems12050163

Ashby, R. W. (1958). Requisite variety and its implications for the control of complex systems. *Cybernetica, 1*(2), 83–99. https://link.springer.com/article/10.1007/s10441-005-7000-7

Balietti, S., Klein, B., & Riedl., C. (2021). Optimal design of experiments to identify latent behavioral types. *Experimental Economics, 24*(3), 772–799. https://doi.org/10.1007/s10683-020-09680-w

Barto, A., Mirolli, M., & Baldassarre, G. (2013). Novelty or surprise? *Frontiers in Psychology, 4*, 907. https://doi.org/10.3389/fpsyg.2013.00907

Berger, J. O. (1985). *Statistical decision theory and Bayesian analysis*. Springer Series in Statistics. Springer. https://doi.org/10.1007/978-1-4757-4286-2

Bilek, E., Zeidman, P., Kirsch, P., Tost, H., Meyer-Lindenberg, A., & Friston, K. (2022). Directed coupling in multi-brain networks underlies generalized synchrony during social exchange. *NeuroImage, 252*, 119038. https://doi.org/10.1016/j.neuroimage.2022.119038

Campbell, J. O. (2016). Universal Darwinism as a process of Bayesian inference. *Frontiers in Systems Neuroscience, 10*, 49. https://doi.org/10.3389/fnsys.2016.00049

Che, Z., Kale, D., Li, W., Bahadori, M. T., & Liu, Y. (2015). Deep computational phenotyping. In L. Cao, C. Zhang, T. Joachims, & G. I. Webb (Eds.), *Proceedings of the 21st ACM SIGKDD International Conference on Knowledge Discovery and Data Mining* (pp. 507–516). ACM. https://doi.org/10.1145/2783258.2783352

Conant, R. C., & Ashby, R. W. (1970). Every good regulator of a system must be a model of that system. *International Journal of Systems Science, 1*(2), 89–97. https://doi.org/10.1080/00207727008920220

Constant, A., Clark, A., Kirchhoff, M., & Friston, K. J. (2022). Extended active inference: Constructing predictive cognition beyond skulls. *Mind & Language, 37*(3), 373–394. https://doi.org/10.1111/mila.12330

Couzin, I. D., Ioannou, C. C., Demirel, G., Gross, T., Torney, C. J., Hartnett, A., Conradt, L., Levin, S. A., & Leonard, N. E. (2011). Uninformed individuals promote democratic consensus in animal groups. *Science, 334*(6062), 1578–1580. https://doi.org/10.1126/science.1210280

DeLanda, M. (2011). *Philosophy and simulation: The emergence of synthetic reason.* Bloomsbury Academic.

Davies, J., & Levin, M. (2022). Synthetic morphology via active and agential matter. *OSF Preprints*. https://doi.org/10.31219/osf.io/xrv8h

Dean, J., & Ghemawat, S. (2008). MapReduce: Simplified data processing on large clusters. *Communication ACM, 51*(1), 107–113. https://doi.org/10.1145/1327452.1327492

Demekas, D., Heins, C., & Klein, B. (2024). An analytical model of active inference in the iterated prisoner's dilemma. In C. L. Buckley (Ed.), *Active inference. IWAI 2023. Communications in computer and information science* (Vol. 1915, pp. 145–172). Springer. https://doi.org/10.1007/978-3-031-47958-8_10

Devaine, M., Hollard, G., & Daunizeau, J. (2014). Theory of mind: Did evolution fool us? *PloS ONE, 9*(2), e87619. https://doi.org/10.1371/journal.pone.0087619

Elias, P. (1955). Predictive coding. I. *IRE Transactions on Information Theory,* 1(1), 16–24. https://doi.org/10.1109/TIT.1955.1055126

Fields, C., Bischof, J., & Levin, M. (2020). Morphological coordination: A common ancestral function unifying neural and non-neural signaling. *Physiology, 35*(1), 16–30. https://doi.org/10.1152/physiol.00027.2019

Fields, C., Friston, K. J., Glazebrook, J. F., & Levin, M. (2022). A free energy principle for generic quantum systems. *Progress in Biophysics and Molecular Biology, 173*, 36–59. https://doi.org/10.1016/j.pbiomolbio.2022.05.006

Fields, C., Glazebrook, J. F., & Levin, M. (2021). Minimal physicalism as a scale-free substrate for cognition and consciousness. *Neuroscience of Consciousness, 2021*(1), niab013. https://doi.org/10.1093/nc/niab013

Fields, C., & Levin, M. (2022). Competency in navigating arbitrary spaces as an invariant for analyzing cognition in diverse embodiments. *Entropy, 24*(6), 819. https://doi.org/10.3390/e24060819

Francis, B. A., & Wonham, W. M. (1976). The internal model principle of control theory. *Automatica, 12*(5), 457–465. https://doi.org/10.1016/0005-1098(76)90006-6

Friston, K. J. (2011). Embodied inference: Or "I think therefore I am, if I am what I think." In W. Tschacher & C. Bergomi (Eds.), *The implications of embodiment: Cognition and communication* (pp. 89–125). Welcome Trust Centre for Neuroimaging. https://psycnet.apa.org/record/2014-14659-005

Friston, K. J. (2013). Life as we know it. *Journal of the Royal Society Interface, 10*(86), 20130475. https://doi.org/10.1098/rsif.2013.0475

Friston, K. J. (2019). A free energy principle for a particular physics. *arXiv*. https://doi.org/10.48550/arXiv.1906.10184

Friston, K. J., Da Costa, L., Hafner, D., Hesp, C., & Parr, T. (2021). Sophisticated Inference. *Neural Computation, 33*(3), 713–763. https://doi.org/10.1162/neco_a_01351

Friston, K. J., Da Costa, L., Sakthivadivel, D. A. R., Heins, R. C., Pavliotis, G. A., Ramstead, M. J. D., & Parr, T. (2023). Path integrals, particular kinds, and strange things. *Physics of Life Reviews, 47*, 1–34.

Friston, K., FitzGerald, T., Rigoli, F., Schwartenbeck, P., & Pezzulo, G. (2017). Active inference: A process theory. *Neural Computation, 29*(1), 1–49. https://doi.org/10.1162/NECO_a_00912

Friston, K. J., Friedman, D. A., Constant, A., Knight, V. B., Fields, C., Parr, T., & Campbell, J. O. (2023). A variational synthesis of evolutionary and developmental dynamics. *Entropy, 25*(7), 964. https://doi.org/10.3390/e25070964

Friston, K. J., Heins, C., Verbelen, T., Da Costa, L., Salvatori, T., Markovic, D., Tschantz, A., Koudahl, M., Buckley C., & Parr, T. (2024). From pixels to planning: Scale-free active inference. *arXiv*. https://doi.org/10.48550/arXiv.2407.20292

Friston, K. J., Levin, M., Sengupta, B., & Pezzulo, G. (2015). Knowing one's place: A free-energy approach to pattern regulation. *Journal of the Royal Society Interface, 12*(105). https://doi.org10.1098/rsif.2014.1383

Friston, K. J., Parr, T., Heins, C., Constant, A., Friedman, D., Isomura, T., Fields, C., Verbelen, T., Ramstead, M., Clippinger, J., & Frith, C. D. (2024). Federated inference and belief sharing. *Neuroscience and Biobehavioral Reviews, 156*, 105500. https://doi.org/10.1016/j.neubiorev.2023.105500

Friston, K. J., Ramstead, M. J., Kiefer, A. B., Tschantz, A., Buckley, C. L., Albarracin, M., Pitliya, R. J., Heins, C., Klein, B., Millidge, B., Sakthivadivel, D. A., St Clere Smithe, T.,

Koudahl, M., Tremblay, S. E., Petersen, C., Fung, K., Fox, J. G., Swanson, S., Mapes, D., & René, G. (2024). Designing ecosystems of intelligence from first principles. *Collective Intelligence, 3*(1). https://doi.org/10.1177/26339137231222481

Friston, K. J., Salvatori, T., Isomura, T., Tschantz, A., Kiefer, A., Verbelen, T., Koudahl, M., Paul, A., Parr, T., Razi, A., Kagan, B., Buckley, C. L., & Ramstead, M. J. D. (2023). Active inference and intentional behaviour. *arXiv*. http://arxiv.org/abs/2312.07547

Goodfellow, I., Pouget-Abadie, J., Mirza, M., Xu, B., Warde-Farley, D., Ozair, S., Courville, A., & Bengio, Y. (2014). Generative adversarial nets. In Z. Ghahramani, M. Welling, C. Cortes, N. Lawrence, & K.Q. Weinberger (Eds), *Advances in Neural Information Processing Systems* (Vol. 27, pp. 1–9).

Guénin-Carlut, A., & Albarracin, M. (2023). On embedded normativity: An active inference account of agency beyond flesh. In C. L. Buckley (Ed.), *Active inference. IWAI 2023. Communications in computer and information science* (Vol. 1915, pp. 91–105). Springer Nature. https://doi.org/10.1007/978-3-031-47958-8_7

Hawkins, J., Ahmad, S., & Cui, Y. (2017). A theory of how columns in the neocortex enable learning the structure of the world. *Frontiers in Neural Circuits, 11*, 81. https://doi.org/10.3389/fncir.2017.00081

Heins, C., Klein, B., Demekas, D., Aguilera, M., Buckley, C. L. (2023). Spin glass systems as collective active inference. In C. L. Buckley (Ed.), *Active Inference. IWAI 2022. Communications in Computer and Information Science* (Vol. 1721). Springer. https://doi.org/10.1007/978-3-031-28719-0_6

Heins, C., Millidge, B., Da Costa, L., Mann, R. P., Friston, K. J., & Couzin, I. D. (2024). Collective behavior from surprise minimization. *Proceedings of the National Academy of Sciences of the United States of America, 121*(17), e2320239121. https://doi.org/10.1073/pnas.2320239121

Heins, C., Millidge, B., Demekas, D., Klein, B., Friston, K., Couzin, I. D., & Tschantz, A. (2022). Pymdp: A Python library for active inference in discrete state spaces. *Journal of Open Source Software, 7*(73), 4098. https://doi.org/10.21105/joss.04098

Hinton, G. E., Sabour, S., & Frosst, N. (2018). Matrix capsules with EM routing. In *International conference on learning representations*. https://openreview.net/forum?id=HJWLfGWRb

Hipólito, I., Ramstead, M. J. D., Convertino, L., Bhat, A., Friston, K., & Parr, T. (2021). Markov blankets in the brain. *Neuroscience and Biobehavioral Reviews, 125*, 88–97. https://doi.org/10.1016/j.neubiorev.2021.02.003

Hohwy, J. (2014). The self-evidencing brain. *Noûs, 50*(2), 259–285. https://doi.org/10.1111/nous.12062

Holland, J. H. (2014). *Signals and boundaries: Building blocks for complex adaptive systems*. MIT Press.

Hyland, D., Gavenčiak, T., Da Costa, L., Heins, C., Kovarik, V., Gutierrez, J., Wooldridge, M. J., & Kulveit, J. (2024). Free-energy equilibria: Toward a theory of interactions between boundedly-rational agents. In *ICML 2024 Workshop on Models of Human Feedback for AI Alignment*. https://openreview.net/pdf?id=4Ft7DcrjdO

Inoue, K., Nakajima, K., & Kuniyoshi, Y. (2020). Designing spontaneous behavioral switching via chaotic itinerancy. *Science Advances*, *6*(46), eabb3989

Jaynes, E. T. (1957). Information theory and statistical mechanics. *Physical Review Journals Archive*, *106*(4), 620–630. https://link.aps.org/doi/10.1103/PhysRev.106.620

Kendall, J. D., Pantone, R. D., Manickavasagam, K., Bengio, Y., & Scellier, B. (2020). Training end-to-end analog neural networks with equilibrium propagation. *ArXiv*. https://api.semanticscholar.org/CorpusID:219260195

Kilteni, K., Groten, R., & Slater, M. (2012). The sense of embodiment in virtual reality. *Presence: Teleoperators and Virtual Environments*, *21*(4), 373–387. https://doi.org/10.1162/PRES_a_00124

Kuchling, F., Friston, K. J., Georgiev, G., & Levin, M. (2020). Morphogenesis as Bayesian inference: A variational approach to pattern formation and control in complex biological systems. *Physics of Life Reviews*, *33*, 88–108. https://doi.org/10.1016/j.plrev.2019.06.001

LeCun, Y. (2023). A path towards autonomous machine intelligence. OpenReview. https://openreview.net/forum?id=BZ5a1r-kVsf

Levin, M. (2019). The computational boundary of a "self": Developmental bioelectricity drives multicellularity and scale-free cognition. *Frontiers in Psychology*, *10*. https://doi.org/10.3389/fpsyg.2019.02688

Levy, P., & Bononno, R. (1997). *Collective intelligence: Mankind's emerging world in cyberspace*. Perseus Books.

Lindley, D. V. (1956). On a measure of the information provided by an experiment. *Annals of Mathematical Statistics*, *27*(4), 86–1005. https://doi.org/10.1214/aoms/1177728069

Lu, S., Harter, D. Wu, G., & Kotturu, P. (2019). The role of effector physicality and risk perception in virtual environments. *Proceedings of the Annual Meeting of the Cognitive Science Society*, *41*, 2208–2213.

Mackay, D. J. C. (1992). Information-based objective functions for active data selection. *Neural Computation*, *4*(4), 590–604. https://doi.org/10.1162/neco.1992.4.4.590

Maull, R. S., & Godsiff, P. (2011). Operationalising and managing variety. Discussion paper, University of Exeter. https://naplesforumonservice.com/wp-content/uploads/2021/02/Godsiff-Maull-Operationalising-and-Managing-Variety-1.pdf

Miller, M., Albarracin, M., Pitliya, R. J., Kiefer, A., Mago, J., Gorman, C., Friston, K. J., & Ramstead, M. J. D. (2022). Resilience and active inference. *Frontiers in Psychology, 13*, 1059117. 21. https://doi.org/10.3389/fpsyg.2022.1059117

Millidge, B., Salvatori, T., Song, Y., Bogacz, R., & Lukasiewicz, T. (2022). Predictive coding: Towards a future of deep learning beyond backpropagation? *arXiv*. https://doi.org/10.48550/arXiv.2202.09467

Murphy, KP., Weiss, Y., & Jordan, MI. (1999). Loopy belief propagation for approximate inference: An empirical study. In K. B. Laskey, & H. Prade (Eds.), *Uncertainty in Artificial Intelligence, Proceedings* (pp. 467–475). Morgan Kaufmann.

Olshausen, B. A., & Field, D. J. (1996). Emergence of simple-cell receptive field properties by learning a sparse code for natural images. *Nature, 381*(6583), 607–609. https://doi.org/10.1038/381607a0

Ondobaka, S., Kilner, J., & Friston, K. J. (2017). The role of interoceptive inference in theory of mind. *Brain and Cognition, 112*, 64–68. https://doi.org/10.1016/j.bandc.2015.08.002

Ororbia, A., & Friston, K. J. (2024). Mortal computation: A foundation for biomimetic intelligence. *arXiv*. https://arxiv.org/abs/2311.09589

Palacios, E. R., Razi, A., Parr, T., Kirchhoff, M., & Friston, K. (2020). On Markov blankets and hierarchical self-organisation. *Journal of Theoretical Biology, 486*, 110089. https://doi.org/10.1016/j.jtbi.2019.110089

Parr, T., Da Costa, L., & Friston, K. J. (2020). Markov blankets, information geometry and stochastic thermodynamics. *Philosophical Transactions of the Royal Society A, 378*(2164), 20190159. https://doi.org/10.1098/rsta.2019.0159

Raju, R. V., Guntupalli, J. S., Zhou, G., Wendelken, C., Lázaro-Gredilla, M., & George, D. (2024). Space is a latent sequence: A theory of the hippocampus. *Science Advances, 10*(31), eadm8470. https://doi.org/10.1126/sciadv.adm8470

Ramstead, M. J. D., Constant, A., Badcock, P. B., & Friston K. J. (2019). Variational ecology and the physics of sentient systems. *Physics of Life Reviews, 31*, 188–205. https://doi.org/10.1016/j.plrev.2018.12.002

Ramstead, M. J. D., Hesp, C., Tschantz, A., Smith, R., Constant, A., & Friston, K. (2021). Neural and phenotypic representation under the free-energy principle. *Neuroscience and Biobehavioral Reviews, 120*, 109–122. https://doi.org/10.1016/j.neubiorev.2020.11.024

Sajid, N., Parr, T., Hope, T. M., Price, C. J., & Friston, K. J. (2020). Degeneracy and redundancy in active inference. *Cerebral Cortex, 30*(11), 5750–5766. https://doi.org/10.1093/cercor/bhaa148

Sakthivadivel, D. A. R. (2022). Towards a geometry and analysis for Bayesian mechanics. *arXiv*. https://arxiv.org/abs/2204.11900

Salvatori, T., Mali, A., Buckley, C. L., Lukasiewicz, T., Rao, R. P. N., Friston, K., & Ororbia, A. (2025). A survey on neuro-mimetic deep learning via predictive coding. *Neural Networks: The Official Journal of the International Neural Network Society, 195*, 108161. https://doi.org/10.1016/j.neunet.2025.108161

Schmidhuber, J. (2006). Developmental robotics, optimal artificial curiosity, creativity, music, and the fine arts. *Connection Science, 18*(2), 173–187. https://doi.org/10.1080/09540090600768658

Schwartenbeck, P., & Friston, K. (2016). Computational phenotyping in psychiatry: A worked example. *eNeuro, 3*(4), ENEURO.0049-16.2016. https://doi.org/10.1523/ENEURO.0049-16.2016

Schwartenbeck, P., Passecker, J., Hauser, T. U., FitzGerald, T. H., Kronbichler, M., & Friston, K. J. (2019). Computational mechanisms of curiosity and goal-directed exploration. *eLife, 8*, e41703. https://doi.org/10.7554/eLife.41703

Sengupta, B., & Friston, K. J. (2018). How robust are deep neural networks? *arXiv*. https://doi.org/10.48550/arXiv.1804.11313

Shannon, C. E. (1948). A mathematical theory of communication. *Bell System Technical Journal, 27*(3), 379–423. https://doi.org/10.1002/j.1538-7305.1948.tb01338.x

Song, Y., Lukasiewicz, T., Xu, Z., & Bogacz, R. (2020). Can the brain do backpropagation? Exact implementation of backpropagation in predictive coding networks. *Advances in Neural Information Processing Systems, 33*, 22566–22579.

Still, S., & Precup, D. (2012). An information-theoretic approach to curiosity-driven reinforcement learning. *Theory in Biosciences, 131*(3), 139–148. https://doi.org/10.1007/s12064-011-0142-z

Strandburg-Peshkin, A., Farine, D. R., Couzin, I. D., & Crofoot, M. C. (2015). Shared decision-making drives collective movement in wild baboons. *Science, 348*(6241), 1358–1361. https://doi.org/10.1126/science.aaa5099

Tissot, T., Levin, M., Buckley, C. L., & Watson, R. (2024). An ability to respond begins with inner alignment: How phase synchronisation effects transitions to higher levels of agency. *bioRxiv*. https://www.biorxiv.org/content/early/2024/02/21/2024.02.16.580248.full.pdf

Toussaint, M., & Goerick, C. (2010). *A Bayesian view on motor control and planning. From motor learning to interaction learning in robots*. Springer. https://api.semanticscholar.org/CorpusID:15852445

Tschantz, A., Seth, A. K., & Buckley, C. L. (2020). Learning action-oriented models through active inference. *PLoS Computational Biology, 16*(4), e1007805. https://doi.org/10.1371/journal.pcbi.1007805

Whitacre, J., & Bender, A. (2010). Degeneracy: A design principle for achieving robustness and evolvability. *Journal of Theoretical Biology, 263*(1), 143–153. https://doi.org/10.1016/j.jtbi.2009.11.008

Winn, J., & Bishop, C. M. (2005). Variational message passing. *Journal of Machine Learning Research, 6*, 661–694. https://www.jmlr.org/papers/volume6/winn05a/winn05a.pdf

Wu, Y., Wayne, G., Graves, A., & Lillicrap, T. (2018). The Kanerva machine: A generative distributed memory. *arXiv*. https://arxiv.org/abs/1804.01756

Yi, Su-in, Kendall, J. D., Williams, R. S., & Kumar, S. (2022). Activity-difference training of deep neural networks using memristor crossbars. *Nature Electronics, 6*, 1–7. https://doi.org/10.1038/s41928-022-00869-w

Zhang, S., Fan, X., Chen, B., & Zhou, M. (2021). Bayesian attention belief networks. *arXiv*. https://arxiv.org/abs/2106.05251

6 Grounded, Embodied, and Situated Intelligence in Humans and Artificial Agents

Carlotta I. Zona, Martin H. Fischer, and Alex Miklashevsky

Introduction

This chapter emphasizes the importance of equipping artificial embodied agents (AEAs) with a perceptuo-motor apparatus that enables them to perceive and act by simulating the cognitive mechanisms of biological agents.

Grounding: The Case of Tactile Sensing

Whether and how biological agents process information from the environment depends on their perceptuo-motor apparatus. While vision has received much attention, tactile sensing is essential for reproducing human perception and is a prerequisite to performing complex motor tasks (Johansson & Flanagan, 2009). Further, the perception of one's body in action as part of oneself has been demonstrated to influence cognitive processes and thus stands as a crucial element of self-awareness and interoception—that is, sensing the body's internal states or signals (AlQallaf & Aragon-Camarasa, 2021; Carvalho & Damasio, 2021; K. Y. Lin et al., 2021; Rodriguez et al., 2017). In turn, self-awareness is a necessary element of consciousness (Reber et al., 2023), and interoception is considered a building block of emotion and abstract concept knowledge (Connell et al., 2018). Clearly, reproducing tactile perception in AEAs presents practical challenges, as tactile sensors are distributed throughout the whole body and comprise various specialized receptors (e.g., for temperature, pressure, pain). Further, tactile sensing/exploring might create disturbances in the AEA's surroundings by colliding with objects.

Robotics has leveraged haptic feedback to enable users to receive tactile information via artificial interfaces. Much of the effort to reproduce tactile sensing so far has been focused on medical use and assistive technologies simulating sensorimotor functions, such as prosthetics and artificial organs (Barontini et al., 2023; Cianchetti et al., 2018; Jadhav et al., 2022). However, some actuators have been developed for dexterous manipulation in everyday tasks (Boivin et al., 2023; Khadivar et al., 2023; Kuppuswamy et al., 2020; She et al., 2021; Yao & Billard, 2023), enabling AEAs to estimate the physical properties of objects, adjust their motor behavior to current environmental conditions, and perform cooperative tasks with humans (e.g., Abiri et al., 2019; Bergholz et al., 2023; Giri et al., 2021; Khodr et al., 2020; Rodríguez-Sedano et al., 2023; Selim et al., 2024; Takahashi et al., 2023; Vouloutsi et al., 2023). Current approaches exploit transducers able to sense pressure and force (Roberts et al., 2021). Crucially, soft materials are not only compliant with objects in the environment but can also be made to use tactile information to infer objects' features. For example, M. A. Lin et al. (2022) designed haptic sensors inspired by feline whiskers that enable AEAs to perceive their surroundings without causing disturbances. When stress resulting from contact between the actuator and an object is detected, the AEA creates a map of the object with no need for visual input. Kuppuswamy et al. (2020) developed a gripper system that uses a vision-based sensor on an inflatable actuator. This gripper can adapt flexibly to objects with different shapes and surfaces and uses shear-stress information for tactile sensing.

A similar approach has used a vision-based sensor on soft-gel actuators (Dong et al., 2017; Donlon et al., 2018; Yuan, Dong, & Adelson, 2017; Yuan, Zhu, et al., 2017) to achieve increasingly fine-grained tactile sensing and dexterous manipulation (Liu, Ma, & Adelson, 2023; Liu, Yañez, & Adelson, 2023; Tippur & Adelson, 2023). For instance, She et al. (2021) developed two coordinated touch-sensitive grippers able to insert cables into jack connectors. One gripper pinches the cable softly while the other holds it more firmly and pulls it to reach its extremity. The continuous tactile feedback loop is a crucial element of this procedure, and the use of soft materials enables collecting information (e.g., about the current orientation of the cable or the attrition generated by the sliding movement), which in turn is used as sensorimotor feedback to correct the AEA's behavior on the fly. K. Y. Lin et al. (2021) presented an approach that combines fiber-based deformation sensors measuring the AEA's physical state (such as bend, twist, and elongation) and sensors measuring overall and local pressure. Furthermore, the properties of conductive fluids have been used to enable proprioceptive and haptic feedback in AEAs in response to bending (Boivin et al., 2023; Helps & Rossiter, 2018).

This body of work shows that "when soft . . . components emerge as capable alternatives to rigid counterparts, they are posed to yield great advances in [. . .] assistive and wearable device applications" (Boivin et al., 2023, p. 1) as well as in human-robot interaction (HRI) (K. Y. Lin et al., 2021). However, while implementing body-based solutions to complex tasks is crucial to model autonomous behavior, the dearth of control parameters implies scarce controllability of AEAs (Mengaldo et al., 2022). In addition, complex morphological computations require extremely complex and expensive hardware, presenting both feasibility (Laschi, 2022) and sustainability concerns (Speck et al., 2017). Thus, because both access to the end products of robotics research and the environmental consequences of developing such products are unevenly distributed, this is likely to further widen the social gap in both wealth and quality of life.

Embodiment: Body-Supported Learning

Recent advances in robotics have stressed that embodiment is as relevant for learning in robots as it is in humans. When learning, evolutionary AEAs have two goals: achieving an optimal body morphology to perform the task (an "evolutionary search" stage) and learning how to perform the task itself (i.e., a "training" stage). While these goals have traditionally been implemented sequentially, recent work has highlighted the crucial contribution of incorporating embodied task learning during rather than after the evolutionary search. Massi et al. (2019) compared the performances of two simulated four-legged AEAs, which adjusted their movement dynamics during the evolutionary search phase. In addition, both AEAs were provided with an "adaptive module", a cerebellum-inspired circuit presenting a modular structure that enables the AEA to learn locomotion. The results showed improved performance when the learning module was involved during the evolutionary search phase compared to when it was available only after the search phase. This result was robust in altered environmental conditions, suggesting (1) that learning was more stable to changes in the environment when it was performed alongside the search phase and (2) that embodiment plays a role not only in humans' but also in AEAs' learning (Basalp et al., 2021; Houbre et al., 2020). In addition, and in contrast to humans, such AEAs can modify their body morphology to solve tasks, accommodate learning, and overcome challenges.

In developmental robotics, the acquisition of concepts is implemented in AEAs via the same principles known from studies on human development, such as joint action, joint attention, and direct instruction (e.g., Di Nuovo & Cangelosi, 2021; Di Nuovo & Jay, 2019; Jacquey et al., 2019). This research has often used a humanoid iCub, whose ability to perceive, integrate, and remember visual and auditory stimuli allows it to recognize and track objects, as well as acquire spoken object names and match them with their referents for later retrieval. What is more, the iCub can acquire higher-order, abstract concepts by aggregating primitive embodied experiences across temporal sequences and combining them conceptually, as observed in human child studies (for reviews, see Cangelosi & Stramandinoli, 2018; Di Nuovo & Cangelosi, 2021). Modeling embodied experiences that support the acquisition of numerical knowledge in AEAs has been a notable proof of concept for embodied cognition (EC) because mathematics has traditionally been considered the abstract knowledge domain *par excellence.*

Challenging this traditional view, effects of sensorimotor experiences (such as finger counting) on the formation of numerical concepts were documented in both behavioral and neurofunctional studies with children (Barrocas et al., 2020; Soylu et al., 2018). Compatible with this evidence, the iCub exhibited improved recognition of spoken digits when the digit's spoken form was linked to an embodied representation of its own fingers during training (Di Nuovo & McClelland, 2019). Interestingly, while the cognitive advantage from embodying finger configurations has also been reported in written-digit recognition (Di Nuovo, 2017), arithmetic biases shown by humans, such as a preference to err by five units from a correct result, also emerge in the iCub due to its ability to reproduce the five-based counting system (Di Nuovo, 2018). In addition to finger configurations, Pecyna et al. (2022) recently used the iCub to explore the role of physically pointing to counted objects for counting accuracy. To do so, the information about the current number in the series was associated with an embodied representation of the angles of arm joints. Similar to prior developmental evidence, counting performance improved when pointing to the objects but not when observing pointing, a control condition lacking the crucial embodied information about joint angles.

A major challenge faced by developmental robotics is that learning requires close supervision and controlled input from the experimenter, making the process resource-consuming. To address this limitation, developing machine-readable databases of embodied experiences is essential (Davies et al., 2021; Hellou et al., 2023). For instance, some approaches have enabled large language models (LLMs) to access perceptual—primarily visual—information. So-called multimodal LLMs (Yin et al., 2023) consist of three elements: data sets of images with corresponding text information (the "eyes" of the model), an LLM itself, and a learnable interface between the two other components. Most recent developments even combine multimodal LLMs with "embodied" physical components, thus allowing the models to actively interact with the external world through robotic hands (Driess et al., 2023).

However, while visual and auditory information can be easily made available to artificial systems, this input still constitutes only a portion of the sensorimotor spectrum routinely processed by human agents. For example, Lynott et al. (2020) found that haptic and interoceptive information is as relevant for conceptual processing as auditory information. Indeed, even olfactory and gustatory information, while negligible for most concepts, can still be crucial for specific semantic categories, such as food items. Following this line of reasoning, making motor information available to an AEA would require providing it with a human-like body supplied with the corresponding sensors and teaching it how to interpret its input throughout development—or, alternatively, using a machine-readable database of sensorimotor experiences. Additionally, a critical issue related to the training of artificial systems involves the bias implicitly encoded in our description of reality. Concerns related to discrimination and representation pose significant challenges in developing these systems and deploying them without human supervision, as they are programmed to perpetuate the patterns found in the data they are trained on.

Situatedness: Predictive Decision-Making

Computing abstract internal states is a possible strategy for AEAs to tackle everyday tasks within their environment. For example, it seems relatively easy to specify explicit control policies for an AEA to catch a ball: With information on launch angle, initial velocity, and gravitational acceleration, the AEA can compute time of flight and horizontally and vertically traveled distance. However, ball-catching success is likely to be low because additional context-dependent factors are not considered—for example, air resistance, which in turn is determined by the ball's size, shape, and material, as well as the properties of the air. Yet, including the relevant parameters in the equation will render the decision-making process less time- and energy-efficient, preventing the AEA from performing the computation in time for action selection and successful execution.

Rather than performing such computations, biological agents rely heavily on current and past sensorimotor processes as pivotal cues in their judgment and decision-making. Crucially, past experiences are used to form expectations about the perceptual parameters associated with specific motor behaviors and events (e.g., Raab, 2017). Comparing current and desired perceptual parameters enables biological agents to predict the world's most likely future state and act on this future state anticipatorily (for reviews, see Clark, 2013, 2022). Such "predictive processing" is crucial in dynamic, time-constrained scenarios where the balance between prediction accuracy and delay in action selection crucially determines the probability of success. For example, biological agents can anticipate and avoid collisions with a looming object by judging the speed of increment in the object's angular size in the visual field. Such a solution is implemented in locusts with only two "movement-detecting" neurons (Rind et al., 2003) and triggers collision avoidance during group flight, providing inspiration for mechanisms enabling swarm-like behavior in AEAs (e.g., Fu et al., 2019).

In artificial agents, interacting with the environment in well-defined ways can lead to developing action-mediated internal states. These do not inherently convey information; instead, they derive meaning through appropriate interactions with the environment. Recent efforts have attempted to exploit such states to inform the AEA's decision-making processes in a highly context-contingent fashion, with a continuous sensorimotor feedback loop enabling the update of such states.

Much research in human-AEA interaction has used deep learning algorithms, recurrent neural networks (e.g., Chen et al., 2022; Jaouedi et al., 2019; Nweke et al., 2018), and 3D skeleton trajectories for action prediction in real-time (Das et al., 2019; Reily et al., 2018; P. Zhang et al., 2019). For

Das, S., Chaudhary, A., Bremond, F., & Thonnat, M. (2019). Where to focus on for human action recognition? In K. Bowyer, S. Dickinson, & G. Medioni (Eds.), *2019 IEEE Winter Conference on Applications of Computer Vision, WACV 2019* (pp. 71–80). IEEE Computer Society. https://doi.org/10.1109/WACV.2019.00015

Davies, S., Lucas, A., Ricolfe-Viala, C., & Di Nuovo, A. (2021). A database for learning numbers by visual finger recognition in developmental neuro-robotics. *Frontiers in Neurorobotics, 15*, 619504. https://doi.org/10.3389/fnbot.2021.619504

Di Nuovo, A. (2017). An embodied model for handwritten digits recognition in a cognitive robot. In D. B. Fogel (Ed.), *2017 IEEE Symposium Series on Computational Intelligence (SSCI)* (pp. 1–6). IEEE. https://doi.org/10.1109/SSCI.2017.8285274

Di Nuovo, A., & Cangelosi, A. (2021). Abstract concept learning in cognitive robots. *Current Robotics Reports, 2*(1), 1–8. https://doi.org/10.1007/s43154-020-00038-x

Di Nuovo, A., & Jay, T. (2019). Development of numerical cognition in children and artificial systems: A review of the current knowledge and proposals for multi-disciplinary research. *Cognitive Computation and Systems, 1*(1), 2–11. https://doi.org/10.1049/ccs.2018.0004

Di Nuovo, A., & McClelland, J. L. (2019). Developing the knowledge of number digits in a child-like robot. *Nature Machine Intelligence, 1*(12), 594–605. https://doi.org/10.1038/s42256-019-0123-3

Dong, S., Yuan, W., & Adelson, E. H. (2017). Improved GelSight tactile sensor for measuring geometry and slip. In *2017 IEEE/RSJ International Conference on Intelligent Robots and Systems (IROS)*, (pp. 137–144). https://doi.org/10.1109/IROS.2017.8202149

Donlon, E., Dong, S., Liu, M., Li, J., Adelson, E., & Rodriguez, A. (2018). GelSlim: A high-resolution, compact, robust, and calibrated tactile-sensing finger. In *2018 IEEE/RSJ International Conference on Intelligent Robots and Systems (IROS), 1*, (pp. 1927–1934). https://doi.org/10.1109/IROS.2018.8593661

Driess, D., Xia, F., Sajjadi, M. S. M., Lynch, C., Chowdhery, A., Ichter, B., Wahid, A., Tompson, J., Vuong, Q., Yu, T., Huang, W., Chebotar, Y., Sermanet, P., Duckworth, D., Levine, S., Vanhoucke, V., Hausman, K., Toussaint, M., Greff, K., Zeng, A., Mordatch, I., & Florence, P. (2023). Palm-e: An embodied multimodal language model. *arXiv*. https://arxiv.org/abs/2303.03378

Fu, Q., Hu, C., Peng, J., Rind, F. C., & Yue, S. (2019). A robust collision perception visual neural network with specific selectivity to darker objects. *IEEE Transactions on Cybernetics, 50*(12), 5074–5088. https://doi.org/10.1109/TCYB.2019.2946090

Giri, G. S., Maddahi, Y., & Zareinia, K. (2021). An application-based review of haptics technology. *Robotics, 10*(1), 29. https://doi.org/10.3390/robotics10010029

Hellou, M., Gasteiger, N., Kweon, A., Lim, J., MacDonald, B. A., Cangelosi, A., & Ahn, H. S. (2023). Development and validation of a motion dictionary to create emotional gestures for the NAO robot. In M. Jang & C. H. Park (Eds.), *2023 32nd IEEE International Conference on Robot and Human Interactive Communication (RO-MAN)* (pp. 897–902). IEEE. https://doi.org/10.1109/RO-MAN57019.2023.10309372

Helps T., & Rossiter. J. (2018). Proprioceptive flexible fluidic actuators using conductive working fluids. *Soft Robot* 5:175–189. https://doi.org/10.1089/soro.2017.0012

Houbre, Q., Angleraud, A., & Pieters, R. (2020). Exploration and exploitation of sensorimotor contingencies for a cognitive embodied agent. In A. P. Rocha, L. Steels, & H. J. van den Herik (Eds.), *Proceedings of the 12th International Conference on Agents and Artificial Intelligence* (pp. 546–554). SCITEPRESS. https://doi.org/10.5220/0008951205460554

Jacquey, L., Baldassarre, G., Santucci, V. G., & O'regan, J. K. (2019). Sensorimotor contingencies as a key drive of development: From babies to robots. *Frontiers in Neurorobotics, 13*, 98. https://doi.org/10.3389/fnbot.2019.00098

Jadhav, S., Majit, M. R. A., Shih, B., Schulze, J. P., & Tolley, M. T. (2022). Variable stiffness devices using fiber jamming for application in soft robotics and wearable haptics. *Soft Robotics, 9*(1), 173–186. https://doi.org/10.1089/soro.2019.0203

Jaouedi, N., Boujnah, N., & Bouhlel, M. S. (2019). Deep learning approach for human action recognition using gated recurrent unit neural networks and motion analysis. *Journal of Computer Science, 15*(7), 1040–1049. https://doi.org/10.3844/jcssp.2019.1040.1049

Johansson, R. S., & Flanagan, J. R. (2009). Coding and use of tactile signals from the fingertips in object manipulation tasks. *Nature Reviews Neuroscience, 10*(5), 345–359. https://doi.org/10.1038/nrn2621

Kaminski, M. E., Rueben, M., Smart, W. D., & Grimm, C. M. (2017). Averting robot eyes. *Maryland Law Review, 76*(4), 983–1024.

Khadivar, F., Yao, K., Gao, X., & Billard, A. (2023). Online active and dynamic object shape exploration with a multi-fingered robotic hand. *Robotics and Autonomous Systems, 166*, 104461. https://doi.org/10.1016/j.robot.2023.104461

Khodr, H., Kianzad, S., Johal, W., Kothiyal, A., Bruno, B., & Dillenbourg, P. (2020). AlloHaptic: Robot-mediated haptic collaboration for learning linear functions. In S. Rossi, F. Ferland, & D. Morocco (Eds.), *2020 29th IEEE International Conference on Robot and Human Interactive Communication (RO-MAN)* (pp. 27–34). IEEE. https://doi.org/10.1109/RO-MAN47096.2020.9223563

Kuppuswamy, N., Alspach, A., Uttamchandani, A., Creasey, S., Ikeda, T., & Tedrake, R. (2020). Soft-bubble grippers for robust and perceptive manipulation. In J.-H. Kim

& J.-B. Song (Eds.), *2020 IEEE/RSJ International Conference on Intelligent Robots and Systems (IROS)* (pp. 9917–9924). IEEE.

Laschi, C. (2022). Embodied intelligence in soft robotics: Joys and sorrows. *IOP Conference Series: Materials Science and Engineering, 1261*(1), 012002. https://doi.org/10.1088/1757-899x/1261/1/012002

Leong, B., & Selinger, E. (2019). Robot eyes wide shut: Understanding dishonest anthropomorphism. In Danah Boyd & J. H. Morgenstern (Eds.), *FAT* 2019—Proceedings of the 2019 Conference on Fairness, Accountability, and Transparency* (pp. 299–308). ACM. https://doi.org/10.1145/3287560.3287591

Li, S., Zhang, L., & Diao, X. (2020). Deep-learning-based human intention prediction using RGB images and optical flow. *Journal of Intelligent and Robotic Systems: Theory and Applications, 97*(1), 95–107. https://doi.org/10.1007/s10846-019-01049-3

Lin, K. Y., Gamboa-Gonzalez, A., & Wehner, M. (2021). Soft robotic sensing, proprioception via cable and microfluidic transmission. *Electronics, 10*(24), 3166. https://doi.org/10.3390/electronics10243166

Lin, M. A., Reyes, E., Bohg, J., & Cutkosky, M. R. (2022). Whisker-inspired tactile sensing for contact localization on robot manipulators. In J.-H. Kim & J.-B. Song (Eds.), *IEEE/RSJ International Conference on Intelligent Robots and Systems (IROS), 2022-Octob*, (pp. 7817–7824). IEEE. https://doi.org/10.1109/IROS47612.2022.9982122

Liu, S. Q., Ma, Y., & Adelson, E. H. (2023). GelSight baby fin ray: A compact, compliant, flexible finger with high-resolution tactile sensing. In R. Yeow, P. Valdivia y Alvarado, J. Paik, & J. Jovanova (Eds.), *2023 IEEE International Conference on Soft Robotics (RoboSoft)* (pp. 1–8). IEEE. https://doi.org/10.1109/RoboSoft55895.2023.10122078

Liu, S. Q., Yañez, L. Z., & Adelson, E. H. (2023). GelSight EndoFlex: A soft endoskeleton hand with continuous high-resolution tactile sensing. In R. Yeow, P. Valdivia y Alvarado, J. Paik, & J. Jovanova (Eds.), *2023 IEEE International Conference on Soft Robotics (RoboSoft)* (pp. 1–6). IEEE. https://doi.org/10.1109/RoboSoft55895.2023.10122053

Lynott, D., Connell, L., Brysbaert, M., Brand, J., & Carney, J. (2020). The Lancaster sensorimotor norms: Multidimensional measures of perceptual and action strength for 40,000 English words. *Behavior Research Methods, 52*(3), 1271–1291. https://doi.org/10.3758/s13428-019-01316-z

Manzi, F., Peretti, G., Di Dio, C., Cangelosi, A., Itakura, S., Kanda, T., Ishiguro, H., Massaro, D., & Marchetti, A. (2020). A robot is not worth another: Exploring children's mental state attribution to different humanoid robots. *Frontiers in Psychology, 11*, 1–12. https://doi.org/10.3389/fpsyg.2020.02011

Maroto-Gómez, M., Bueno-Adrada, M., Malfaz, M., Castro-González, Á., & Salichs, M. Á. (2024). Human–robot pair-bonding from a neuroendocrine perspective: Modeling the effect of oxytocin, arginine vasopressin, and dopamine on the social behavior of an autonomous robot. *Robotics and Autonomous Systems, 176*, 104687. https://doi.org/10.1016/j.robot.2024.104687

Massi, E., Vannucci, L., Albanese, U., Capolei, M. C., Vandesompele, A., Urbain, G., Sabatini, A. M., Dambre, J., Laschi, C., Tolu, S., & Falotico, E. (2019). Combining evolutionary and adaptive control strategies for quadruped robotic locomotion. *Frontiers in Neurorobotics, 13*, 1–19. https://doi.org/10.3389/fnbot.2019.00071

McGinn, C. (2020). Why do robots need a head? The role of social interfaces on service robots. *International Journal of Social Robotics, 12*(1), 281–295. https://doi.org/10.1007/s12369-019-00564-5

Mengaldo, G., Renda, F., Brunton, S. L., Bächer, M., Calisti, M., Duriez, C., Chirikjian, G. S., & Laschi, C. (2022). A concise guide to modelling the physics of embodied intelligence in soft robotics. *Nature Reviews Physics, 4*(9), 595–610. https://doi.org/10.1038/s42254-022-00481-z

Mollahosseini, A., Abdollahi, H., Sweeny, T. D., Cole, R., & Mahoor, M. H. (2018). Role of embodiment and presence in human perception of robots' facial cues. *International Journal of Human Computer Studies, 116*, 25–39. https://doi.org/10.1016/j.ijhcs.2018.04.005

Nazir, T. A., Lebrun, B., & Li, B. (2023). Improving the acceptability of social robots: Make them look different from humans. *PLoS ONE, 18*(11), e0287507. https://doi.org/10.1371/journal.pone.0287507

Nweke, H. F., Teh, Y. W., Al-garadi, M. A., & Alo, U. R. (2018). Deep learning algorithms for human activity recognition using mobile and wearable sensor networks: State of the art and research challenges. *Expert Systems with Applications, 105*, 233–261. https://doi.org/10.1016/j.eswa.2018.03.056

Pecyna, L., Cangelosi, A., & Di Nuovo, A. (2022). A robot that counts like a child: A developmental model of counting and pointing. *Psychological Research, 86*(8), 2495–2511. https://doi.org/10.1007/s00426-020-01428-8

Raab, M. (2017). Motor heuristics and embodied choices: How to choose and act. *Current Opinion in Psychology, 16*, 34–37. https://doi.org/10.1016/j.copsyc.2017.02.029

Reber, A. S., Miller, W. B., & Baluška, F. (2023). Consciousness: Unicellular organisms know the secret. *Nature, 620*(7972), 37. https://doi.org/10.1038/d41586-023-02437-4

Reily, B., Han, F., Parker, L. E., & Zhang, H. (2018). Skeleton-based bio-inspired human activity prediction for real-time human–robot interaction. *Autonomous Robots, 42*(6), 1281–1298. https://doi.org/10.1007/s10514-017-9692-3

Rind, F. C., Santer, R. D., Blanchard, J. M., & Verschure, P. F. M. J. (2003). Locust's looming detectors for robot sensors. In F. G. Barth, J. A. C. Humphrey, & T. W. Secomb (Eds.), *Sensors and sensing in biology and engineering* (pp. 237–250). Springer Vienna. https://doi.org/10.1007/978-3-7091-6025-1_17

Roberts, P., Zadan, M., & Majidi, C. (2021). Soft tactile sensing skins for robotics. *Current Robotics Reports, 2*(3), 343–354. https://doi.org/10.1007/s43154-021-00065-2

Rodriguez, I., Martinez-Otzeta, J. M., Lazkano, E., Ruiz, T., & Sierra, B. (2017). On how self-body awareness improves autonomy in social robots. In *2017 IEEE International Conference on Robotics and Biomimetics (ROBIO)* (pp. 1688–1693). https://doi.org/10.1109/ROBIO.2017.8324661

Rodríguez-Sedano, F. J., Conde, M. Á., Rodríguez-Lera, F. J., & Chaparro-Peláez, J. (2023). Measuring the impact of haptic feedback in collaborative robotic scenarios. *Universal Access in the Information Society*. https://doi.org/10.1007/s10209-023-01040-8

Selim, M., Dresscher, D., & Abayazid, M. (2024). A comprehensive review of haptic feedback in minimally invasive robotic liver surgery: Advancements and challenges. *The International Journal of Medical Robotics and Computer Assisted Surgery, 20*(1), e2605. https://doi.org/10.1002/rcs.2605

She, Y., Wang, S., Dong, S., Sunil, N., Rodriguez, A., & Adelson, E. (2021). Cable manipulation with a tactile-reactive gripper. *The International Journal of Robotics Research, 40*(12–14), 1385–1401. https://doi.org/https://doi.org/10.1177/02783649211027233

Soylu, F., Lester, F. K., & Newman, S. D. (2018). You can count on your fingers: The role of fingers in early mathematical development. *Journal of Numerical Cognition, 4*(1), 107–135. https://doi.org/10.5964/jnc.v4i1.85

Speck, O., Speck, D., Horn, R., Gantner, J., & Sedlbauer, K. P. (2017). Biomimetic bio-inspired biomorph sustainable? An attempt to classify and clarify biology-derived technical developments. *Bioinspiration and Biomimetics, 12*(1), 011004. https://doi.org/10.1088/1748-3190/12/1/011004

Takahashi, M., Nagano, H., Tazaki, Y., & Yokokohji, Y. (2023). Effective haptic feedback type for robot-mediated material discrimination depending on target properties. *Frontiers in Virtual Reality, 4*, 1070739. https://doi.org/10.3389/frvir.2023.1070739

Tippur, M. H., & Adelson, E. H. (2023). GelSight360: An omnidirectional camera-based tactile sensor for dexterous robotic manipulation. In R. Yeow, P. Valdivia y Alvarado, J. Paik, & J. Jovanova (Eds.), *2023 IEEE International Conference on Soft Robotics (RoboSoft)* (pp. 1–8). IEEE. https://doi.org/10.1109/RoboSoft55895.2023.10122097

Vouloutsi, V., Cominelli, L., Dogar, M., Lepora, N., Zito, C., & Martinez-Hernandez, U. (2023). Towards living machines: Current and future trends of tactile sensing, grasping, and social robotics. *Bioinspiration & Biomimetics, 18*(2), 025002. https://doi.org/10.1088/1748-3190/acb7b9

Wang, B., & Rau, P. L. P. (2019). Influence of embodiment and substrate of social robots on users' decision-making and attitude. *International Journal of Social Robotics, 11*(3), 411–421. https://doi.org/10.1007/s12369-018-0510-7

Wang, Z., Boularias, A., Mülling, K., Schölkopf, B., & Peters, J. (2017). Anticipatory action selection for human–robot table tennis. *Artificial Intelligence, 247*, 399–414. https://doi.org/10.1016/j.artint.2014.11.007

Xia, C., Weng, C. Y., Zhang, Y., & Chen, I. M. (2020). Vision-based measurement and prediction of object trajectory for robotic manipulation in dynamic and uncertain scenarios. *IEEE Transactions on Instrumentation and Measurement, 69*(11), 8939–8952. https://doi.org/10.1109/TIM.2020.2994602

Yao, K., & Billard, A. (2023). Exploiting kinematic redundancy for robotic grasping of multiple objects. In *IEEE Transactions on Robotics* (pp. 1–21). https://doi.org/10.1109/TRO.2023.3253249.

Yin, S., Fu, C., Zhao, S., Li, K., Sun, X., Xu, T., & Chen, E. (2023). A survey on multimodal large language models. *arXiv*. 2306.13549. https://doi.org/10.48550/arXiv.2306.13549

Yuan, W., Dong, S., & Adelson, E. (2017). GelSight: High-resolution robot tactile sensors for estimating geometry and force. *Sensors, 17*(12), 2762. https://doi.org/https://doi.org/10.3390/s17122762

Yuan, W., Zhu, C., Owens, A., Srinivasan, M. A., & Adelson, E. H. (2017). Shape-independent hardness estimation using deep learning and a GelSight tactile sensor. In *Proceedings—IEEE International Conference on Robotics and Automation* (pp. 951–958). https://doi.org/10.1109/ICRA.2017.7989116.

Zhang, K., Cao, Z., Liu, J., Fang, Z., & Tan, M. (2018). Real-time visual measurement with opponent hitting behavior for table tennis robot. *IEEE Transactions on Instrumentation and Measurement, 67*(4), 811–820. https://doi.org/10.1109/TIM.2017.2789139

Zhang, P., Lan, C., Xing, J., Zeng, W., Xue, J., & Zheng, N. (2019). View adaptive neural networks for high performance skeleton-based human action recognition. *IEEE Transactions on Pattern Analysis and Machine Intelligence, 41*(8), 1963–1978. https://doi.org/10.1109/TPAMI.2019.2896631.

7 Embodied Cognition and Strong AI

Lawrence Shapiro and Francesco Bianchini

Introduction

What does embodied cognition (EC) tell us about the prospects for strong artificial intelligence (AI)? Our use of the word "strong" in this context is meant to invoke the now familiar distinction between strong and weakly embodied artificial intelligence (WEAI). WEAIs, like Siri, Alexa, medical diagnostics, navigation guidance systems, and countless other "smart" devices, make use of computational processes to perform, typically, narrowly defined tasks. A navigation guidance system is wonderful for calculating a route to the grocery store that can avoid highways, warn you away from speed traps, and maximize your fuel efficiency, but it is hopeless for planning a dinner party. Strong AIs, on the other hand, should exhibit all the flexibility, creativity, and versatility of a human mind, being able to achieve the same complex results as human beings. Whereas WEAIs do not claim to have human-like intelligence, a strong AI will be no less of a thinker than a human being.

Searle introduced the notion of strong AI more than forty years ago (Searle, 1980). This is relevant because the psychological framework at that time was the classical, computational, and symbolic cognitivist framework. The contrast between weak AI and strong AI was between mere modeling of the mind (programs used as tools for studying the mind) and something that could be defined as a mind in all respects (programs that are, in fact, minds). In strong AI, programs are minds because, in Searle's words, "the programmed computer has cognitive states, the programs are not mere tools that enable us to test psychological explanations; rather, the programs are themselves the explanations" (Searle, 1980, p. 418).

The ensuing debate characterized this distinction in different ways for various reasons. First, intentionality, consciousness—even in a subjective sense—or a general form of intelligence have been included among the requirements that characterize the mind and that a strong AI system must have. Furthermore, concerning the latter aspect, the initial distinction has been superimposed on that between narrow AI (relating to task-specific systems) and artificial general intelligence (AGI), a distinction that is not in Searle's original discussion.

Second, AI approaches have undergone a broad and radical evolution, influenced by the embodied turn in cognitive science and the neural turn in AI. These new approaches, including robotics, have blurred previously clear-cut distinctions in the field, raising new questions. For example, can we discuss strong AI without considering the embodiment of the artificial system? Does it still make sense to talk about consciousness, phenomenal states, or computational cognitive states in relation to strong AI? Are the human-like results that advanced neural networks produce, especially those developed since 2010, indicative of strong AI?

Initially, one of the requirements for strong AI was for artificial systems to perform complex, flexible, and creative tasks *in the same way* as humans, using similar mechanisms, processes, and cognitive states. Today, this constraint seems less relevant. Instead, having a physical body situated in a context and exhibiting dynamic behavior in interaction with the environment has become more important for strong AI over the past twenty-five years. Yet, the AI systems that now come closest to human-like performance are generative AI, which are neither cognitive in the traditional sense nor embodied in the, by now, classic sense.

The most recent innovation in AI is the idea of large language models (LLMs). LLMs certainly appear to be cognitively sophisticated. Prompting some of the excitement surrounding their appearance is their capacity to pass a test that Alan Turing thought sufficient for demonstrating human-like intelligence. According to this test, an AI that could convince a human being that it was in conversation with another human being, rather than an AI, is genuinely intelligent. By Turing's lights, LLMs are strong AIs. One might react to this conclusion with pride and awe: teams of human engineers have achieved something extraordinary. They have created a thinker. An alternative reaction, our reaction, goes in the opposite direction. Turing's test for intelligence is vulnerable to false positives, for LLMs are not strong AIs.

Why doubt that LLMs are intelligent? They are based on a particular type of neural network that implements a statistical-predictive procedure using a mechanism originally designed to improve the performance of automatic translators. This mechanism, later utilized in linguistic generative AI, is known technically as attention (Vaswani et al., 2017). In broad terms, this particular field of machine learning (ML) focuses on implementing a process that encodes and decodes inputs (words and texts) in a progressively contextualized manner. More generally, the concept of meaning does not seem present, even though these systems, at a certain level of abstraction, are brain-like and network-like. Some scholars have defined LLMs as stochastic parrots due to their intrinsic characteristics (Bender et al., 2021). The label "parrot" is apt. Just as parrots can "speak" without knowing what they are saying, LLMs excel at producing coherent text with no idea of the meaning of the words within. Intelligence, however, we understand the term, involves more than the sort of mechanical pattern matching that is the basis for the LLM's conversational aptitude.

In the first part of this chapter, we will describe how researchers in the field of EC conceive of the "more" that is needed to turn a weak AI into a strong one. Machines like LLMs lack intelligence because, these researchers say, they simply recognize correlations between words without ever attaching meanings to these words. Such machines are limited to syntactic operations, never aware of the semantics of the words they manipulate. According to a popular view in EC, real intelligences like ourselves traffic in *modal* symbols: Our thoughts, unlike the objects of pure syntax that LLMs recognize, are encoded in a sensorimotor format. This format contains sensorimotor information arising from the manner in which an embodied organism encounters the world.

In the sections to follow we explain where this response to the "syntax problem" goes awry. The first difficulty concerns the claim that sensorimotor reactivation grounds *concept* meaning. We believe that proponents of this claim confuse concepts with what Rey (1983) called *conceptions*. It may be true that human conceptions are informed by sensorimotor reactivation even if the concept's meaning is not. The second difficulty concerns the idea that sensorimotor reactivation can provide a grounding for meaning. Such a thought merely pushes the problem down the road. We can also ask how sensorimotor activity is in itself meaningful, and a theory that answers this question might also suffice to clarify how concepts become meaningful.

Once we get clear on these issues, we will see that the opposition to strong AI that some EC theorists have expressed requires refinement. If strong AI refers to an AI that shares human conceptions of the world, then perhaps it is impossible. On the other hand, if an AI is strong in virtue of possessing human concepts, it may not be too far away.

Interestingly, an aspect that seems to unite the old and new visions of AI is intentionality, or in more contemporary terms, the question of conceptual meaning. This is not trivial, given the many differences that separate the symbolic approach from the embodied approach to cognition and AI. Finding a line of continuity in the question of conceptual meaning is an almost unexpected result, considering this has been one of the main problems, if not *the* main problem, of classic AI. This problem has been disruptive and yet remains unresolved. That an AI system can have *truly* meaningful concepts seems to be as impossible a challenge today as it was in the past. From this point of view, the evolution of AI has remained unchanged, except in its approaches. So, what does embodiment really contribute to the production of AI?

Concepts as Units of Thought

Psychologists have long been interested in understanding the nature of *concepts*, where concepts are typically glossed as being the "units" of thought. Consider the concept of *jewelry*: Concepts, like that of *jewelry*, have meanings. On a first pass, *jewelry* means something like "a precious, elegant, wearable piece of metal."[1] The *jewelry* concept acts as a unit of thought in the following senses. First, because its meaning remains relatively fixed, it makes possible thoughts about the same thing over time: the jewelry you are dreaming about on Monday is the same thing you dream about on Friday. Second, the stability of concepts underlies the capacity for inference, ensuring that the contents of premises "carry through" to the conclusions that follow. If the concepts involved in premises differed in meaning from those in conclusions, every argument would end up as an equivocation. Third, concepts are essential for communication. Jack and Jill can discuss jewelry, disagree about which kinds of jewelry are preferable, describe their favorite ways to wear jewelry, and so on because they share the concept *jewelry*. Finally, the existence of concepts allows for the consistency of category judgments over time. Thalia, a jewelry-lover, never fails to find the best jewelry at an estate sale. Her skill at doing so depends on her deploying the same *jewelry* concept on each occasion that she goes to such a sale. Because the concept's meaning remains unchanged, Thalia succeeds in categorizing the best jewelry at the estate sales she visits.

That concepts play these essential roles makes sense of their characterization as the units of thought. Thought takes place in a medium of concepts. Inference, communication, and categorization require concepts. We'd hardly be thinking things without such capacities, and we'd have no such capacities without concepts. Going further, if strong AI is possible, then it must also be possible to endow machines with concepts. Without concepts, a machine would lack the units of thought. It could not think, infer, communicate, or categorize. It could not think like a human being and this could not satisfy the condition for strong AI.

Concepts and Concept Labels

We mentioned earlier that LLMs could pass the Turing test while not possessing intelligence. Intelligence, at least of a human sort, demands concepts. How might a machine be endowed with concepts? One common approach involves creating networks of connections between concepts. The meaning of a concept emerges from its connections to other concepts. For instance, below is the network that provides meaning to the concept *jewelry*.

This network defines *jewelry* in terms of its relationship to other concepts. The number and length of the connections between these concepts reveals something about their significance for the meaning of *jewelry*. For example, the concepts *precious, wearable, metal,* and *elegant* cluster together, suggesting that jewelry is the kind of thing that, if exhibiting the property of being precious, probably exhibits some of the properties to which the other concepts refer. The concept *ring* is less associated with other concepts central to *jewelry* than is the concept *elegant*. And what it is to be a piece of jewelry—the meaning of *jewelry*—does not involve the concept *plastic* at all.

Whatever the virtues of this network-style approach to concepts, an obvious distinction suffices to show its limitations. It would be an easy matter to program a computer—whether with a standard symbolic program or a connectionist one—to replicate the network of connections that define *jewelry*. Doing so, however, would hardly produce a thinking machine. This is because the connections in the machine's network are not between actual

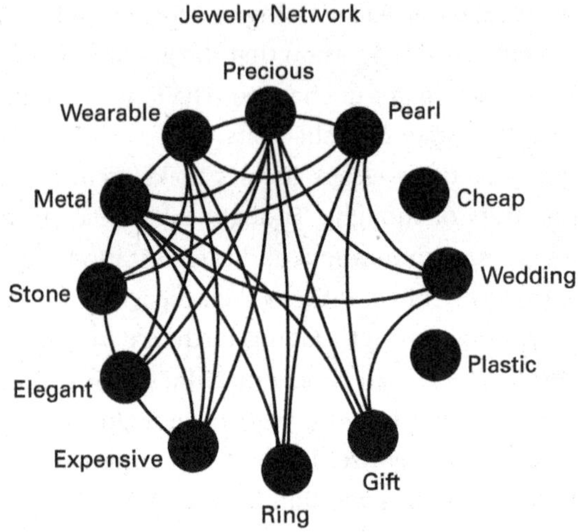

Figure 7.1
A network characterization of the concept *jewelry*. *Source*: the author.

concepts but only between concept *labels*. Like the network that figure 7.1 depicts, which draws lines between *words* rather than concepts, the networks that computers construct are similarly constrained. The networks involve no actual concepts but only labels for concepts. Thinking involves the creation and manipulation of concepts, not concept labels.

This insight lies behind the criticism that EC theorists raise against traditional theories of concepts. According to this criticism, these traditional theories confuse concepts with concept labels.[2] The psychologists Art Glenberg and David Robertson have been especially clear about how, in their view, this confusion dooms traditional theories. They ask whether theories of concepts, like the network one we have been examining, can be "theories of human meaning" (2000, p. 381).[3] Motivating a negative answer, they believe, is the fact that the symbols in such networks are abstract and "arbitrarily" related to their referents. In essence, they say, the symbols are like labels or words. Words are abstract insofar as a word like "jewelry" can apply to a variety of substances that might differ in various ways. The connection between the word and its referent is also arbitrary, in the sense that there is nothing special about "jewelry" that makes it the correct word for referring to jewelry—other words could have served as well.

The symbols in the jewelry network are like words in also being abstract and arbitrary. Consequently, Glenberg and Robertson believe it makes sense to ask whether "the meaning of an abstract symbol (a word) can arise from the conjunction of relations to other undefined abstract symbols" (2000, p. 381). Citing Searle's famous Chinese room thought experiment (Searle, 1980), they offer a negative answer to this question on the grounds that a network of abstract and arbitrary symbols fails a condition for meaning: "abstract, arbitrary symbols, such as words, need to be grounded in something other than relations to more abstract arbitrary symbols if any of those symbols are to be meaningful" (Glenberg & Robertson, 2000, p. 381). Because the symbols in the jewelry network are abstract and arbitrary, the connections between them cannot produce meaning. It's as if, Glenberg and Robertson claim (following Harnad, 1990), you hope to learn a foreign language using a dictionary that defines a foreign word only in terms of other foreign words that you do not understand. "Obviously, no matter how many words you look up, that is, no matter how many structural relations you determine among the arbitrary symbols, you will never figure out the meaning of any of the words" (Glenberg & Robertson, 2000, p. 381–382). Necessary for meaning is that the abstract and arbitrary symbols in the network no longer be abstract and arbitrary. They must be in some sense *grounded*.

In the remainder of this chapter, we will address the following issues that the discussion so far has raised. First, we will consider how EC theorists seek to ground concepts. We will then turn to a purported implication of this method of grounding for the possibility of strong AI: some embodied researchers have claimed that AIs could never be human-like unless they shared with human beings similar sensorimotor systems. To conclude, we will argue that this implication requires refinement. While AIs may not conceptualize the world as human beings do, they might nonetheless have human-like concepts. We will also see that the theory of meaning that EC theorists adopt is at best incomplete.

Embodied Concepts

How does a symbol acquire meaning? In the present context, this question concerns symbols that stand for concepts. Presumably, these symbols exist in our brain and would exist in computers as well if a strong AI were instantiated in a computer. If Glenberg and Robertson are correct, then the wrong answer to our question is this: symbols acquire meaning by standing in relation to other symbols. The problem with this answer is that the symbols in this network of connections stand in relation only to each other. Therefore, we are left wondering how they ever come to be about the *meaning* of symbols. To be meaningful, symbols must at some point, and somehow, become connected not just to each other but also to the objects or properties about which they aim.

The EC researcher Lawrence Barsalou has developed and defended a well-received (within the EC field) theory about how concepts are grounded. Central to his view is the idea that the symbols constituting our conceptual system are grounded, and thus meaningful, by virtue of being *modality-specific*. Among the modalities involved in conceptual thinking are our various sensory modalities as well as motor activations.

Consider once more the concept *jewelry*. On the embodied theory of concepts, our concept *jewelry* arises from our sensorimotor interactions with it. In our encounters with jewelry we have touched, manipulated, seen, and heard it. We have held it in our fingers, hung it from our ears, and heard its jangling. We have felt its smoothness and its weight. Barsalou proposes that our sensory and motor experiences with jewelry become encoded in the nervous system, packaged together in "conjunctive" neurons (Barsalou et al., 2003).

When thinking about jewelry in its absence, when talking about jewelry, making inferences about jewelry, or daydreaming about jewelry, the conjunctive neuron that holds information about previous sensorimotor interactions with jewelry activates the very sensorimotor systems through which one's experiences with jewelry were filtered. The concept *jewelry*, in this view, acquires its meaning because it is grounded in the sensorimotor experiences one has when in contact with jewelry. Similarly, as Glenberg remarks, the meaning of "hammer" consists of "(a) how a hammer looks, (b) how it sounds when used, (c) how it feels when hefted, and (d) how to use it" (Glenberg, 2015, p. 182, response to Mahon).

In contrast to the abstract and arbitrary symbols on which traditional theories of concepts rest, symbols that we might characterize as *amodal*, the modal symbols at the core of the embodied account of concepts are neither abstract nor arbitrary. They consist in reenactments of the very sensorimotor activity that birthed them in the first place. They acquire their meaning not through their connections to other symbols, which in turn are connected only to other symbols, but rather through reenacting the sensorimotor interactions with the objects and properties they are about.[4]

A wealth of behavioral and neurophysiological data support the existence of embodied concepts.[5] For instance, Barsalou et al. (1999) found that subjects tend to rely on imagery, which involves sensorimotor reactivation, when answering questions about an object's features. Similarly, neuroscientists have found that when a subject reads a word like "kick," motor areas of the subject's brain associated with kicking are active (Pulvermüller, 2005). Such a finding would be expected if the subject's thoughts about kicking consisted in part in the motor activity that corresponds to kicking.

To summarize, EC researchers reject traditional theories of concepts because, they believe, such theories cannot account for how concepts acquire their meanings. As they understand traditional theories, they deal in abstract and arbitrary symbols, and they assume that meaning can arise from connections between abstract and arbitrary symbols. But, charge those in the EC community, abstract and arbitrary symbols are ungrounded. They never make contact with the world, and connections between them bring them no closer to the world. Embodied concepts, on the other hand, are grounded in the very sensorimotor activities that come from contact with the objects and properties that concepts represent. Embodied concepts acquire their meaning not through their connections to other, ungrounded, concepts but through the sensorimotor activity that comes from experience with the world.

Embodied Concepts and Strong AI

Let us grant, for now, the coherence of embodied concepts. We shall suppose that concepts acquire their meaning by virtue of coding for a simulation of sensorimotor activity. The meaning of a concept consists in the patterns of sensorimotor activity that contact with the concept's referent typically cause. Thus, as Glenberg claims, the meaning of our concept *hammer* is something like "object that makes a banging sound, that is metallic-looking on one end and wooden-looking on the other, that is gripped with the hand and swung in a downward motion." Likewise, the meaning of *jewelry* is an "object that is wearable, shiny, and jangly." These are the meanings of *hammer* and *jewelry* because these are the sensorimotor processes that are reactivated when having *hammer* or *jewelry* thoughts.

From this theory of concepts, an argument against the possibility of strong AI follows directly. The argument begins with the claim that semantic networks like the one that defines *jewelry* fail to endow concepts with meaning. Consequently, an AI that is genuinely capable of thought rather than just a mere simulation of thought must be programmed with something more than a semantic network. The next premise asserts the adequacy of embodied concepts as an account of how concepts become meaningful. However, the argument continues, computers either have no sensorimotor capacities at all, in which case they also lack meaningful concepts and thus cannot think, *or* they do have sensorimotor capacities, but not like those that human beings possess, so even if they can think, they would not think like human beings. Either way, strong AI, AI that is indistinguishable from human intelligence, is unattainable.

This argument is explicit in some discussions of embodied concepts. Here, for instance, is Barsalou declaiming the possibility of strong AI:

> Because peripheral devices in computers differ considerably from human sensory-motor systems, implementations of perceptual symbol systems[6] in computers should have a distinctly nonhuman character. Contrary to the claims of functionalism, computers should not be capable of implementing a human conceptual system, because they do not have the requisite sensory-motor systems for representing human concepts. Although it is intriguing to consider how a perceptual symbol system might be implemented technologically, it is probably naïve to believe that such a system would correspond closely to human intelligence. Such correspondence awaits the development of artifacts that are much more biological in nature. (Barsalou, 1999, pp. 607–608)

This passage suggests that Barsalou takes the second route in the dilemma with which the above argument closes. That is, he allows that computers might be equipped with sensorimotor systems that suffice for producing meaningful concepts and, correspondingly, intelligence. However, insofar as these sensorimotor systems differ from a human being's, the meanings of the concepts that they ground will also differ from the meaning of human concepts.

What would *jewelry* mean for an AI whose visual color perception is much more fine-grained than our own, so that it distinguishes between pieces of jewelry that we'd label all as gold? Lacking ears, this AI cannot experience the feel of jewelry as it hangs from its earlobes, bumping against its neck. We grip jewelry with our fingers, but the AI uses vacuum suction to pick up and hold objects the size of a ring. Given these many differences between human and AI sensorimotor systems, human and AI concepts would differ in meaning. Consequently, although the AI might be intelligent, its intelligence would not be human intelligence. It would not be a strong AI.

Concepts versus Conceptions

We believe the argument against strong AI that we sketched above, and that Barsalou appears to endorse, fails. It runs into difficulty because the theory of concepts on which it rests is not actually a theory of concepts at all; and the theory of concepts it rejects, a network theory, is not prone to the objections that it assumes. Once we sort out these issues, we will see that AIs can have concepts, even human-like concepts, although they may diverge sharply from human beings in their *conceptions* of the world. This difference between having a concept and having a conception reopens the possibility of strong AI, although it also points to a vagueness in the definition of strong AI that deserves more attention.

Consider the concept *bachelor*. This is a favorite example among philosophers because the meaning of "bachelor" is easy to specify with necessary and sufficient conditions. *Bachelor* means "unmarried man." A first point to make about this concept is that it hardly seems one for which an analysis in terms of sensorimotor reenactment is appropriate. What sensorimotor activity does the property of being unmarried invoke? Making this problem especially difficult is that many things—surely most—are unmarried. In addition to men, many women are unmarried. So too are most children. Books, automobiles, and camels are similarly unmarried. If concepts acquire their meaning in virtue of activation in sensorimotor brain regions that respond to the concept's referent, we need an account of the sensorimotor activity that being unmarried triggers. It hardly seems likely, however, that unmarried males, females, children, books, automobiles, and camels all cause in us sensorimotor activations of the same kind—the "unmarried" kind.

More generally, the theory of embodied concepts faces a challenge when seeking to explicate the meaning of abstract concepts, such as *unmarried* or *justice* or *radical*. It's in the nature of abstract concepts to abstract away from the very properties of objects with which we have sensorimotor contact—their visual, auditory, and tactile properties, as well as the properties that make our physical interactions with them possible. Despite this difficulty, we nevertheless make frequent use of abstract concepts. Of course, this point is not lost on members of the embodied community, and Barsalou has himself offered thoughts about how abstract concepts might be embodied (Barsalou, 1999, 2005; Barsalou et al., 2003).[7] We raise the difficulty of abstraction at this point to motivate the idea that the sensorimotor activity that co-occurs with conceptual thought may not be constitutive of concepts after all. It might serve some other role in thought, as we will soon suggest.

Let's return to *bachelor*. The meaning of this concept is "unmarried man." But thoughts about bachelors, thoughts in which the *bachelor* concept appears, often involve more than thoughts about unmarried men. Thoughts about bachelors "bring to mind" a variety of associated thoughts. Many of these associated thoughts might be useful for identifying bachelors when a man's marital status is unknown. For instance, a common stereotype, perhaps dated, depicts bachelors as lotharios, as living in "pads" that feature expensive sound systems and extensive liquor collections, as being youthful rather than elderly, as being vain about their appearances. These features of bachelors are not, strictly speaking, part of the meaning of the concept *bachelor*. After all, some bachelors might exhibit none of these features, and some individuals who meet the *bachelor* stereotype might not be bachelors at all; perhaps they are married, or perhaps they are women.

Following Rey (1983), we wish to distinguish a concept, which represents something like the conditions for membership in some category, from a conception, which provides information about features typically associated with a concept's referent.[8] In the example just discussed, the features typically associated with bachelors, those that fit the bachelor stereotype, constitute the *bachelor* conception. Having such a conception often serves an epistemological function, helping one to justify beliefs about who is and who is not a bachelor. Yet, importantly, a conception is not itself a concept. As we have seen, individuals can satisfy a *bachelor* conception while not being a bachelor and can fail to satisfy a *bachelor* conception despite being a bachelor. Concepts and conceptions are different things.

Granting the coherence of this distinction between concepts and conceptions, we now submit that the sensorimotor activity that embodied theorists identify with concepts in fact constitutes a conception. Recall Glenberg's description of the meaning of "hammer." He said that it consists of "(a) how a hammer looks, (b) how it sounds when used, (c) how it feels when hefted, and (d) how to use it" (Glenberg 2015, p, 182). But (a)–(d) are much better construed as a conception of hammers than as the meaning of *hammer*. In the first place, one could know how hammers look, sound, feel, and are used without knowing what a hammer *is*, without knowing that a hammer is a tool used for driving nails into hard surfaces. Likewise, one could know that a hammer is a tool used for driving nails into hard surfaces without ever having any sensorimotor contact with hammers. Moreover, just as the bachelor conception is useful for helping to identify bachelors, so too, a conception of a hammer can be valuable when trying to locate a hammer; one looks for objects that satisfy (a)–(d).

We believe that the sensorimotor reactivations that accompany thought more likely create conceptions than they do the meaning of concepts. Our conception of bachelors includes the imagery, sounds, odors, and physical interactions that exposure to bachelors elicit. EC teaches us that thoughts about these features of our *bachelor* conception come by way of reactivation of sensorimotor regions of the brain. This is an interesting finding, but not one that supports the conclusion that the *meaning* of *bachelor* is anything other than "unmarried man."

Implications for Strong AI

With the distinction between concepts and conceptions now available, we can revisit Barsalou's claim that an AI could not have a human conceptual system unless it also possesses a human sensorimotor system. More likely true, we believe, is that an AI with a sensorimotor system that differs from our own would fail to conceptualize the world as a human being does; its conceptions would diverge from our own. This claim, however, is hardly surprising. Even prior to the work on which the theory of embodied concepts is based, psychologists had evidence that judgments about the world—conceptions—were often body-based. For instance, the ecological psychologist William Warren showed that subjects' judgments about whether a staircase is climbable—whether they conceptualized the staircase as climbable—depended on their height (Warren, 1984). Similarly, in later work, Warren and Whang (1987) showed that subjects' judgments of whether they could pass through a doorway—their conception of a doorway's "passability"—is a function of their shoulder breadth.

More generally, it seems obvious that how a person conceptualizes the world, how they think about jewelry, or bachelors, or staircases, or birds, depends on the kinds of experiences with these things that they have had. The interesting claim emerging from the field of EC is that these experiences leave a trace in our sensorimotor systems and that this trace initiates reactivation of those same sensorimotor brain regions when later thinking about the objects that originally created the trace. EC thus offers an explanation of how our conceptions are formed and maintained. But this is not a theory of *concept meaning*, as it is typically billed.

We see now that the appellation "strong AI" needs refining. If a strong AI is one whose conceptions of the world are human-like, then insofar as an AI's sensorimotor systems differ from our own, they will not be human-like and thus will not be strong AIs. On the other hand, if sensorimotor reactivation informs conceptions and not concepts, so that concept meaning arises independently of sensorimotor activity, and strong AI requires similarity in concepts and not conceptions, then the possibility of strong AI remains open.

At the very least, this observation adds to the growing realization that the concept of strong AI has evolved over time, particularly with the embodied turn in cognitive science and AI. Defining strong AI today is more complex, and the issue is closely linked to recent debates on the relationship between artificial and natural intelligence (Andler 2023; Hernández-Orallo, 2017). The challenge of artificially simulating or re-creating the general capabilities of intelligence and cognition raises the question: What constitutes strong AI? Is it an artificial system capable of reasoning about everything? One that exhibits human language skills to talk about anything and answer all types of questions? Or is it a system that can apply general cognitive skills to solve a wide range of problems?

The question of concept meaning remains a crucial unifier, extending beyond the problem of symbol manipulation (as highlighted by Searle's Chinese room argument) and grounding (Harnad, 1990) to include embodied approaches to AI and robotics and the criticisms they have raised. For instance, situated robotics (Brooks, 1991) risks being too distant from replicating the organization of living systems, failing to capture the construction and coevolution of conceptual meaning essential for achieving *actual* intelligence in an artificial system (Di Paolo, 2003). These approaches have been criticized for remaining within a mechanistic framework, offering a functional explanation of cognition but failing to grasp the true nature of intentionality and meaning (Ziemke, 2016). Similarly, the most radical embodied approaches, such as enactivist, nonrepresentationalist, and dynamicist perspectives, have been critiqued as insufficiently explanatory regarding the problem of meaning in AI (Froese & Taguchi, 2019). We believe that maintaining the possibility of strong AI, in light of the evolution of this notion following the embodied turn, not only highlights an unresolved problem but also offers a way to reconsider the conceptual framework for explaining concept meaning in intelligent artificial systems. In this context, the proposal of embodied intelligence (EI) and efforts to define its requirements and boundaries are particularly relevant.

If EC provides the wrong answer to the question about how concepts acquire their meaning, what is the right answer? If concept meaning is not grounded in sensorimotor activity, how does it arise; and could it arise in an AI? Here, we must be more speculative. The first point to make is that the concern Glenberg raises about network analyses of concept meaning is not really apt. Glenberg worries that symbols do not acquire meaning merely through their connections to other meaningless symbols. This is the point of Searle's Chinese room thought experiment and Harnad's foreign dictionary example. But traditional cognitive science has never been committed to such a theory of concept meaning. Consider once more figure 7.1, which depicts the network of concepts that determine the meaning of *jewelry*. Glenberg assumes that the nodes in this network are *labels* for concepts, rather than standing for concepts themselves. If they were just labels—abstract and arbitrary symbols without any meaning of their own—then he is doubtless correct that the connections between them would not suffice for meaning.

Alternatively, however, suppose that the nodes in the network are not labels but represent actual concepts. It's no longer far-fetched that the meaning of *jewelry* might arise from connections between concepts like *precious*, *metal*, *elegant*, and *expensive*. Glenberg mischaracterizes traditional theories of concept acquisition when he assumes that the network analyses they adopt trade in labels rather than concepts.

Of course, we still want to know how the concepts that participate in these networks acquire their meaning and whether a theory of human concept meaning might as well apply to an AI. We close this chapter with two points about these questions. First, we assume that the patterns of neural activity that instantiate concepts, that represent meanings, do so by virtue of their relation to things in the world or to other concepts that are related to things in the world. Philosophers have offered any number of theories about which sort of relation to the world might suffice to endow a pattern of neural activity with meaning. The relation might be causal or it might involve a lawful correlation of some sort (Dretske, 1988; Fodor, 1987). Insofar as these theories are correct, they support two claims. First, concept meaning does not require sensorimotor reenactment. Second, AIs can possess meaningful concepts. Such AIs would only need sensory states that, modulo various conditions, carry information about properties in the world.

The second point we wish to make about concept meaning is that it is simply a special case of a more general problem about the origins of meaning. Those endorsing the embodied account of symbol grounding believe that concept meaning arises from sensorimotor activity that is already meaningful. *Hammer* means an object that looks, sounds, feels, *like so*. And because the sensorimotor regions of the brain that represent how things look, sound, and feel are *already* meaningful, they can do the job of grounding concept meaning. But just as we can ask how a concept acquires its meaning, we can also ask how sensorimotor regions of the brain come to represent visual, auditory, and tactile properties in the world. For instance, suppose activity in visual cortex represents the visual appearance of a hammer. How does it do so? What makes neural activity in the back of the brain mean "visual appearance of a hammer"? It makes no sense to say that the meaning of this activity is grounded in sensorimotor reactivation because we are wondering how sensorimotor activity becomes meaningful in the first place.

Conclusion

We conclude with a summary of our main points. First, EC researchers have confused concepts with conceptions. What they identify, for example, as the *concept* of a hammer, is more likely the *conception* of a hammer, where this distinction is between what *hammer* means and what we typically associate with hammers—that is, the sensorimotor activity that accompanies contact with hammers. Second, doubts that EC researchers have expressed about whether an AI could ever think like a human being are more accurately construed as doubts that an AI would share a human being's conceptions about the world, not whether an AI could share a human being's concepts. If "strong AI" refers to an AI that is endowed with human-like concepts rather than conceptions, then strong AI is possible despite likely differences between human and AI sensorimotor systems. Third, the embodied theory of concept meaning is at best incomplete, for it assumes the meaningfulness of sensorimotor activity that it takes to ground concept meaning.

Notes

1. Philosophers typically distinguish the referent of a concept from what Frege (1893) called its sense. More recently, Prinz (2002) drew a similar distinction between a concept's intentional content and its cognitive content. These distinctions help to make sense of the idea that the concepts *morning star* and *evening star* refer to the same object (they have the same reference or intentional content) while meaning different things (they differ in their senses or cognitive contents). It's because the concepts differ in their senses that one can be surprised to learn that they refer to the same object. We understand the embodied theory of concepts to be mainly about a concept's sense or cognitive content and will largely ignore questions about how concepts refer.

2. Dijkstra (2024) makes this point in her defense of embodied approaches to concepts.

3. The theories that Glenberg and Robertson target are the hyperspace analogue to language (Burgess & Lund, 1997) and latent semantic analysis (Landauer & Dumais, 1997); however, their criticisms of these theories apply just as well to the network account of concepts we described above.

4. Why do we not actually see, feel, and hear jewelry when thinking about it? The suggestion is that sensorimotor reenactments are "offline," meaning that they do not result in conscious awareness of an object's properties when calling the concept to mind.

5. Of course, the evidence faces challenges on several fronts. Some argue that it can be explained in ways that do not support the existence of embodied concepts (Mahon 2015; Mahon & Caramazza, 2005, 2008; Mahon & Hickcock, 2016). Others point to repeated failures to replicate crucial experimental findings (Machery, 2024).

6. Barsalou uses the label "perceptual symbol" for what we have been calling "embodied concepts."

7. For an excellent critical discussion of the issues surrounding the abstraction challenge that embodied concepts face, see Dove (2016).

8. See Shapiro (2019, chapter 4) for additional discussion.

References

Andler, D. (2023). *Intelligence artificielle, intelligence humaine: La double énigme.* Gallimard.

Barsalou, L. W. (1999). Perceptual symbol systems. *Behavioral and Brain Sciences, 22*(4), 577–660. https://doi.org/10.1017/S0140525X99002149

Barsalou, L. W. (2005). Abstraction as dynamic interpretation in perceptual symbol systems. In L. Gershkoff-Stowe & D. H. Rakison (Eds.), *Building object categories in developmental time* (pp. 407–450). Psychology Press.

Barsalou, L., Simmons, W., Barbey, A., & Wilson, C. (2003). Grounding conceptual knowledge in modality-specific systems. *Trends in Cognitive Sciences, 7*, 84–91. https://doi.org/10.1016/S1364-6613(02)00029-3

Barsalou, L., Solomon, K., & Wu, L. (1999). Perceptual simulation in conceptual tasks. In M. K. Hiraga, C. Sinha, & S. Wilcox (Eds.), *Cultural, typological, and psychological perspectives in cognitive linguistics* (pp. 209–228). John Benjamins B.V.

Bender, E. M, Gebru, T., McMillan-Major, A., & Shmitchell, S. (2021). On the dangers of stochastic parrots: Can language models be too big? In *Proceedings of the 2021 ACM conference on fairness, accountability, and transparency* (pp. 610–623). ACM. https://doi.org/10.1145/3442188.3445922

Brooks, R. A. (1991). Intelligence without representation. *Artificial Intelligence, 47*(1–3), 139–159. https://doi.org/10.1016/0004-3702(91)90053-M

Burgess, C., & Lund, K. (1997). Modelling parsing constraints with high-dimensional context space. *Language and Cognitive Processes, 12*(2–3), 177–210. https://doi.org/10.1080/016909697386844

Dijkstra, K. (2024). Memory and action. In L. Shapiro & S. Spaulding (Eds.), *The Routledge handbook of embodied cognition* (2nd ed., pp. 431–441). Routledge.

Di Paolo, E. A. (2003). Organismically-inspired robotics: Homeostatic adaptation and teleology beyond the closed sensorimotor loop. In K. Murase & T. Asakura (Eds.), *Dynamical systems approach to embodiment and sociality* (pp. 19–42). Advanced Knowledge International.

Dove, G. (2016). Three symbol ungrounding problems: Abstract concepts and the future of embodied cognition. *Psychonomic Bulletin and Review, 23*, 1109–1121. https://doi.org/10.3758/s13423-015-0825-4

Dretske, F. (1988). *Explaining behavior: Reasons in a world of causes.* MIT Press.

Fodor, J. (1987). *Psychosemantics.* MIT Press.

Froese, T., & Taguchi, S. (2019). The problem of meaning in AI and robotics: Still with us after all these years. *Philosophies, 4*(2), 14. https://doi.org/10.3390/philosophies4020014

Glenberg, A. (2015). Response to Mahon: Unburdening cognition from abstract symbols. *Canadian Journal of Experimental Psychology, 69*(2), 181–182. https://doi.org/10.1037/cep0000057

Glenberg, A., & Robertson, D. (2000). Symbol grounding and meaning: A comparison of high-dimensional and embodied theories of meaning. *Journal of Memory and Language, 43*(3), 379–401. https://doi.org/10.1006/jmla.2000.2714

Harnad, S. (1990). The symbol grounding problem. *Physica D, 42*(1–3), 335–346. https://doi.org/10.1016/0167-2789(90)90087-6

Hernández-Orallo, J. (2017). *The measure of all minds. Evaluating natural and artificial intelligence.* Cambridge University Press.

Landauer, T., & Dumais, S. (1997). A solution to Plato's problem: The latent semantic analysis theory of acquisition, induction, and representation of knowledge. *Psychological Review, 194*, 211–240.

Machery, E. (2024). The replication crisis in embodied cognition research. In L. Shapiro & S. Spaulding (Eds.), *The Routledge handbook of embodied cognition* (2nd ed., pp. 475–484). Routledge.

Mahon, B. (2015). What is embodied about cognition? *Language, Cognition and Neuroscience, 30*(4), 420–429. https://doi.org/10.1080/23273798.2014.987791

Mahon, B., & Caramazza, A. (2005). The orchestration of the sensory-motor systems: Clues from neuropsychology. *Cognitive Neuropsychology, 22*, 480–494. https://doi.org/10.1080/02643290442000446

Mahon, B., & Caramazza, A. (2008). A critical look at the embodied cognition hypothesis and a new proposal for grounding conceptual content. *Journal of Physiology, 102*, 59–70. https://doi.org/10.1016/j.jphysparis.2008.03.004

Mahon, B., & Hickcock, G. (2016). Arguments about the nature of concepts: Symbols, embodiment, and beyond. *Psychonomic Bulletin and Review, 23*, 941–958. https://doi.org/10.3758/s13423-016-1045-2

Prinz, J. (2002). *Furnishing the mind: Concepts and their perceptual basis.* MIT Press.

Pulvermüller, F. (2005). Brain mechanisms linking language and action. *Nature Reviews: Neuroscience, 6*(7), 1–8. https://doi.org/10.1038/nrn1706

Rey, G. (1983). A reason for doubting the existence of consciousness. In *Consciousness and Self-Regulation: Volume 3: Advances in Research and Theory* (pp. 1–39). Springer.

Searle, J. (1980). Minds, brains, and programs. *Behavioral and Brain Sciences, 3*, 417–424.

Vaswani, A., Shazeer, N., Parmar, N., Uszkoreit, J., Jones, L., Gomez, A. N., Kaiser, L., & Polosukhin, I. (2017). Attention is all you need. In *Proceedings of 31st international conference on neural information processing systems* (pp. 6000–6010). Curran Associates.

Warren, W. (1984). Perceiving affordances: Visual guidance of stair climbing. *Journal of Experimental Psychology: Human Perception and Performance, 10*(5), 683–703. https://doi.org/10.1037/0096-1523.10.5.683

Warren, W. & Whang, S. (1987). Visual guidance of walking through apertures: Body-scaled information for affordances. *Journal of Experimental Psychology: Human Perception and Performance, 13*(3), 371–383. https://doi.org/10.1037/0096-1523.13.3.371

Ziemke, T. (2016). The body of knowledge: On the role of the living body in grounding embodied cognition. *Biosystems, 148*, 4–11. https://doi.org/10.1016/j.biosystems.2016.08.005

8 Embodied AI in Machine Learning: Is It Really Embodied?

Matej Hoffmann and Shubhan Parag Patni

Introduction

Embodied artificial intelligence (EAI) is gaining momentum in the machine learning (ML) communities with the goal of leveraging current progress in artificial intelligence (AI) (deep learning, transformers, large language and visual-language models (VLMs)) to empower robots. In this chapter we put this work in the context of Good Old-Fashioned Artificial Intelligence (GOFAI) (Haugeland, 1989) and the behavior-based or embodied alternatives (R. A. Brooks, 1991; Pfeifer & Scheier, 2001). We claim that the AI-powered robots are only weakly embodied and inherit some of the problems of GOFAI. Moreover, we review and critically discuss the possibility of cross-embodiment learning (O'Neill et al., 2024). We identify fundamental roadblocks and propose directions on how to make progress.

GOFAI-Powered Robots Never Really Worked

AI in the 1950s and 1960s, later called Good Old-Fashioned Artificial Intelligence (GOFAI) by Haugeland (1989), held that the key to intelligence is computation with symbols that represent the world. The keywords were algorithmic nature, symbolic computation, and representation. GOFAI was very successful in formal domains (like chess), where the state of the world is discrete and directly accessible, and standard AI techniques (like search) can be applied. While the focus has been on abstract "thinking," when entering the real world, a relationship had to be established between the dynamic, continuous, partially accessible reality out there and the internal world representation. That is, reality had to be sensed and mapped onto the internal world model, in which the "thinking" was performed. Finally, whatever action was selected, it had to be executed in the real world. The approach thus became known as the *sense-think-act* architecture. The "interfaces" with the real world—previously uninteresting and underestimated—became fundamental and practical challenges. The "frame problem" (keeping the internal representation of the world consistent with the real world outside) and the "symbol-grounding problem" (concerned with the relationship of the symbolic representation and the outside world) are the most serious of the fundamental problems. The initial hopes of applying, for example, theorem solving (e.g., Fikes & Nilsson, 1971) to real-world robot tasks have never really materialized, and AI and robotics have developed largely independently, meeting when complex high-level plans, and hence reasoning, were required (Beetz et al., 2016; Russell & Norvig 2021).[1]

Robotics before the Learning Era

It is not within the scope of this chapter to provide an overview of the field of robotics and its history. Before the "robot learning era" in recent years, robotics research focused on developing the foundations (kinematics, dynamics, motion planning, and control), sensing and perception, localization, and mapping, as well as industrial robotics and, more recently, human-robot interaction (HRI) (see the corresponding parts in Siciliano & Khatib, 2016). Robotics had more intersections with control theory than with AI. The impressive robot agility demonstrated by the robots from Boston Dynamics, like Atlas, have been achieved through careful engineering and the use of model predictive control rather than learning.

Embodiment, Behavior-Based, and Soft Robotics

During the GOFAI era, the computer was not only a tool but also a metaphor for the mind in the paradigm known as cognitivism (Fodor, 1975). The response to this in cognitive science was embodied cognition (EC), which holds that cognitive processes are constitutively shaped by the interaction with the world through the agent's body, emphasizing the role of action rather than world modeling and representations (e.g., Engel et al., 2013). In order to demonstrate that behaviors that would be considered intelligent or cognitive by many do not have to come from internal world models and computation over them, machines with minimalistic controllers were built, and their interaction with the environment was observed. Grey Walter was the pioneer of this approach, building electronic machines with a minimal "brain" that displayed phototactic-like behavior (Walter, 1953). This was picked up by Valentino Braitenberg (1986), who built a series of two-wheeled vehicles of increasing complexity. Already, the most primitive ones, in which sensors were directly connected to motors (exciting or inhibiting them), displayed sophisticated behaviors. Although the driving mechanisms are simple and entirely deterministic, the interaction with the real world gives rise to complex behavioral patterns.

Motivated by practical considerations (i.e., to build robots that can interact with the world in real time), Rodney Brooks (1991) openly attacked the GOFAI position in the seminal article "Intelligence without Representation." Through building robots that interact with the real world, such as insect robots, he realized that "when we examine very simple level intelligence, we find that explicit representations and models of the world simply get in the way. It turns out to be better to use the world as its own model" (Brooks, 1991).

Inspired by biological evolution, Brooks created a decentralized control architecture consisting of different layers; every layer is a more or less simple coupling of sensors to motors. The levels operate in parallel but are built in a hierarchy (hence the term subsumption architecture) (Brooks, 1986).

Embodiment means more than only "intelligence requires a body." The behavior of any system is not merely the outcome of an internal control structure (such as the central nervous system); it is also affected by the ecological niche in which the system is physically embedded, by its morphology (the shape of its body and limbs, as well as the type and placement of sensors and effectors), and by the material properties of the elements composing the morphology (Pfeifer & Bongard, 2006). Taking embodiment seriously is mandatory for the rapidly developing field of soft robotics (e.g., Yasa et al., 2023), where the complexity and nonlinearity of the robot bodies makes it hard to employ traditional control schemes and the dynamics of the material substrate needs to be exploited.

Current Weakly Embodied AI (WEAI)

Large-scale, so-called foundation models in natural language processing (NLP) and computer vision can enable capable AI systems by providing general-purpose pretrained models that can often outperform their narrowly targeted counterparts trained on smaller but more task-specific data. The idea of applying the same strategy to control robots is appealing. The community originating in computer vision, ML, and NLP, but now connecting their models to robots, uses the label EAI (e.g., Deitke et al., 2022; Liu et al., 2024; Vanhoucke, 2024). The idea is to leverage the reasoning ("common sense") capabilities of such models as well as their capability to understand visual scenes (images) to produce plans that a robot can execute.

In this chapter, we argue that the embodiment in this work is weak or shallow and discuss the implications this has on the field—what are principled roadblocks and where are productive future research directions.

Implications of Embodiment for Making Better Robots

The implications of embodiment for any agent (biological or artificial) are illustrated graphically in figure 8.1 (Pfeifer et al., 2007).

Below, we list selected premises of embodiment to serve as a checklist to gauge how embodied WEAI really is. The *body* or body *morphology* refers to the shape of the body and limbs, as well as the type and placement of sensors and effectors. The list draws on the principles of intelligent systems from Pfeifer and Scheier (2001) and Pfeifer and Bongard (2006). Behavior is not in the brain/controller but in the closed-loop interaction of the controller, the body, and the environment.

1. **Body morphology facilitates control.** Interaction of the body with the environment can facilitate the production of behavior with simpler control by the nervous system. In extreme cases, only physics (i.e., no sensing, control, or even actuation) may suffice to produce behavior like walking (McGeer, 1990). Mechanical feedback loops (see the movement and mechanical feedback arrows in figure 8.1) can even serve to stabilize the system (see self-stabilization in Pfeifer et al., 2007 or physical implications of embodiment in Hoffmann & Pfeifer, 2012). Complex nonlinear bodies offer a bigger potential for their intrinsic dynamics to be exploited to produce desired behavior as studied in soft robotics (Hauser et al., 2023).
2. **Sensor morphology facilitates perception.** The placement, shape, and mechanical properties of sensors, importantly, determine what will be sensed. This is certainly true for tactile sensors, for example, but also for

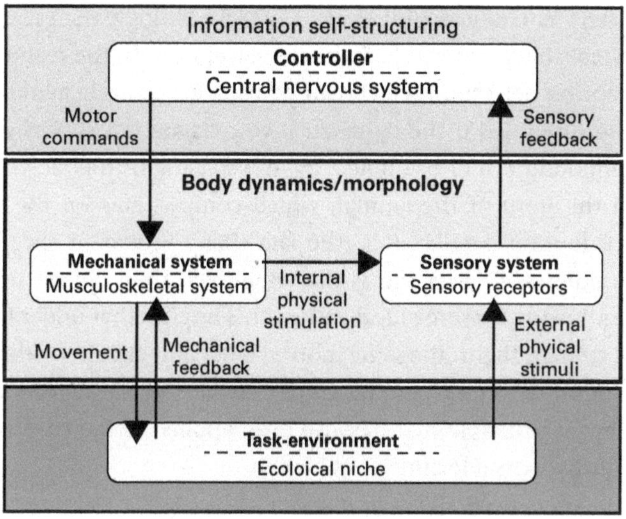

Figure 8.1
Implications of embodiment (the interplay of information and physical processes). Driven by motor commands, the musculoskeletal system (mechanical system) of the agent acts on the external environment (task environment or ecological niche). The action leads to rapid mechanical feedback characterized by pressure on the bones, torques in the joints, and passive deformation of skin tissue. In parallel, external stimuli (pressure, temperature, and electromagnetic fields) and internal physical stimuli (forces and torques developed in the muscles and joint-supporting ligaments, as well as accelerations) impinge on the sensory receptors (sensory system). The patterns induced thus depend on the physical characteristics and morphology of the sensory systems and on the motor commands. Especially if the interaction is sensorimotor coordinated, as in foveation, reaching, or grasping movements, information structure is generated. The effect of the motor command strongly depends on the tunable morphological and material properties of the musculoskeletal system, where by "tunable" we mean that properties such as shape and compliance can be changed dynamically: during the forward swing of the leg in walking, the muscles should be largely passive, whereas when hitting the ground, high stiffness is required, so that the materials can take over some of the adaptive functionality on impact, such as the damped oscillation of the knee joint. *Source*: Figure and caption from Pfeifer et al., 2007. The figure has been redrawn by the author.

visual sensors (Franceschini et al., 1992). In biology, the eyes of different animals have evolved to facilitate perception. In the realm of insect eyes, motion detection, which can be used for obstacle avoidance, can often be simplified if the light-sensitive cells are not spaced evenly. In the compound eye of the house fly, the spacing of the facets is denser toward the front of the animal, which compensates for the phenomenon of motion parallax (i.e., the fact that at constant speed, objects on the side travel faster across the visual field than objects toward the front). Allowing for some idealization, this implies that under the condition of straight flight, the same motion detection circuitry—the elementary motion detectors—can be employed for motion detection for the entire eye, a principle that has also been applied to the construction of navigating robots (Hoshino et al., 2000).

3. **Sensorimotor coordination and active perception.** Closed-loop interaction with the environment and the choice of specific actions may significantly transform what is perceived and make a perceptual task like discrimination much easier compared to passive perception or random movements of the agent. Such *information self-structuring* in simple robots was quantified by Lungarella and Sporns (2006) using information theoretic measures. The work being done on active perception (in particular, active vision) (Bajcsy et al., 2018) and the principle of sensorimotor coordination (Pfeifer & Bongard, 2006) are relevant here.
4. **Parallel loosely coupled processes.** Intelligence—at least as we know it from biology—is emergent from a large number of parallel processes[2] coordinated through the embodied interaction with the environment (Pfeifer & Bongard, 2006). The subsumption architecture (Brooks, 1986) makes a similar point but adds a hierarchy to the parallel processes.
5. **Principle of ecological balance.**
   a. Given a certain task environment, there has to be a match between the complexities of the agent's sensory, motor, and neural systems.
   b. There is a certain balance or task distribution between morphology, materials, control, and environment.

Embodiment somewhat overlaps with the notion of *morphological computation* (MC), but we want to emphasize the role of the body in facilitating behavior and hence simplifying control and in perception. We do not regard these processes as computational (in line with Müller & Hoffmann, 2017).

Representative Works of Weakly Embodied AI

A generalist agent (Reed et al., 2022) can be regarded as a natural extension of the ML and NLP models, as it retains the capabilities of the image- and language-based foundation models like image caption generation and chat, combines them with playing Atari games, and also brings a real robotics task into the same model—the RGB stacking benchmark, where a robot arm stacks colored blocks using input from an RGB camera.

There is an explosion of work in EAI (here, WEAI)—that is, the use of large models in robotics. To give the discussion concrete contours, we briefly describe the evolution of this young field: we discuss work from 2022 to 2024 inspired by the perspective of Vanhoucke (2024). The deployment of large language models (LLMs) into robotics has naturally started from planning, which was lifted into "semantic space" from geometric space (see figure 8.2A). An example is the "say-can" model (Ahn et al., 2022). The reasoning and planning capabilities traditionally handled by automatic inference systems in the GOFAI era now leverage the "common sense" power of LLMs.

The world state, now represented in language, had to be extracted from the world (perception or state estimation module). The next logical step

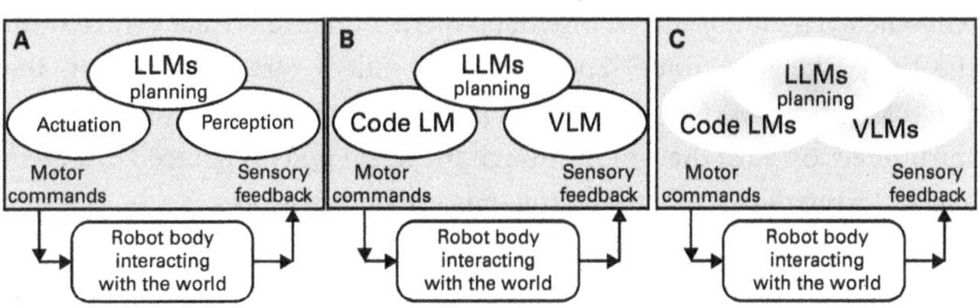

Figure 8.2
Evolution of approaches in EAI. The next modules to be "AI-powered" were the perception and actuation modules, through visual language models (VLMs) and code language models (code LMs), respectively, as shown in panel B. *Source*: Redrawn and adapted from Vanhoucke (2024).

was thus to deploy VLMs, which can convert images into language representations, as the perception module (e.g Plato, ca. 380–347 B.C.E./1997; Zeng et al., 2022). To keep the representation in sync with the real world (addressing the frame problem), periodic state reestimation and replanning is necessary, as exemplified in the "inner monologue" (Huang et al., 2022).

The final bastion of traditional robotics that was given LLM treatment in the words of Vanhoucke (2024) was the action. This was achieved by having the language model generate code to be executed by the robot "code LM," as in the article "Code as Policies" (Liang et al., 2023), Text2Motion (Lin et al., 2023), ProgPrompt (Singh et al., 2023), or ChatGPT for Robotics (Vemprala et al., 2024).

The interfaces between the modules—LLM as planner, VLM as state estimator, and Code LM as action generator—have become a bottleneck in these architectures. Therefore, the next step was to "blend these modules in" as schematically illustrated in figure 8.2C. PALM-E (Driess et al., 2023) blended the LLM and VLM; RT-1 (Zitkovich et al., 2023) blended the perception and actuation (VLM and Code LM). The final logical next step was to reason jointly about the entire problem, blending all the module boundaries while still leveraging "Internet data." Examples of this approach are RT-2 (Zitkovich et al., 2023) and VC-1 (Yokoyama et al., 2023).

How Embodied Is Weakly Embodied AI?

Comparing figures 8.1 and 8.2, it is apparent that WEAI is really weakly embodied. The implications of embodiment in figure 8.1 are centered on the body dynamics (blue) and task-environment (green) blocks. In the WEAI architectures (figure 8.2), what matters is the controller or brain, and the interaction with the world through the agent body is reduced to a mere arrow closing the loop between the sensors and actuators.

Let us look through the "embodiment checklist" and see where WEAI stands.

0. Behavior is not in the brain/controller but in the closed-loop interaction of the controller, the body, and the environment.
   - In WEAI, the behavior is in the controller—in the large model. Complete long-horizon tasks can be planned in the model, and the state in the model has to be kept in sync with the world. Embodied interaction with the environment is not exploited.

1. Body morphology facilitates control.
    - In WEAI, there is little room for the agent body to be exploited to fulfill the task. The robot platforms used—like mobile robots or robot arms with two-finger grippers—do not feature rich body dynamics that could be exploited for the task. Moreover, the details of the sensory and motor interfaces of the robots are abstracted away. For example, if the action space is the position and orientation of the end effector and the open/close gripper command, it leaves little room for the model to make interesting use of the robot morphology.
2. Sensor morphology facilitates perception.
    - WEAI leverages the power of deep neural networks to find patterns in static images (e.g., object recognition) and how what is on the images can be translated into text (VLMs). The inputs need to be more or less standard RGB images. The power of convolutional neural networks and visual transformers on tasks like recognition will hardly transfer to particular sensor morphologies like with non-uniform spacing of light sensitive cells. Thus, sensor morphologies designed for a particular task environment are not compatible with the WEAI approach.
3. Sensorimotor coordination and active perception.
    - Typical WEAI architectures aim to maximally exploit the common sense of LLMs to solve tasks without the need for additional training—also called "zero shot." In this case, there is no possibility to exploit the closed-loop embodied interaction with the environment. When additional training (fine-tuning) of the model takes place, it typically uses offline data sets collected when, for example, a human was teleoperating the robot to solve the task. The operator may exploit sensorimotor coordination to solve the task, and this would be transferred to the robot controller. However, there has to be a match in the *situatedness*;[3] that is, the human and the robot have to see through similar "eyes" "An agent is 'situated' if it can acquire information about the current situation through its sensors in interaction with the environment" (Pfeifer & Scheier, 2001, p. 72). The human and the robot have to see through similar "eyes." If the operator drives the robot using, for example, a third-person view but the robot has an egocentric camera, the transfer will not work.

4. Parallel loosely coupled processes.
   - WEAI architectures may consist of several blocks—VLM → LLM → code LM—with parallel processing inside, but there is typically a single loop connecting the sensors and motors. Hence, there is only one process and one timescale.
5. Principle of ecological balance.
   - WEAI is strongly susceptible to ecological imbalance, as the architectures feature a gigantic brain. The sensory inputs can be text and vision, with some developments in adding other modalities. The robots typically have rather simple morphology and, hence, action possibilities. However, the main source of ecological imbalance is the action representation—typically a low-dimensional discrete action space.

Active Embodied Interaction versus Offline Learning

The impressive progress in NLP and computer vision has been enabled by two main factors: (1) large data sets of text and images were available; (2) passive supervised learning was appropriate for tasks like predicting the next word in a sequence or labeling the object in an image. A big part of the WEAI community thinks that getting enough data will "solve robotics" as well. Data sets are typically collected using teleoperation. From such data sets, imitation learning or offline reinforcement learning methods are applied to train a controller (e.g., Walke et al., 2023). However, successful behavior in the real world inherently requires sensorimotor coordination and active perception (see also a recent article on the limits of passive AI [Pezzulo et al., 2024]). If the robot cannot sample the actions and their consequences itself, learning is bound to be inefficient. We illustrate this with a cartoon in figure 8.3. Panel A shows the setup from the seminal work of Held and Hein (1963). Kittens were raised in darkness from birth until they entered the experimental setup and were split into two groups. The kittens from the first group (on the right in figure 8.3A) were allowed to actively move around the arena, while the kittens from the second group were passively moved around on a trolley, driven by the movement of the first kittens. After this "training phase" kittens from both groups were presented with behavioral tests (visually guided paw placement, avoidance of visual cliff, blink to an approaching object). Importantly, only the kittens from the actively moving group succeeded in the tests of visually guided behavior. Panel B illustrates how robot data sets are created.

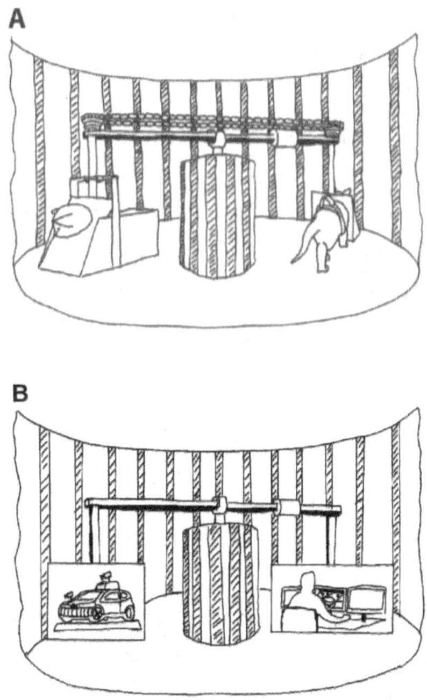

Figure 8.3
Active embodied interaction versus being driven around. (A) Setup from Held and Hein (1963). The kitten on the right is actively walking in the apparatus; the kitten on the left is being passively carried around but receiving similar stimulation. (B) A cartoon representation of data set generation in WEAI. A human teleoperates a robot, which only passively receives sensory stimulation. *Source*: the author. Panel A adapted from Held & Hein, 1963.

A human (on the right) is remotely controlling a robot (a car equipped with sensors in this case), and the data are recorded. The robot has one important advantage. While the passive kitten did not have access to the motor commands of the active kitten, the robot data set may contain both sensory and motor signals (if the operator uses the same actuation interface). However, it is still not possible for the robot to actively sample the actions and see their effects and associated rewards, as in standard online reinforcement learning.

Natural WEAI Extensions

WEAI researchers are aware of the gap that still exists between "high-level language knowledge" and "low-level robotics knowledge," and there is work under way to bridge it. For example, there are several extensions to the robotic transformer family of models that bring motion-centric representations (RT-trajectory; Gu et al., 2023), prompt with visual iterative optimization (PIVOT—iterative visual prompting; Nasiriany et al., 2024), or introduce hierarchies to increase the granularity of the language instructions (e.g., for "put the coke on the table," there would be another more granular description starting with "move arm forward," etc.) (RT-hierarchy; Belkhale et al., 2024). Szot et al. (2024) developed a multimodal LLM and an action space adapter for continuous action spaces.

Touch is a sensory modality that is especially relevant for manipulation and that researchers are trying to add to the models (e.g., Sferrazza et al., 2023; F. Yang et al., 2024).

Cross-Embodiment Embodied AI: An Oxymoron?

The need for large data sets to train these models triggered several cross-embodiment initiatives—that is, "crowd-sourcing" data from different labs and different robots performing various tasks in order to train models that can master different tasks, environments, and embodiments (see Jaquier et al., 2024, for a review of transfer learning in robotics). Open X-Embodiment is a large initiative across many institutions focusing primarily on developing a large robot data set (O'Neill et al., 2024). Currently, sixty data sets have been brought into a common format, featuring twenty-two different robot embodiments (mostly robot arms). Building on previous models, RT-1 and RT-2, the new model, RT-X, receives a history of recent RGB images from a camera placed over the robot's shoulder and language instructions (e.g., "pick up the orange fruit") as input and produces (predicts) a seven-dimensional action vector controlling the end effector x, y, z, roll, pitch, yaw, and gripper opening or the rates of these quantities. The action is tokenized into 256 bins uniformly distributed along each dimension. The results show some positive transfer such that cotraining on data collected on multiple robots improves performance on a training task. However, generalization to new robots was not yet studied.

J. Yang et al. (2024) have put together a data set featuring a heterogeneous set of robots (cars, quadrupedal robots, wheeled robots, and manipulator arms) aggregated from eighteen manipulation, navigation, and driving data sets. Here, no language instructions are part of the model. The input to the network is a history of RGB images and the goal expressed as an egocentric image (object being grasped by the manipulator; mobile robot reaching a waypoint). The output is similar to the RT-X model: a seven-dimensional discrete action space. A single goal-conditioned policy was trained, capable of controlling robotic arms, quadcopters, quadrupeds, and mobile bases. They found that cotraining with navigation data can enhance robustness and performance in goal-conditioned manipulation with a wrist-mounted camera. Zero-shot generalization to a new embodiment was evaluated.

All Robots in One (Wang et al., 2024) is another cross-embodiment initiative focusing solely on acquiring a data set and extending the typical sensory inputs (text and vision) with audio and touch.

At first glance, EAI across embodiments seems contradictory. A key implication of EC is that our bodies constitutively shape the way we think (Pfeifer & Bongard, 2006), and hence different bodies require different brains. How can a single large model (or "brain") command bodies as diverse as cars, legged robots, and robot arms?

The reason why the efforts reviewed above have had some success is that the *embodiment* there is *shallow*. These projects have essentially created an abstraction layer that reduces every robot to the Cartesian coordinates of its gripper, or the coordinates of the robot's center of mass in case of a mobile robot. The models were trained to learn a mapping from camera images to a low-dimensional action space, conditioned on a goal expressed in language (e.g., "pick apple from . . . and place on . . ."; O'Neill et al., 2024) or as a goal image (J. Yang et al., 2024). In this formulation, the fact that the robot has joints—a defining feature of a robot arm—is abstracted away.

The control has a coarse spatial (discretized low-dimensional action space) and temporal resolution (3–10 Hz). Interaction forces are not considered. Thus, one could easily conclude that this is not "real robotics" or "real embodiment."

Fundamental Roadblocks: Do Foundation Models Stand in the Way?

Where WEAI Is Not Like GOFAI

While GOFAI and cognitivism rested on the manipulation of symbols that represent the world, WEAI could be classified as emergentism or connectionism, harnessing the power of neural networks.[4] WEAI representations are learned from data and are much more flexible than the largely handcrafted rule-based systems. Large models can also be more flexibly connected to the real world through VLMs for perception (mitigating the frame problem) and code LMs for action. Finally, learning can be employed throughout the whole pipeline—end-to-end when appropriate.

Where WEAI Inherits GOFAI Problems

1. **Symbol grounding problem.** LLMs are praised for having "common sense," which can be leveraged not only to chat but also to make action plans for the real world. This common sense comes from the large text corpora that the models have digested. This provides much better grounding than GOFAI has provided. In fact, Harnad (1990) proposes connectionism as a possible remedy to the symbol grounding problem. However, grounding derived solely from text corpora has intrinsic limitations. An LLM may hold a conversation on riding a bicycle, skiing, or fishing, but proper grounding on performing this in the real world when embodied in a robot is largely lacking. If language is the "common currency" and the LLM modules are used for reasoning, the problem of grounding is inevitable. Human cognition is grounded (Barsalou, 2008) through internal multimodal sensorimotor simulations. Using multimodal LLMs (MLLMs) and grounding them into different embodiments and their associated action spaces will be necessary (Szot et al., 2024).
2. **Embodiment and situatedness.** In WEAI, intelligence is in the LLM (or multiple LLM) Knowledge extracted from texts, images, and videos constitutes a *world model*, which is largely disembodied or indirectly shallowly embodied in a human-like embodiment (since the text and images on the Internet are human-centered). As the details of the body morphology are not represented, they cannot be meaningfully exploited by the "reasoning engine" for facilitating control or perception. The large model also somehow stands in the way of exploiting sensorimotor coordination or direct perception (Gibson, 1979)—shortcuts exploiting closed-loop interaction with the environment.
3. **"Soft real-time" interaction.** A long time has passed since the Stanford Cart of Hans Moravec, where long "thinking" periods separated every

movement (Moravec, 1983). Now we have much more computational power, but also, the models have grown considerably. Closing the loop at 3–10 Hz (O'Neill et al., 2024) is not fast enough for real tasks in robotics.

Where WEAI Brings Its Own New Challenges

The large models are truly large, bringing enormous computational and energy costs. Even inference—that is, only "running" the model—requires considerable computational resources. For example, training GPT-3 (now a "small and old" model) on only text data required 29,000 GPUs and consumed 1,287 MWh, compared to an average household, which consumes 10–12 MWh annually (De Vries, 2023). EAI agents dealing with multimodal data, especially visual data, will require more storage and computation and consume more energy. Further, each inference request was calculated to consume as much as 3–4 Wh, as much as an LED bulb running for one hour (De Vries, 2023). Each planning sequence would require multiple requests, and physical bodies need to carry enough computational power so that the model can be deployed in real-time.

Let us reexamine Brooks's objections (1991) to GOFAI: "When we examine very simple level intelligence we find that explicit representations and models of the world simply get in the way. It turns out to be better to use the world as its own model" (Brooks, 1991, p. 140).

Does this apply to WEAI as well? It largely does. The WEAI representations are not explicit. However, the power of WEAI is exactly in leveraging the world models. Exploiting direct interaction with the world is in theory possible if learning in a particular robot embodiment is used, but it would essentially have to bypass the LLM.

Outlook: The Future of WEAI in Robotics

Let us sketch possible directions of future developments around WEAI.

WEAI for Flexible Visual Perception and Planning

The conservative scenario is that the power of LLMs and VLMs, leveraging what can be learned from data already available on the Internet, will be exploited for robotics for flexible planning and for (passive) visual perception, along the lines of the natural WEAI extensions described above. The interface to the robot will still rely on a robot control API taking care of real-time interaction, like in the current models that only "reason" about the end effector position or robot center of mass.

Making WEAI More Embodied

Some developments within WEAI naturally extend it but at the same time have deeper implications. One such development is the use of egocentric views in the visual inputs as demonstrated by J. Yang et al. (2024). Unlike the models that use third-person views and over-the-shoulder cameras that see the world but also the robot in it, egocentric views may open the door to direct or active perception and sensorimotor coordination. We also argue that when data sets are collected by teleoperating robots, the view of the robot operator and the view that is recorded for subsequent training should be an egocentric view.

Another important concept is that of object affordances—what one can do with an object (e.g., a chair has a "sit on" affordance) (Gibson, 1979). Affordances have also been discussed in work on WEAI (Ahn et al., 2022; Singh et al., 2023), where they are treated as properties of the objects alone. However, affordances depend on the perceptual and motor capabilities of the agent (a robot arm cannot sit on a chair), and the ability to perceive them is acquired by the agent through a long sensorimotor learning process (Jamone et al., 2018). This perspective should find its way into the models, introducing robot-aware affordances (Schiavi et al., 2023).

New Foundation Models for Robotics?

Perhaps new foundation models will have to be created for robotics. However, that is associated with both practical and fundamental roadblocks. The obvious practical complication is that such data are hard to come by, although several efforts are underway (e.g., Walke et al., 2023); Leaving cross-embodiment aside for now, the quantity of the data collected from real robots is nothing compared to the text and image data available on the internet. Data augmentation techniques or realistic robot simulators could constitute a partial remedy.

However, there are also fundamental problems with foundation models for robotics. In natural language processing, massive text corpora themselves provide all the training data needed: models can be trained to predict the next word or sentence and to capture longer-range dependencies through the attention mechanisms of transformer architectures (Vaswani et al., 2017). Image databases like ImageNet (Deng et al., 2009) typically need labels (describing what is in the picture) in addition. Robot data sets, collected by driving a sensorized robot in an environment and teleoperating it to perform a task, would comprise a time series of multimodal data that can be used to train models. Driving around a real robot with cameras in environments with varying lighting conditions already introduces more variety in the collected images than in typical image data sets. With different robots, the variability of the multimodal data sets would grow dramatically, substantially reducing what can be transferred across the different platforms. There is thus a fundamental trade-off: as the robot embodiments become more diverse and more of their different sensory and action spaces are represented by the model, the individual robots profit less from the shared gigantic "brain." This is a principled limitation of foundation models for robotics, and only partial workarounds exist: (1) an abstraction layer hiding the robot embodiment, like in the low-dimensional discrete action space (O'Neill et al., 2024; J. Yang et al., 2024), and (2) standardized robots (Walke et al., 2023)—meaning only one robot type is used.

Another fundamental limitation is the passive mode of data collection (see "Active Embodied Interaction versus Offline Learning," above). Imitation learning allows the model to copy how a human operating a robot solves a task, but the situation resembles the one in figure 8.3. To exploit embodiment, active learning through closed-loop embodied interaction with the environment is necessary.

Finally, if foundation models for robotics continue focusing on the "robot brain" only like in figure 8.3C, they will still constitute a modern subsymbolic sense-think-act architecture with a single loop through the robot and environment, which essentially means there is a single timescale for the interaction (cf. Iida & Giardina, 2023). To make proper use of all the implications of embodiment, multiple interaction loops at different timescales should be considered, and more details about the mechanical and sensory system added to the model such that the model can exploit them (see figure 8.1). Alternatively, these loops may not need to be rerepresented (modeled—"Sometimes the world is its own best model." [Brooks, 1991]), but active learning will be necessary to take advantage of them.

Active Learning in a Single Embodiment (Reinforcement Learning Instead of LLM)

Robot learning (Peters et al., 2016) has relied on deep neural networks (like deep reinforcement learning), but not on foundation models. The learning has also typically been active and in a single robot embodiment, often with a big part of it done in simulation (e.g., FastGrasp'D) (Turpin et al., 2023). This approach is inherently much more embodied and will not hit the fundamental roadblocks described above.

Touch Is Not Like Vision

Tactile inputs have been added to some MLLMs recently. The approach that worked so well for vision can extend to touch, provided that there is a unified representation for touch. Yang et al. (2024) developed UniTouch: A unified tactile model for vision-based touch sensors. However, there are several roadblocks: (1) there is no touch on the Internet; (2) tactile data are strongly dependent on the robot hand or gripper used and the type of sensor (see also Pliska et al., 2024); and (3) the way haptic (tactile and proprioceptive) data are collected is intrinsically active—the explorative movements are key for recognition.

Cross-Embodiment Outlook Though contradictory at first glance, the cross-embodiment idea is going to continue to receive attention. The bigger the embodiment gap, the less positive transfer there can be. However, at least in the motor domain, progress appears possible. For example, Bohlinger et al. (2024) developed a single controller that generalizes across highly diverse robot morphologies. In biology, there likewise appears to be a shared organizational design for motion controllers across species (Ijspeert & Daley, 2023).

Low-dimensional descending commands from the brain work in concert with low-level spinal control (feedforward and feedback) and could be attractive for robotics foundation models; the foundation models would learn to supply these descending inputs (like locomotion speed and direction), and the rest will be up to low-level robot controllers.

Truly Embodied AI Similar to GOFAI, for WEAI, the robot embodiment constitutes the uninteresting periphery that should be abstracted or standardized. This form of "Cartesian dualism" will in the end hinder progress. For optimal performance, codesign or coevolution of brains and bodies will be necessary. This is of course a big challenge, but work being done in this area is approaching the problem from a modeling and control perspective (Pekarek, 2010; Zardini et al., 2021) or Deep Evolutionary Reinforcement Learning (DERL) (Gupta et al., 2021). Gupta et al. made agents undergo both lifetime learning and evolution simultaneously but asynchronously, more in line with how species of biological organisms actually evolve. They reproduced the Baldwin effect, which states that evolutionary selection prefers not only the morphologies with the highest fitness but also the morphologies that can learn and adapt the quickest.

True embodiment will inevitably need to exploit physics. Physical interaction of robot bodies with the environment can be made part of the existing models through differentiable physics (de Avila Belbute-Peres et al., 2018). Using transformers for modeling physical systems may be possible (Geneva & Zabaras, 2022). Alternatively, the controller should be given the chance to learn to exploit the interaction with the environment directly—without a model.

Acknowledgments

This work was cofunded by the European Union under project ROBOPROX (registration number CZ.02.01.01/00/22_008/0004590).

Notes

1. In Russell & Norvig (2021), there is only one chapter (26) on Robotics. Conversely, in Siciliano & Khatib (2016), there is only one chapter (Beetz et al. 2016) on AI reasoning methods for robotics.

2. These are parallel interaction loops—mechanical or informational—operating at different timescales. It has nothing to do with the possibility of parallelization on a graphical processing unit (GPU).

3. "An agent is 'situated' if it can acquire information about the current situation through its sensors in interaction with the environment" (Pfeifer & Scheier 2001, p. 72).

4. The fact that LLMs use tokens, which can also be regarded as symbols, does not play an important part here.

References

Ahn, M., Brohan, A., Brown, N., Chebotar, Y., Cortes, O., David, B., Finn, C., Fu, C., Gopalakrishnan, K., Hausman, K., Herzog, A., Ho, D., Hsu, J., Ibarz, J., Ichter, B., Irpan, A., Jang, E., Ruano, R. J., Jeffrey, K., . . . Zeng, A. (2022). Do as I can, not as I say: Grounding language in robotic affordances. *arXiv*. https://doi.org/10.48550/arXiv.2204.01691.

Bajcsy, R., Aloimonos, Y., & Tsotsos, J. K. (2018). Revisiting active perception. *Autonomous Robots*, *42*(2), 177–196. https://doi.org/10.1007/s10514-017-9615-3

Barsalou, L. W. (2008). Grounded cognition. *Annual Review of Psychology*, *59*, 617–645. https://doi.org/10.1146/annurev.psych.59.103006.093639

Beetz, M., Chatila, R., Hertzberg, J., & Pecora, F. (2016). AI reasoning methods for robotics. In B. Siciliano & O. Khatib (Eds.), *Springer handbook of robotics* (pp. 329–356). Springer International Publishing. https://doi.org/10.1007/978-3-319-32552-1_14

Belbute-Peres, F. A., Smith, K., Allen, K., Tenenbaum, J., & Kolter, J. Z. (2018). End-to-end differentiable physics for learning and control. In S. Bengio et al. (Eds.), *Advances in Neural Information Processing Systems 31* (pp. 7178–7189). Curran Associates, Inc.

Belkhale, S., Ding, T., Xiao, T., Sermanet, P., Vuong, Q., Tompson, J., Chebotar, Y., Dwibedi, D., & Sadigh, D. (2024). RT-H: Action hierarchies using language. *arXiv*. https://doi.org/10.48550/arXiv.2403.01823

Bohlinger, N., Czechmanowski, G., Krupka, M., Kicki, P., Walas, K., Peters, J., & Tateo, D. (2024). One policy to run them all: An end-to-end learning approach to multi-embodiment locomotion. *arXiv*. http://arxiv.org/abs/2409.06366

Braitenberg, V. (1986). *Vehicles: Experiments in synthetic psychology*. MIT Press.

Brooks, R. A. (1986). A robust layered control system for a mobile robot. *IEEE Journal on Robotics and Automation*, *2*(1), 14–23. https://doi.org/10.1109/JRA.1986.1087032

Brooks, R. A. (1991). Intelligence without representation. *Artificial Intelligence*, *47*(1), 139–159. https://doi.org/10.1016/0004-3702(91)90053-M

Deitke, M., Batra, D., Bisk, Y., Campari, T., Chang, A. X., Chaplot, D. S., Chen, C., Perez-D'Arpino, C., Ehsani, K., Farhadi, A., L. Fei-Fei, L., Francis, A., Gan, C., Grauman, K., Hall, D., Han, W., Jain, U., Kembhavi, A., Krantz, J, . . . Wu, J. (2022). Retrospectives on the embodied AI workshop. *arXiv*. https://doi.org/10.48550/arXiv.2210.06849

Deng, J., Dong, W., Socher, R., Li, L.-J., Li, K., & Fei-Fei, L. (2009). ImageNet: A large-scale hierarchical image database. In G. Hua, M. J. Black, & J. Matas (Eds.), *2009 IEEE Conference on Computer Vision and Pattern Recognition* (pp. 248–255). IEEE. https://doi.org/10.1109/CVPR.2009.5206848

De Vries, A. (2023). The growing energy footprint of artificial intelligence. *Joule*, *7*(10), 2191–2194. https://doi.org/10.1016/j.joule.2023.09.004

Driess, D., Xia, F., Sajjadi, M. S., Lynch, C., Chowdhery, A., Ichter, B., Wahid, A., Tompson, J., Vuong, Q., Yu, T., Huang, W., Chebotar, Y., Semanet, P., Duckworth, D., Levine, S., Vanhoucke, V., Hausman, K., Toussant, M, . . . Florence, P. (2023). PaLM-E: An embodied multimodal language model. *arXiv*. http://arxiv.org/abs/2303.03378

Engel, A. K., Maye, A., Kurthen, M., & König, P. (2013). Where's the action? The pragmatic turn in cognitive science. *Trends in Cognitive Sciences, 17*(5), 202–209. https://doi.org/10.1016/j.tics.2013.03.006

Fikes, R. E., & Nilsson, N. J. (1971). Strips: A new approach to the application of theorem proving to problem solving. *Artificial Intelligence, 2*(3), 189–208. https://doi.org/10.1016/0004-3702(71)90010-5

Fodor, J. A. (1975). *The language of thought*. Harvard University Press.

Franceschini, N., Pichon, J. M., Blanes, C., & Brady, J. M. (1992). From insect vision to robot vision [and discussion]. *Philosophical Transactions: Biological Sciences, 337*(1281), 283–294. https://doi.org/10.1098/rstb.1992.0106

Geneva, N., & Zabaras, N. (2022). Transformers for modeling physical systems. *Neural Networks, 146*, 272–289. https://doi.org/10.1016/j.neunet.2021.11.022

Gibson, J. J. (1979). *The ecological approach to visual perception*. Houghton Mifflin.

Gu, J., Kirmani, S., Wohlhart, P., Tan, A., Jampani, V., Soricut, R., Hsu, K., Kew, J., Rao, A., Tsao, K., Xu, C., Finn, C., Bousmalis, K., Levine, S., & Vanhoucke, V. (2023). RT-trajectory: Robotic instruction following with predicted trajectory plans. *arXiv*. https://doi.org/10.48550/arXiv.2311.01977

Gupta, A., Savarese, S., Ganguli, S., & Fei-Fei, L. (2021). Embodied intelligence via learning and evolution. *Nature Communications, 12*(1), 5721. https://doi.org/10.1038/s41467-021-25874-z

Harnad, S. (1990). The symbol grounding problem. *Physica D, 42*(1–3), 335–346. https://doi.org/10.1016/0167-2789(90)90087-6

Haugeland, J. (1989). *Artificial intelligence: The very idea*. MIT Press.

Hauser, H., Nanayakkara, T., & Forni, F. (2023). Leveraging morphological computation for controlling soft robots: Learning from nature to control soft robots. *IEEE Control Systems Magazine, 43*(3), 114–129. https://doi.org/10.1109/MCS.2023.3253422

Held, R., & Hein, A. (1963). Movement-produced stimulation in the development of visually guided behavior. *Journal of Comparative and Physiological Psychology, 56*(5), 872–876. https://doi.org/10.1037/h0040546

Hoffmann, M., & Pfeifer, R. (2012). The implications of embodiment for behavior and cognition: Animal and robotic case studies. *arXiv*. https://doi.org/10.48550/arXiv.1202.0440

Hoshino, K., Mura, F., & Shimoyama, I. (2000). Design and performance of a micro-sized biomorphic compound eye with a scanning retina. *Journal of Microelectromechanical Systems*, *9*(1), 32–37. https://doi.org/10.1109/84.825774

Huang, W., Xia, F., Xiao, T., Chan, H., Liang, J., Florence, P., Zeng, A., Tompson, J., Mordatch, I., Chebotar, Y., Sermanet, P., Brown, N., Jaskson, T., Luu, L., Levine, S., Hausman, K., & Ichter, B. (2022). Inner monologue: Embodied reasoning through planning with language models. *arXiv*. http://arxiv.org/abs/2207.05608

Iida, F., & Giardina, F. (2023). On the timescales of embodied intelligence for autonomous adaptive systems. *Annual Review of Control, Robotics, and Autonomous Systems*, *6*, 95–122. https://doi.org/10.1146/annurev-control-063022-094301

Ijspeert, A. J., & Daley, M. A. (2023). Integration of feedforward and feedback control in the neuromechanics of vertebrate locomotion: A review of experimental, simulation and robotic studies. *Journal of Experimental Biology*, *226*(15), jeb245784. https://doi.org/10.1242/jeb.245784

Jamone, L., Ugur, E., Cangelosi, A., Fadiga, L., Bernardino, A., Piater, J., & Santos-Victor, J. (2018). Affordances in psychology, neuroscience, and robotics: A survey. *IEEE Transactions on Cognitive and Developmental Systems*, *10*(1), 4–25. https://doi.org/10.1109/TCDS.2016.2594134

Jaquier, N., Welle, M. C., Gams, A., Yao, K., Fichera, B., Billard, A., Ude, A., Asfour, T., & Kragic, D. (2024). Transfer learning in robotics: An upcoming breakthrough? A review of promises and challenges. *The International Journal of Robotics Research*, 02783649241273565. https://doi.org/10.1177/02783649241273565

Liang, J., Huang, W., Xia, F., Xu, P., Hausman, K., Ichter, B., Florence, P., & Zeng, A. (2023). Code as policies: Language model programs for embodied control. In K. Althoefer, J. Taylor, & H. Qiao (Eds.), *2023 IEEE International Conference on Robotics and Automation (ICRA)* (pp. 9493–9500). IEEE. https://doi.org/10.1109/ICRA54114.2023.10160591

Lin, K., Agia, C., Migimatsu, T., Pavone, M., & Bohg, J. (2023). Text 2Motion: From natural language instructions to feasible plans. *Autonomous Robots*, *47*(8), 1345–1365. https://doi.org/10.1007/s10514-023-10131-7

Liu, Y., Chen, W., Bai, Y., Liang, X., Li, G., Gao, W., & Lin, L. (2024). Aligning cyber space with physical world: A comprehensive survey on embodied AI. *arXiv*. http://arxiv.org/abs/2407.06886

Lungarella, M., & Sporns, O. (2006). Mapping information flow in sensorimotor networks. *PLoS Computational Biology*, *2*(10), e144. https://doi.org/10.1371/journal.pcbi.0020144

McGeer, T. (1990). Passive dynamic walking. *The International Journal of Robotics Research*, *9*(2), 62–82. https://doi.org/10.1177/027836499000900206

Moravec, H. P. (1983). The Stanford cart and the CMU Rover. *Proceedings of the IEEE, 71*(7), 872–884.

Müller, V. C., & Hoffmann, M. (2017). What is morphological computation? On how the body contributes to cognition and control. *Artificial Life, 23*(1), 1–24. https://doi.org/10.1162/ARTL_a_00219

Nasiriany, S., Xia, F., Yu, W., Xiao, T., Liang, J., Dasgupta, I., Xie, A., Driess, D., Wahid, A., Xu, Z., Vuong, Q., Zhang, T., Lee, T-W. E., Lee, K-H., Xu, P., Kirmani, S., Zhu, Y., Zeng, A., . . . Ichter, B. (2024). PIVOT: Iterative visual prompting elicits actionable knowledge for VLMs. *arXiv*. https://doi.org/10.48550/arXiv.2402.07872

O'Neill, A., Rehman, A., Gupta, A., Maddukuri, A., Gupta, A., Padalkar, A., Lee, A., Pooley, A., Gupta, A., Madlekar, A., Jain, A., Tung, A., Bewlet, A., Herzog, A., Irpan, A., Jhazatsky, A., Rai, A., Gupta, A., Wang. A. . . . Lin, Z. (2024). Open X-embodiment: Robotic learning datasets and RT-X models. *arXiv*. https://doi.org/10.48550/arXiv.2310.08864

Pekarek, D. N. (2010). *Variational methods for control and design of bipedal robot models*. California Institute of Technology. https://search.proquest.com/openview/b20b505eda84786a0cf678df894452e0/1?pq-origsite=gscholar&cbl=18750

Peters, J., Lee, D. D., Kober, J., Nguyen-Tuong, D., Bagnell, J. A., & Schaal, S. (2016). Robot learning. In B. Siciliano & O. Khatib (Eds.), *Springer handbook of robotics* (pp. 357–398). Springer International Publishing. https://doi.org/10.1007/978-3-319-32552-1_15

Pezzulo, G., Parr, T., Cisek, P., Clark, A., & Friston, K. (2024). Generating meaning: Active inference and the scope and limits of passive AI. *Trends in Cognitive Sciences, 28*(2), 97–112. https://doi.org/10.1016/j.tics.2023.10.002

Pfeifer, R., & Bongard, J. (2006). *How the body shapes the way we think: A new view of intelligence*. MIT Press.

Pfeifer, R., Lungarella, M., & Iida, F. (2007). Self-organization, embodiment, and biologically inspired robotics. *Science, 318,* 1088–1093. https://doi.org/10.1126/science.1145803

Pfeifer, R., & Scheier, C. (2001). *Understanding intelligence*. MIT Press.

Pliska, M., Patni, S., Mares, M., Stoudek, P., Straka, Z., Stepanova, K., & Hoffmann, M. (2024). Single-grasp deformable object discrimination: The effect of gripper morphology, sensing modalities, and action parameters. *IEEE Transactions on Robotics, 40,* 4414–4426. https://doi.org/10.1109/TRO.2024.3463378

Reed, S., Żołna, K., Parisotto, E., Colmenarejo, S. G., Novikov, A., Barth-Maron, G., Giménez, M., Sulsky, Y., Kay, J., Springenberg, J. T., Eccles, T., Bruce, J., Razavi, A., Edwards, A., Heess, N., Chen, Y., Hadsell, R., Vinyals, O., Bordbar, M., & de Freitas, N.

(2022). A generalist agent. *Transactions on Machine Learning Research.* https://openreview.net/forum?id=1ikK0kHjvj

Russell, S. J., & Norvig, P. (2021). *Artificial intelligence: A modern approach.* Pearson.

Schiavi, G., Wulkop, P., Rizzi, G., Ott, L., Siegwart, R., & Chung, J. J. (2023). Learning agent-aware affordances for closed-loop interaction with articulated objects. In K. Althoefer, J. Taylor, & H. Qiao (Eds.), *2023 IEEE international conference on robotics and automation (ICRA)* (pp. 5916–5922). IEEE. https://doi.org/10.1109/ICRA48891.2023.10160747

Sferrazza, C., Seo, Y., Liu, H., Lee, Y., & Abbeel, P. (2023). The power of the senses: Generalizable manipulation from vision and touch through masked multimodal learning. OpenReview. https://openreview.net/forum?id=FMsmo01TaI.

Siciliano, B., & Khatib, O. (2016). *Springer handbook of robotics.* Springer International Publishing. https://doi.org/10.1007/978-3-319-32552-1

Singh, I., Blukis, V., Mousavian, A., Goyal, A., Xu, D., Tremblay, J., Fox, D., Thomason, J., & Garg, A. (2023). ProgPrompt: Generating situated robot task plans using large language models. In K. Althoefer, J. Taylor, & H. Qiao (Eds.), *2023 IEEE International Conference on Robotics and Automation (ICRA)* (pp. 11523–11530). IEEE. https://doi.org/10.1109/ICRA48891.2023.10161317

Szot, A., Mazoure, B., Agrawal, H., Hjelm, D., Kira, Z., & Toshev, A. (2024). Grounding multimodal large language models in actions. *arXiv.* https://doi.org/10.48550/arXiv.2406.07904

Turpin, D., Zhong, T., Zhang, S., Zhu, G., Heiden, E., Macklin, M., Tsogkas, S., Dickinson, S., & Garg, A. (2023). Fast-Grasp'D: Dexterous multi-finger grasp generation through differentiable simulation. In *2023 IEEE international conference on robotics and automation (ICRA)* (pp. 8082–8089). https://doi.org/10.1109/ICRA48891.2023.10160314

Vanhoucke, V. (2024, March 4). The state of robot learning. *Medium* (blog). https://vanhoucke.medium.com/the-state-of-robot-learning-639dafffbcf8

Vaswani, A., Shazeer, N., Parmar, N., Uszkoreit, J., Jones, L., Gomez, A. N., Kaiser, Ł., & Polosukhin, I. (2017). Attention is all you need. In I. Guyon, U. V. Luxburg, S. Bengio, H. Wallach, R. Fergus, S. Vishwanathan, & R. Garnett (Eds.), *Advances in Neural Information Processing Systems 30* (pp. 5998–6008). Curran Associates, Inc. https://proceedings.neurips.cc/paper/2017/file/3f5ee243547dee91fbd053c1c4a845aa-Paper.pdf

Vemprala, S., Bonatti, R., Bucker, A., & Kapoor, A. (2024). ChatGPT for robotics: Design principles and model abilities. *IEEE Access, 12,* 56306–56321. https://doi.org/10.1109/ACCESS.2024.3389088

Walke, H. R., Black, K., Zhao, T. Z., Vuong, Q., Zheng, C., Hansen-Estruch, P., He, A. W., Yan, V., Abhishek, A., Kan, J., Goel, K., Finn, C., Kumar, A., & Levine, S. (2023). BridgeData V2: A dataset for robot learning at scale. In *Proceedings of the 7th Conference on Robot Learning* (pp. 1723–1736). PMLR. https://proceedings.mlr.press/v229/walke23a.html

Walter, G. W. (1953). *The living brain* (G. W. Walter, Ed.). Norton & Co.

Wang, Z., Zheng, H., Nie, Y., Xu, W., Wang, Q., Ye, H., Li, Z., Zhou, G., Wang, T., Zhang, S., Han, L., He, B., Wang, Y., Sun, S., Wang, Z., & Luo, P. (2024). All robots in one: A new standard and unified dataset for versatile, general-purpose embodied agents. *arXiv*. http://arxiv.org/abs/2408.10899

Yang, F., Feng, C., Chen, Z., Park, H., Wang, D., Dou, Y., Zeng, Z., Chen, X., Gangopadhyay, R., Owens, A., & Wong, A. (2024). Binding touch to everything: Learning unified multimodal tactile representations. *arXiv*. http://arxiv.org/abs/2401.18084

Yang, J., Glossop, C., Bhorkar, A., Shah, D., Vuong, Q., Finn, C., Sadigh, D., & Levine, S. (2024). Pushing the limits of cross-embodiment learning for manipulation and navigation. *arXiv*. https://doi.org/10.48550/arXiv.2402.19432

Yasa, O., Toshimitsu, Y., Michelis, M. Y., Jones, L. S., Filippi, M., Buchner, T., & Katzschmann, R. K. (2023). An overview of soft robotics. *Annual Review of Control, Robotics, and Autonomous Systems*, *6*, 1–29. https://doi.org/10.1146/annurev-control-062322-100607

Yokoyama, N., Clegg, A., Truong, J., Undersander, E., Yang, T. Y., Arnaud, S., Ha, S., Batra, D., & Rai, A. (2023). ASC: Adaptive skill coordination for robotic mobile manipulation. *IEEE Robotics and Automation Letters*, *9*(1), 779–786. https://doi.org/10.1109/LRA.2023.3336109

Zardini, G., Milojevic, D., Censi, A., & Frazzoli, E. (2021). Co-design of embodied intelligence: A structured approach. In *2021 IEEE/RSJ international conference on intelligent robots and systems (IROS)* (pp. 7536–7543). IEEE. https://doi.org/10.1109/IROS51168.2021.9636513

Zeng, A., Attarian, M., Ichter, B., Choromanski, K., Wong, A., Welker, S., Tombari, F., Purohit, A., Ryoo, M., Sindhwani, V., Lee, J., Vanhoucke, V., & Florence, P. (2022). Socratic models: Composing zero-shot multimodal reasoning with language. *arXiv*. http://arxiv.org/abs/2204.00598

Zitkovich, B., Yu, T., Xu, S., Xu, P., Xiao, T., Xia, F., Wu, J., Wolhart, P., Welker, S., Wahid, A., Vuong, Q., Vanhoucke, V., Tran, H., Soricut, R., Singh, A., Singh, J., Sermant, P., Sanketi, P.R., Salazar, G. . . .Han, K. (2023). RT-2: Vision-language-action models transfer web knowledge to robotic control. *arXiv*. http://arxiv.org/abs/2307.15818

9 The Embodied Future of Robotics

David Howard

Introduction

It's fair to say that, for vast swathes of the robotics community, embodied artificial intelligence (EAI) is synonymous with uploading a neural network onto a robot and seeing what it can do. Not that there's anything wrong with that, of course—it's utilitarian (in that it *works*) as well as impressive (in that the robots can do backflips, cook your breakfast, and execute many other complex tasks). The robots themselves are embodied, in that they possess a physical presence and can affect the world, and they contain all the items typically associated with EAI (per this book's definition)—a means to elicit behavior (a brain, sensors, actuators, and the aforementioned body), an environment in which to behave, and some measure of adaptation or optimization to that environment. Ultimately, robots need controllers, and under certain constraints, a machine-learned policy running on a generalist robot—anything from Boston Dynamics, Unitree, Anybotics, and the like, is the best answer. We do this a lot in our research program; it's interesting research and industry wants it.

At the culmination of this chapter, I'll share some thoughts on what the key roadblocks are to achieving EAI, specifically for robotics, and moreover, what we can do as a community to make some progress. I'll propose some research directions, together with a background of some of the problems that have been encountered across evolutionary, soft, and other robotics disciplines that have prevented the broadscale uptake of EAI. Reflecting on the stated goal of this collection—to formulate and promote an inclusive and holistic view of EAI—I'll conclude with a view of a cross-disciplinary future of the field.

More Embodiment in Robotics

When it comes to embodied intelligence (EI) in robotics, it feels like we are missing a trick. And let me be up-front about my personal feelings that the characterization of EAI meaning *anything on a robot* is a little confusing and at worst a hype-fueled exercise in branding. It hearkens back to the 1980s, when everything was *cybernetics*, or now when everything is *AI*. Maybe I'm becoming too cynical in my advancing years. Maybe it's good that EI is how researchers wish to classify their work; it could be, after all, a way of broadening the research community. Maybe this leads to a future where stronger flavors of EI are more strongly pursued by more of the research community. We shall wait and see.

My specific issue is that it reduces EAI—a wonderful, rich, and inspirational concept, something to be strived for—into a rather mundane and, for some definitions, a "solved" problem. My lingering feeling is that many research communities—not just robotics—would benefit from exposure to the core philosophical concepts around embodiment and what it might *really* mean for their fields of research.

Let's stick to robotics for now, as I am hardly an expert in some of the other fields that may benefit from such exposure. Put simply, the trick we are missing for robotics is a lack of intertwinement. We have a body and a brain and an environment—the raw materials for EAI, but as (predominantly) engineers, we are taught to cut problems up into pieces following reductionist principles, solve the easier problems, and then bolt everything back together afterward. And yet, the theory of embodied cognition (EC) (Gallagher, 2006) postulates that entities display performant behavior that emerges via a tight coupling between their environment, their body, and their brain. See the issue? EAI encourages us to employ systems thinking—to consider the whole rather than the parts and to encourage interactions between the three subcomponents as the predominant problem-solving mechanism (Pfeifer & Bongard, 2006). However, that's not typically how we are trained to build robotic systems.

It seems, then, that in addition to the development of novel materials, novel computing methodologies, novel engineering and design principles, and novel paradigms for realizing increasingly strongly embodied robots, we also need to realize some fundamental shift in thinking that permeates all the way to educational curricula. Moreover, we must continue to see evidence that the ends justify the means and that adopting "EAI robotics" over standard robotic systems is worthwhile. We can open the door to roboticists working across myriad different problems and domains, but we cannot force them through. A daunting task, to be sure.

Yet the rewards may be immense. We know that current robotic systems have significant limitations, and while recent technological developments have made impressive leaps to address some of these—for example, the use of large language models (LLMs) to allow humans and robots to more naturally interact and potentially understand each other (Zeng et al., 2023)—other limitations remain critically unaddressed. We want robots to survive outdoors, to tackle unstructured and unknown environments, to learn and adapt like natural life does, and to overcome extreme conditions. We want robots that can recover from severe damage—robots that can heal. As engineers (in the most inclusive possible sense of the word), we want the option of methodologies that provide these characteristics, that take robotic performance to the limits of what is possible. We want to invent novel design methodologies that coordinate material, morphology, control, and environment holistically to generate emergent behaviors, unlike anything we've ever seen from a robot. And finally, we want to inspire new ways of thinking about fundamental questions—*What is a robot? What should a robot be capable of?* It's hard to argue that such goals are not worthwhile.

To frame this chapter, it might be useful to assess where various parts of robotics research fall to the diagram provided in figure 9.1. We can see "traditional robotic learning" encompassing the brain (controller) and environment, for example, with an immutable morphology (Leal et al., 2024; Soori et al., 2023). Soft robotics and evolutionary robotics are arguably the most direct implementations of EAI in robotics. The more unique parts of evolutionary robotics (Eiben, 2014)—namely, the ability to co-optimize morphology and controller, albeit typically with a static environment—are represented in the body-brain overlap. Morphological computation (MC) (Nygaard et al., 2021) is chiefly the domain of soft robotics (Rus & Tolley, 2015) and exploits interactions between the body and environment (Liow et al., 2024; Xavier et al., 2022).

In terms of popularity, traditional robotic learning is significantly ahead of the other two types, as it encompasses all the "put a neural network on a robot" work previously described (Soori et al., 2023), as well as modern approaches based on foundation models (e.g., Leal et al., 2024; O'Neill

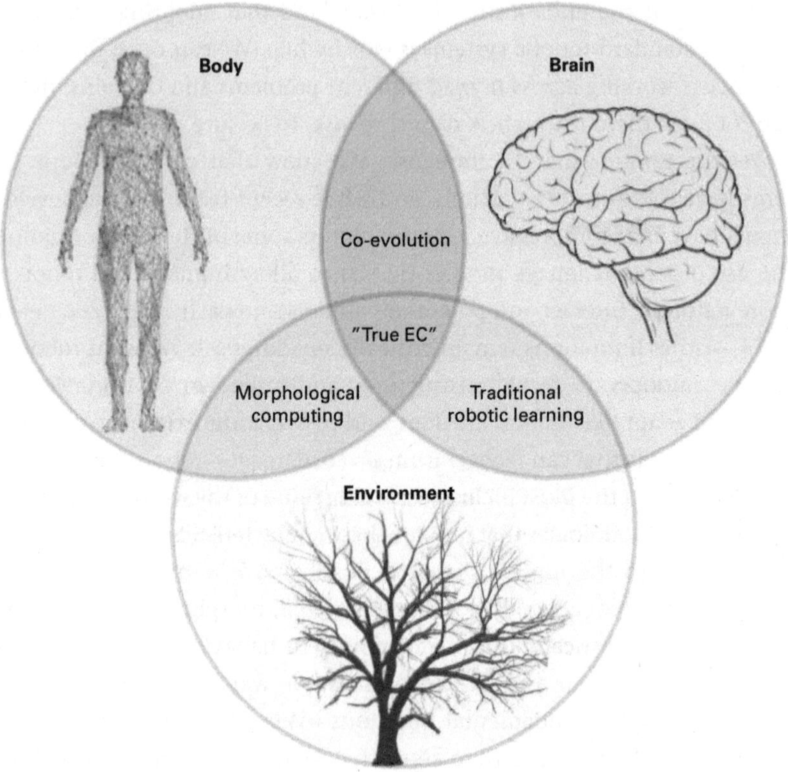

Figure 9.1
Delineation of the body-brain-environment triad called EC. *Source*: With permission from Howard et al., 2022.

et al., 2023). As well as being the oldest subfield of the three, it has also benefitted the most from the continued rise of deep learning, GPU-accelerated simulation, and, more recently, foundation models.

Soft robotics (active from the 2010s onward) is a relatively nascent field, yet it contains a huge amount of potential as soft, compliant structures more naturally interact with complex environments and naturally drives research to consider morphology as well as controller. It also considers EAI-adjacent phenomena including morphing (Shah et al., 2021), multitask embodiments, and multienvironment transitions (Yao et al., 2024). Furthermore, it is a natural testing ground for all kinds of work in soft sensing, multifunction materials, and advanced fabrication—all potential precursors for "true EAI" robotics.

Evolutionary robotics is represented by a small but dedicated community that has been around since the 1990s. It is home to numerous works related to EI (Howard, 2022), including environmental mediation (Kriegman, Walker, & Bongard, 2018), the interplay between ontogenetic and phylogenetic change (Bongard & Pfeifer, 2003), plasticity in genotype and phenotype (Moczek et al., 2011), and function recovery (Cully et al., 2015). Furthermore, couching the field in the trappings of Darwinism, specifically the heightened role of the environment in driving the emergence of behavior, naturally lends itself to being a tool for the study of EI and furthering our understanding of EI principles (Kriegman, Walker, et al., 2018). Evolutionary robotics includes delineation of an adaptive body, coupled brain, and environmental assessment, and there is a high degree of shared terminology (e.g., *emergence*, *coupling*, and *behavior*).

Through these similar but somewhat distant fields, we have some of the required elements, at various levels of sophistication and maturity. We also have an emerging suite of frameworks with which to consider these fields in the context of various definitions of embodiment (Füchslin et al., 2013; Howard et al., 2019).

A few things are missing, however. Notably, few works belong in the central section in figure 9.1. The closer we get, the more we may see the benefits of EAI. The crux of the argument is that each element of the EI triad must be balanced against the others—body, brain, and environment—much as each must be properly considered, represented, and empowered to drive the emergence of behavior. The complexity of the brain must match the behavioral freedom of the body and the challenge imposed by the environment (Pfeifer & Bongard, 2006). Similarly, the body must be sensitive enough to derive meaningful learning signals from the environment, adaptable enough to handle the environment, and flexible and controllable enough to cogenerate useful behaviors with the brain. The environment must be represented in enough fidelity, and contain enough challenge, to drive those behaviors to emerge. Raising the complexity and fidelity of each element promotes the capabilities of the EI system to a higher level, with the potential to realize robots with higher-order embodied competencies.

This must also be reflected in the way we teach and in calls for specific tools to be developed and made available. Many researchers in the field, including Pfeifer, Bongard, and Ciocarlie, advocate for a holistic approach to designing robots with EI, arguing for a deeply integrated approach where the morphology, sensing capabilities, and computational aspects are designed in tandem to promote harmonious functionality between the different elements. This integrated perspective is essential for advancing the field of EI and is one that I would add a small caveat to: the environment must also be properly represented and considered at all stages of this process, potentially in more fidelity than the standard techniques, including finite elements method (FEM) and vibrational analysis (again, conducted in isolation), allow. Of course, an open question is how we actually alter curricula—and the most direct way may be to disperse ideas around embodiment broadly across academia, so that lecturers want to teach it and students are aware of its existence and want to learn it.

Interesting Research for Embodied Intelligence in Robotics

Robots provide a reasonably direct analogue to natural creatures when implementing EI; both possess a well-defined brain and physical body, and both act in an environment. However, many aspects of EI robotics are in flux. Here, I will introduce some of these open questions and attempt to provide some reasonable research pathways to push the field forward.

How Bioinspired Should We Be?

Bioinspiration permeates to the very core of EI robots. Researchers seek to grant animal-like abilities to their robots, together with a level of adaptability and robustness that mimics natural life. Significant bodies of work exist in bioinspired robotics (e.g., robotic birds replete with artificial feathers, dog-like quadrupeds with flexible spines, soft octopuses), as well a push to develop new materials (Mengüç et al., 2017) that closely mimic their biological counterparts—for example, gecko-inspired materials exploiting Van der Waals forces for adhesion.

The other main use of bioinspiration is as a design algorithm for the robots themselves, most directly implemented by evolutionary robotics that allow for, among other things, automatic optimal sensor placement for proprioception, high-level morphology design for performance across a range of environments, and in the codevelopment of matching controllers for a given embodiment.

Saying that most evolutionary robotics is bioinspired is a little charitable. Given its roots in evolutionary computing—primarily a software optimization methodology paying only lip service to very high-level and abstract representations of biological reality—it is perhaps unsurprising that this is so. Divorced from any physical concerns, evolutionary computing is mainly driven by things like computational complexity, runtime, and search efficiency. There are exceptions—for example, Cartesian genetic programming (Miller & Turner, 2015) implements an algorithmic version of noncoding DNA—but the main algorithms favor mathematics over biology. To give an example, differential evolution (Das & Suganthan, 2010) is one of the most popular evolutionary algorithms. It uses more than two parents for mutation and is based on the mathematics of efficient vector operations.

The argument is that because evolutionary robotics relates strongly to EI in a way that the rest of evolutionary computing does not, there is significant potential in becoming more bioinspired. For a start, evolutionary robotics is inspired by evolutionary biology and genetics, yet it hasn't updated to match current theory in many decades. So much has been discovered in this time that could be useful for practitioners yet has not permeated into the field of evolutionary robotics. Even phenomena that were well-known many decades ago are rarely studied in the field. This is a clear case for closer collaboration across domains, as the benefits of self-organizing and emergent representations/discovery algorithms cannot be understated. In some cases, we are seeing a slide into parallel universes where Lamark was right (Jelisavcic et al., 2017)—more efficient, for sure, but approaching more of a mathematical construct than related to biological reality.

One issue is finding representations that are both evolvable (can be discovered during machine learning (ML)) and expressive (provide enough phenotypic diversity to include "good" solutions). Typically, evolved robots develop environmentally dependent morpho-functional features but, due to algorithmic complexity, are typically defined at a level that prohibits truly performant behavior. In the most common case, the forms (the body part of the body-brain-environment trio) are too simple—made of sticks or blocks—and the choice of form too limited. This then limits the level of EI that can be attained.

One solution is to combine evolution with gradient-based algorithms to further refine and detail the final phenotypes (Pinskier et al., 2024), but we can also investigate more bioinspired ways to achieve this possibility.

Doursat (Delile et al., 2017) has a catalogue of work using cell growth as a motif, and there is much more to be done along these lines, harnessing the latest theory from the biology.

Algorithms may be tailored to produce biological form and function; Hockings and Howard (2020) provide a wish-list selection of mechanisms, including cell lines, signaling, and mechanisms of limb development, to promote the emergence of bio-mimicking embodiments and discover new design spaces that encourage useful embodied behaviors. The key feature of these algorithms is *emergence*, in that once the (relatively simple) rules are set, complexity can naturally emerge. This is how nature does it, and it directly addresses computational complexity issues that prevail in evolutionary robotics, especially in coevolutionary (body-brain) learning.

One final important facet of the bioinspiration question is asking what we should use as our substrate. Originally, we were limited to bits and bytes in a simulator; now (although simulation remains a crucial tool) we are seeing more diverse and freeform methods for material instantiation of our embodied creations, which gives us the benefit of real environmental interactions to study and drive our embodiment loops. The "evolution of things" (Eiben & Smith, 2015) succinctly summarizes how real-world robotic morphology development has been revitalized by 3D printing that allows the forms themselves to be easily fabricated, yet significant challenges remain in terms of, for example, integrating sensing, actuation, and processing, miniaturization, and creating bio-adjacent materials with which to build our robots.

Of course, we could go one step further and start making our robots out of biological material such as cells (Blackiston et al., 2021). The appeal of this is that it brings us level in some sense to nature in terms of having a substrate that intrinsically provides many of the properties we desire—for example, self-organizing structure across levels, developmental growth, repair, and of course the ability of cells to naturally multiply. Our design algorithm could theoretically use plasticity mechanisms, for example, rather than our coding them into digital simulacra, which synergizes nicely with the idea of setting up a process and letting the embodied complexity emerge. This research is still nascent, however, and it is unclear how programmable the process is, as well as how to scale it into more "robot-sized" morphologies and how to deal with the ethical and social perception of such a proposition.

In Silico or in Materio?

A longstanding research question centers around how we design, fine-tune, and assess our creations. Physically building everything is slow and costly. Simulation is fast and flexible but due to reality gap effects is unlikely to properly represent reality. This is especially the case for EI, where we may be dealing with multipart multimaterial bodies, soft bodies (Mengaldo et al., 2022), and complex, contact-rich interactions with environments (Collins et al., 2021). Furthermore, the materials that continue to be created are unlikely to come with simulator-friendly models. This is clearly a problem, especially as EI places the onus on environmental accuracy to instantiate rich information flows between agent and environment (Nygaard et al., 2021).

Much recent work advocates for a hybrid approach, combining the speed of simulation with accuracy from injections of real-world data. Accuracy can be improved through system identification of hard-to-model phenomena (Hwangbo et al., 2019), creating neural networks that can be embedded into the simulator. Physics-informed neural networks have shown promise for accelerating both accuracy and speed in soft simulation (Wang et al., 2024). This is particularly relevant as it speeds up soft FEM, which despite being a simulation technique, is still rather slow and inflexible. Finally, differentiable simulation (Newbury et al., 2024) reduces computational complexity by providing gradient information over part of the design problem (Hu et al., 2019). In summary, then, we need real data, but we need ways to work that data into fast, cheap, and flexible embodiment-designing tools.

One-Upping Nature

As the de facto substrate of embodiment, nature usually holds the advantage over artificial systems when implementing EI. Computational approaches do have the upper hand in two ways, however: in *providence* and in *control*. Simulation provides an omniscient view of the information streams coming into and out of the robot-env loop, giving us an unprecedented view into the works of EAI, in a way that no natural system can. We can analyze these information flows to tweak them specifically to encourage useful EAI loops to develop, yet how to effectively perform this analysis is an open question. Computational approaches similarly allow us to easily repeat experiments, perform ablation studies, and so on, allowing us to perform meaningful scientific analysis on these complex systems.

Natural environments come with their own dynamism, which life must overcome (or at least learn to live with). In simulation, we can exert direct environmental control to encourage the emergence of beneficial embodied interactions. Shaping the learning process through curricula (Bengio et al., 2009) allows incremental learning of increasingly complex behaviors, which in reality would be somewhere between cumbersome and impossible.

Education, Standardization, and Collaboration

Finally, it's worth considering some steps the combined research community around EI can make to move forward.

Reduce the Overheads through Standardization

Many of the suggestions listed above come with some form of overhead—be it added mechanical complexity, modeling difficulty, or the time spent coding the software frameworks that allow us to study EAI. At what point, and for what situations, do the overheads become worthwhile? The cost-benefit assessment relates to the value of the problem being addressed and the value of the insights gleaned from the specific study. It also relates to the level of education in EAI—to what extent it is taught as an alternative engineering paradigm—and the availability of readily accessible engineering and design tools that lower the barrier to entry for EI research and promote education and participation.

In terms of research, we are starting to see the codification of some of the more ephemeral ideas around embodiment and vital discussion on how we implement EI in robotic systems in the future. Accessible, codified, and standardized design tools—including plug-ins, software packages, and simulators—that allow us to create holistic entities and program and analyze embodiment loops will go a long way to opening the field to interested researchers and encouraging further innovations without increasing their "EI overheads." Similarly, the standardization and availability of both data sets and experimental equipment will ease some of this burden. The opportunity for reuse may also encourage researchers to engage in more open, longer-term, and impactful embodiment studies. Metrics for measuring and assessing specific properties of a broadened notion of EI will also help and, ideally, can be shared across both natural and artificial systems; the differences don't necessarily preclude us from (inclusively and generally) describing and studying these embodied systems in a unified way as per this volume.

Reach Out across Communities

EI is intriguing as researchers across diverse disciplines are studying the same problem (in the general systems theory sense) in different ways, creating opportunities for deep collaboration that are virtually unparalleled across the scientific disciplines. This volume is evidence of such. There is a tremendous opportunity for EI to be a vehicle for inclusivity, which in turn can generate an overarching, holistic, and cross-disciplinary field. We are already seeing this in some areas—for example, robotics has a view of EI that is separate, but not at odds with, a human-robot interaction (HRI) viewpoint borne of social sciences and social robotics. It is exciting to think about EI and the crossovers among disparate disciplines that are yet unexplored, and I for one am excited about the future.

References

Bengio, Y., Louradour, J., Collobert, R., & Weston, J. (2009). Curriculum learning. In L. Bottou & M. Littman (Eds.), *Proceedings of the 26th Annual International Conference on Machine Learning* (pp. 41–48). ACM. https://doi.org/10.1145/1553374.1553380

Blackiston, D., Lederer, E., Kriegman, S., Garnier, S., Bongard, J., & Levin, M. (2021). A cellular platform for the development of synthetic living machines. *Science Robotics, 6*(52), eabf1571. https://doi.org/10.1126/scirobotics.abf1571.

Bongard, J. C., & Pfeifer, R. (2003). Evolving complete agents using artificial ontogeny. In R. Hara Fumio & Pfeifer (Ed.), *Morpho-functional machines: The new species* (pp. 237–258). Springer Japan. https://doi.org/10.1007/978-4-431-67869-4_12.

Ciocarlie, M. (2022). What is embodied intelligence? *Columbia Engineering News*, December 13, 2021. https://www.engineering.columbia.edu/about/news/what-embodied-intelligence

Collins, J., Chand, S., Vanderkop, A., & Howard, D. (2021). A review of physics simulators for robotic applications. *IEEE Access, 9*, 51416–51431. https://doi.org/10.1109/ACCESS.2021.3068769

Cully, A., Clune, J., Tarapore, D., & Mouret, J.-B. (2015). Robots that can adapt like animals. *Nature, 521*(7553), 503–507. https://doi.org/10.1038/nature14422

Das, S., & Suganthan, P. N. (2010). Differential evolution: A survey of the state-of-the-art. *IEEE Transactions on Evolutionary Computation, 15*(1), 4–31. https://doi.org/10.1109/TEVC.2010.2059031

Delile, J., Herrmann, M., Peyriéras, N., & Doursat, R. (2017). A cell-based computational model of early embryogenesis coupling mechanical behaviour and gene regulation. *Nature Communications*, *8*(1), 13929. https://doi.org/10.1038/ncomms13929

Eiben, A. E. (2014). Grand challenges for evolutionary robotics. *Frontiers in Robotics and AI*, *1*, 4. https://doi.org/10.3389/frobt.2014.00004

Eiben, A. E., & Smith, J. (2015). From evolutionary computation to the evolution of things. *Nature*, *521*(7553), 476–482. https://doi.org/10.1038/nature14544

Füchslin, R. M., Dzyakanchuk, A., Flumini, D., Hauser, H., Hunt, K. J., Luchsinger, R. H., Reller, B., Scheidegger, S., & Walker, R. (2013). Morphological computation and morphological control: Steps toward a formal theory and applications. *Artificial Life*, *19*(1), 9–34. https://doi.org/10.1162/ARTL_a_00079

Gallagher, S. (2006). *How the body shapes the mind*. Clarendon Press. http://dx.doi.org/10.1093/0199271941.001.0001

Hockings, N., & Howard, D. (2020). New biological morphogenetic methods for evolutionary design of robot bodies. *Frontiers in Bioengineering and Biotechnology*, *8*, 621. https://doi.org/10.3389/fbioe.2020.00621

Howard, D. (2022). From the lab to the field with evolutionary field robotics. *Frontiers in Robotics and AI*, *9*, 1027389. https://doi.org/10.3389/frobt.2022.1027389

Howard, D., Collins, J., & Robinson, N. (2022). Taking shape: A perspective on the future of embodied cognition and a new generation of evolutionary robotics. *IOP Conference Series: Materials Science and Engineering*, *1261*(1), 12018. https://doi.org/10.1088/1757-899X/1261/1/012018

Howard, D., Eiben, A. E., Kennedy, D. F., Mouret, J.-B., Valencia, P., & Winkler, D. (2019). Evolving embodied intelligence from materials to machines. *Nature Machine Intelligence*, *1*(1), 12. https://doi.org/10.1038/s42256-018-0009-9

Hu, Y., Liu, J., Spielberg, A., Tenenbaum, J. B., Freeman, W. T., Wu, J., Rus, D., & Matusik, W. (2019). ChainQueen: A real-time differentiable physical simulator for soft robotics. In G. Dudek & B. Boulet (Eds.), *2019 International Conference on Robotics and Automation (ICRA)* (pp. 6265–6271). IEEE. https://doi.org/10.1109/ICRA.2019.8794333

Hwangbo, J., Lee, J., Dosovitskiy, A., Bellicoso, D., Tsounis, V., Koltun, V., & Hutter, M. (2019). Learning agile and dynamic motor skills for legged robots. *Science Robotics*, *4*(26), eaau5872. https://doi.org/10.1126/scirobotics.aau5872

Jelisavcic, M., Kiesel, R., Glette, K., Haasdijk, E., & Eiben, A. E. (2017). Analysis of Lamarckian evolution in morphologically evolving robots. *Artificial Life Conference Proceedings 14*, 214–221. https://doi.org/10.7551/ecal_a_038

Kriegman, S., Cheney, N., & Bongard, J. (2018). How morphological development can guide evolution. *Scientific Reports*, *8*(1), 13934. https://doi.org/10.1038/s41598-018-31868-7

Kriegman, S., Walker, J., Shah, D., Kramer, R., & Bongard, J. (2018). Interdependence of morphology and control in the evolution of soft robots. In P. Corke & J. Roberts (Eds.), *Proceedings of the 2018 IEEE International Conference on Robotics and Automation (ICRA)* (pp. 7832–7837). IEEE. https://doi.org/10.1109/ICRA.2018.8460544

Kriegman, S., Walker, J., & Bongard, J. (2018). Morphological evolution of soft robots through material properties and environmental mediation. *Memetic Computing, 10*(3), 267–277. https://doi.org/10.1007/s12293-018-0255-0

Leal, I., Choromanski, K., Jain, D., Dubey, A., Varley, J., Ryoo, M., Lu, Y., Liu, F., Sindhwani, V., Vuong, Q., Sarlos, T., Oslund, K., Hausman, K., & Roa, K. (2024). SARA-RT: Scaling up robotics transformers with self-adaptive robust attention. In Z. Wang, H. Asama, & W.-H. Liao (Eds.), *2024 IEEE International Conference on Robotics and Automation (ICRA)* (pp. 6920–6927). IEEE. https://doi.org/10.1109/ICRA57147.2024.10611411

Liow, L., Brett, J., Pinskier, J., Hanson, L., Tidswell, L., Kottege, N., & Howard, D. (2024). A compliant robotic leg based on fibre jamming. *IEEE Transactions on Robotics, 40*, 4578–4597. https://doi.org/10.1109/TRO.2024.3443714

Mengaldo, G., Renda, F., Brunton, S. L., Bächer, M., Calisti, M., Duriez, C., Chirikjian, G. S., & Laschi, C. (2022). A concise guide to modelling the physics of embodied intelligence in soft robotics. *Nature Reviews Physics, 4*(9), 595–610. https://doi.org/10.1038/s42254-022-00481-z

Mengüç, Y., Correll, N., Kramer, R., & Paik, J. (2017). Will robots be bodies with brains or brains with bodies? *Science Robotics, 2*, eaar4527. https://doi.org/10.1126/scirobotics.aar4527

Miller, J. F., & Turner, A. J. (2015). Cartesian genetic programming. In S. Silva & A. I. Esparcia-Alcázar (Eds.), *Proceedings of the Companion Publication of the 2015 Annual Conference on Genetic and Evolutionary Computation* (pp. 179–189). ACM. https://doi.org/10.1145/2739482.2756571

Moczek, A. P., Sultan, S., Foster, S., Ledón-Rettig, C., Dworkin, I., Nijhout, H. F., Abouheif, E., & Pfennig, D. W. (2011). The role of developmental plasticity in evolutionary innovation. *Proceedings of the Royal Society B: Biological Sciences, 278*(1719), 2705–2713. https://doi.org/10.1098/rspb.2011.0971

Newbury, R., Collins, J., He, K., Pan, J., Posner, I., Howard, D., & Cosgun, A. (2024). A review of differentiable simulators. *IEEE Access, 12*, 96985–97010. https://doi.org/10.1109/ACCESS.2024.3425448

Nygaard, T. F., Martin, C. P., Torresen, J., Glette, K., & Howard, D. (2021). Real-world embodied AI through a morphologically adaptive quadruped robot. *Nature Machine Intelligence, 3*(5), 410–419. https://doi.org/10.1038/s42256-021-00320-3.

O'Neill, A., Rehman, A., Gupta, A., Maddukuri, A., Gupta, A., Padalkar, A., Lee, A., Pooley, A., Gupta, A., & Mandlekar, A. (2023). Open x-embodiment: Robotic learning datasets and rt-x models. *ArXiv* https://doi.org/10.48550/arXiv.2310.08864

Pfeifer, R., & Bongard, J. (2006). *How the body shapes the way we think: A new view of intelligence*. MIT Press. https://api.semanticscholar.org/CorpusID:25705913

Pinskier, J., Wang, X., Liow, L., Xie, Y., Kumar, P., Langelaar, M., & Howard, D. (2024). Diversity-based topology optimization of soft robotic grippers. *Advanced Intelligent Systems, 6*(4), 2300505. https://doi.org/10.1002/aisy.202300505

Rus, D., & Tolley, M. T. (2015). Design, fabrication, and control of soft robots. *Nature, 521*(7553), 467–475. https://doi.org/10.1038/nature14543

Shah, D., Yang, B., Kriegman, S., Levin, M., Bongard, J., & Kramer-Bottiglio, R. (2021). Shape changing robots: Bioinspiration, simulation, and physical realization. *Advanced Materials, 33*(19), 2002882. https://doi.org/10.1002/adma.202002882

Soori, M., Arezoo, B., & Dastres, R. (2023). Artificial intelligence, machine learning and deep learning in advanced robotics, a review. *Cognitive Robotics, 3*, 54–70. https://doi.org/10.1016/j.cogr.2023.04.001

Wang, X., Dabrowski, J. J., Pinskier, J., Liow, L., Viswanathan, V., Scalzo, R., & Howard, D. (2024). PINN-Ray: A physics-informed neural network to model soft robotic fin ray fingers. *ArXiv*. https://doi.org/10.48550/arXiv.2407.08222

Xavier, M. S., Tawk, C. D., Zolfagharian, A., Pinskier, J., Howard, D., Young, T., Lai, J., Harrison, S. M., Yong, Y. K., Bodaghi, M., & Fleming, A. J. (2022). Soft pneumatic actuators: A review of design, fabrication, modeling, sensing, control and applications. *IEEE Access, 10*, 59442–59485. https://doi.org/10.1109/access.2022.3179589

Yao, D. R., Kim, I., Yin, S., & Gao, W. (2024). Multimodal soft robotic actuation and locomotion. *Advanced Materials*, 2308829. https://doi.org/10.1002/adma.202308829

Zeng, F., Gan, W., Wang, Y., Liu, N., & Yu, P. S. (2023). *Large language models for robotics: A survey. ArXiv*. https://arxiv.org/abs/2311.07226.

III Adaptation, Morphology, and Emergence

Part III shifts the focus to adaptation and morphology and emergence and the potential of bioinspired approaches in technology and artificial intelligence (AI). This part explores how biological organisms can internally "rehearse" risky actions or explore potential behaviors within their physical structure before committing to engagement with the external world, a "morphological pretraining" that stands in stark contrast to machine learning (ML) pretraining. Biohybrid systems could inherit morphological pretraining capabilities directly from their living biological constituents, offering novel pathways for adaptation. The conventional view of cognition is challenged by describing the morphological info-computational (ICON) framework, based on morphological computing (MC), which emphasizes the physical body's role in shaping computation. Embodied artificial intelligence (EAI) based on a human-computer interaction that can effectively interact with humans in the real world is proposed. While acknowledging the remarkable recent successes of generative models like large language models (LLMs), this part proposes that the development of artificial social agents should prioritize the nature of shared existence, emphasizing fluid interaction and nonlinear dynamic engagement with the environment rather than the accumulation of abstract representational knowledge.

Common Themes

MC, which utilizes the physical properties of organisms to produce and control behavior, suggests that the form (morphology) of an organism or system plays a crucial role in its computational processes. The ICON perspective incorporates self-organization as a key process in the framework and aligns with Friston's ideas on self-organized dynamics minimizing free energy. Both approaches propose a hierarchical model: the ICON framework uses a multiscale hierarchy of info-computational processes, and Friston's uses hierarchical predictive coding (Dodig-Crnkovic, 2014; chapter 4). In the ICON view, a body determines the organism's behavior through its intrinsic physical properties with causal powers (Dodig-Crnkovic, 2014; chapter 11).

Evolutionary and morphological pretraining rely on evolutionary algorithms (in the former) and physical structure (in the latter) to aid how an organism interacts with the environment through simplifying control and information processing (Pfeifer & Bongard, 2006). The fact that different physical systems can achieve similar cognitive outcomes is central to evolutionary biology and suggests a complex interplay between the physical form and the solution (see Rouleau & Levin, 2023). Even before animals developed brains to execute goal-directed behavior, cells had ways to coordinate their activity to survive through regeneration, embryogenesis, and metamorphosis (Baluška & Levin, 2016). During periods of reduced selection pressure, organisms can afford to practice "risky moves" in various spaces (genetic, metabolic, behavioral) (chapter 10).

Soft robotics represents one way to engineer bioinspired robots or to create biobots—robots composed of biological rather than technological components (Xenobots and anthrobots) (chapter 10). Soft robotics allows the development of technologies that are not purely bioinspired. Exciting developments have been shown in manipulation, locomotion, and application-orientated capabilities such as manipulation of food items. Soft mimicry allows the creation of intelligent systems by emulating nature's adaptations. However, there are still unresolved challenges in the scalability of sensing, actuation, and control in comparison to biological systems (Hawkes et al., 2021).

Nonlinear dynamics may offer insight into the conditions of human-machine systems to solve problems together, as it has been a productive theoretical framework in the human movement sciences (Chemero, 2009; Dreyfus, 2007; Kelso, 1995). Thus, an agent with a rich sensorimotor repertoire can act in the world successfully simply by letting itself fall into stable patterns of interaction with the environment. Current artificial systems have difficulty enabling *being with others* since they don't share the same world with humans. It is possible to use "reverse self-engineering" to rebuild biological functions so that machines share actions and attitudes toward the same existential possibilities as their human partners (chapter 12).

References

Baluška, F., & Levin, M. (2016). On having no head: Cognition throughout biological systems. *Frontiers in Psychology, 7*. https://doi.org/10.3389/fpsyg.2016.00902

Chemero, A. (2009). *Radical embodied cognitive science*. MIT Press.

Dodig-Crnkovic, G. (2014). Modeling life as cognitive info-computation. In A. Beckmann, E. Csuhaj-Varjú, & K. Meer (Eds.), *Computability in Europe* (pp. 153–162). Springer. https://doi.org/10.1007/978-3-319-08019-2_16

Dreyfus, H. L. (2007). Why Heideggerian AI failed and how fixing it would require making it more Heideggerian. *Philosophical Psychology, 20*(2), 247–268. https://doi.org/10.1080/09515080701239510

Hawkes, E. W., Majidi, C., & Tolley, M. T. (2021). Hard questions for soft robotics. *Science Robotics, 6*(53). https://doi.org/10.1126/scirobotics.abg6049

Kelso, J. A. S. (1995). *Dynamic patterns: The self-organization of brain and behavior*. MIT Press.

Pfeifer, R., & Bongard, J. (2006). *How the body shapes the way we think: A new view of intelligence*. MIT Press.

Rouleau, N., & Levin, M. (2023). The multiple realizability of sentience in living systems and beyond. *eNeuro 10*(11) 1–7. https://doi.org/10.1523/ENEURO.0375-23.2023

During these periods of reduced selection pressure, evolution, and/or development, the organism itself is free to practice risky moves in genetic space, metabolic space, morphospace, or the space of possible behaviors. These can all be considered forms of pretraining: biologicals can explore where various behaviors fall within a risk/opportunity spectrum. This can help them learn to avoid high-risk/low-opportunity behaviors and how to carefully enter into high-risk/high-opportunity ones. Some of these concepts have been directly or indirectly incorporated into machine learning (ML) methods. These are reviewed in the next section. However, the additional paths via which embodied systems can pretrain themselves to grapple with the danger and uncertainty of their external environments by simulating parts of that world within themselves, has yet to be explored in robotics. This new set of opportunities for robotics is explored in the "Pretraining in Robotics" section. By building robots out of biological rather than technological components, the resulting "biobots" may be able to inherit morphological pretraining mechanisms that have already been installed into those biological components by natural selection. This opportunity is explored in the "Pretraining in Biobotic" section. Concluding remarks about the future of morphological pretraining are provided in the final section.

Pretraining in ML

ML has, since its inception, acknowledged that a careful trade-off must be struck between risky behavior and behavior likely to yield useful behavior. This is commonly referred to as the exploration/exploitation trade-off. Intuitively, and in practice, machine learning methods are designed to favor exploration early and exploitation later.

Pretraining in Premodern Machine Learning

This can be seen in early ML methods such as simulated annealing (Bertsimas & Tsitsiklis, 1993), in which a "temperature" hyperparameter is gradually reduced during training: high temperatures allow candidate solutions to a given problem to make large jumps in solution space; as the temperature falls, the magnitude of solution changes is gradually reduced, relying on the assumption that solutions are increasingly likely to have found a slope leading up to a global optimum as training continues. Other norms within the ML community that point to its awareness that practicing risky play within a protected space is important is the separation of training and testing regimes. Freezing artificial intelligence (AI) model parameters during testing limits the model to exploiting what it has learned during training.

Evolutionary computation is a branch of ML, and evolution strategies are themselves a particular form of evolutionary computation (Hansen et al., 2015). In ES, each genetic parameter that encodes some part of a candidate solution to a given problem has a step size associated with it: this denotes the amount that mutation may vary that parameter when it produces offspring solutions. This early form of automatically tuning the amount of adaptation that occurs during optimization inspired adaptive learning rates (Liu et al., 2019), a subject of study that continues to be of interest within the ML community.

Pretraining in Modern ML

Finally, pretraining of foundation models has risen to prominence in the deep learning community. Once again, these views indirectly acknowledge the value of enabling risky play before exposing a learner to a real-world challenge. Pretraining methods have followed the observation that it is often easier to provide a relatively small amount of domain-specific training data to a foundation model previously trained on a large amount of domain-agnostic training data, compared to training a foundation model for a given domain ab initio.

However, all of the above examples operate on passive, nonembodied learners, like bit strings in evolutionary algorithms or neural networks in ML. Embodied machines offer literal internal spaces within which risk play can occur to prepare it for risky encounters with the outside world. The beginnings of such exploitation in embodied machines like robots is explored in the next section.

Pretraining in Robotics

All current robots, including autonomous vehicles, drones, and semiautonomous machines, like energy harvesters, have limited internal mechanical complexity in the sense of a large number of semi-independent components. As a simple example, if the wheel of an autonomous vehicle blows out, the wheel itself does not contain semi-independent components that can migrate to the site of damage and heal the wound, thus saving the vehicle as a whole from having to adapt to the injury. However, much progress is being made on self-healing materials (e.g., Tan et al., 2021; Wen et al., 2021). This example is but one of many that suggests developing machines within machines could yield robots that are more useful and safer than current machines: if the machine as a whole is overwhelmed by a surprise from within, it would have no recourse but to fail unexpectedly because nothing within can provide a partial or complete compensation.

Nonembodied Rehearsal: Rigid Robots

Grappling with surprise is a primary concern for roboticists, as real-world deployment of physical machines is likely to bring that machine into contact with environments very different from those in which it was trained (see figure 10.1a). Roboticists have therefore considered incorporating internal rehearsal into autonomous machines. The author participated in one such study in which internal rehearsal was cognitive rather than physical: a legged robot constantly trained physical simulations to increasingly capture its own body plan (Bongard et al., 2006). When the robot suffered an injury such as the loss of a leg, the robot updated its "self-models" to reflect this fact. The robot could then use these self-models to rehearse potentially risky actions before attempting them in reality. Subsequent work with a different legged robot (Cully et al., 2015) demonstrated that sufficient rehearsal with a self-model before deployment into real worlds, in which damage did occur, could confer adaptation onto the machine with it having to adapt self-models during deployment. This latter approach thus, in retrospect, can be viewed as an important step toward morphological pretraining, even if no physical components within the body of the robot participated in rehearsals.

Embodied Rehearsal: Soft Robots

Rigid robots, however, even if filled with semi-independent components, would by definition resist internal dynamics ever influencing how the robot exerts force on its external environment due to its rigid shell. Soft robots, on the other hand, by their very nature propagate internal forces to their surfaces. If designed properly, they could modulate which internal forces propagate to their surface, as illustrated in figure 10.1b. Finally, they may be designed or learn to suppress forces arising from risky internal actions and allow vetted internal actions to escape and guide whole-body actions. One initial step in this direction is particle robots (Savoie et al., 2019; S. Li et al., 2019; Ma et al., 2024). This class of autonomous machine is composed of a number of semi-independent robots surrounded by a deformable container. The actions of the internal robots are indirectly aggregated and directed by their interactions with each other and their container. The most well-known exemplar from this class (S. Li et al., 2019) was composed of carefully designed cylindrical components that could dynamically attach and detach from one another. This enabled the overall robot that contained them to achieve locomotion and other simple sensorimotor behaviors, but it also provides the opportunity for subsets of mostly or completely independent subassemblies within the body of the robot to practice risky behaviors. If successful behaviors are found internally, such subassemblies could recruit more or all of the remaining components within the particle robot's body to magnify this successful behavior to the robot as a whole.

A Richer Inner Life: Fluid Robots

If the size of independent components within a particle robot can be sufficiently reduced and the number of them sufficiently increased, not only do more opportunities for internal rehearsal become possible but the robot as a whole can begin to act like a fluid, as shown in figure 10.1c. S. Li et al. (2019) explored scaling of internal particles, and Bielawski et al. (2024) explored optimizing the behavior of "fluid robots" in a differentiable simulator using gradient descent.

Considering a robot as a fluid rather than a collection of a few rigid or soft components may offer even more opportunities for morphological pre-training than nonfluid soft robots do. The actions of one or a few particles within a fluid or granular material can trigger waves that propagate through the material. Careful orchestration of where and when waves are triggered can cause them to reinforce or negate each other via positive or negative interference. One can then envisage prospective actions being encoded as vibrations within a fluid robot. These vibratory prospective actions could be interrogated by other waves that encode environmental features: prospective actions that increasingly survive interrogatory waves would propagate through more of the robot's body. Prospective actions overwhelmed by interrogatory vibrations are failing internal tests and do not propagate to the robot's surface. This hypothetical process would recapitulate motor imagery, in which candidate actions arising in the motor cortex descend to muscle groups, where they become actualized. It would also embody the concept of neural Darwinism (Edelman, 1993), in which ideas putatively compete in the brain for control of the motor system.

Exploiting vibrational dynamics within a granular or fluid material to simulate actions and test them using self-models or world models that are also encoded as vibration may seem counterintuitive. However, a growing body of work shows that vibration may be an attractive modality with which to compute because multiple computations can be performed in the same place at the same time. F. Li et al. (2014) showed that results of Boolean operations can be observed arriving at the same particle but at different points on the frequency spectrum. Parsa et al. (2022a, 2022b, 2024) showed that evolutionary algorithms can be employed to tune the features of grains within a granular material such that two specific Boolean operations arrive at the same place at the same time but at different points on the frequency spectrum. Beaulieu et al. (2024) took this concept of "polycomputing" one step further by showing how AI can further exploit vibrations within a granular material to further increase the material's computational density.

Levin and Bongard (2023) have speculated about how and why polycomputing may exist in biological materials. Finally, there is evidence that neurons may employ soliton waves in addition to electricity for transporting information within neural tissue (Heimburg & Jackson, 2005), and new evidence shows that neurons mix frequencies (Luff et al., 2024). Although this biological evidence does not prove that organisms use vibration to internally rehearse risky actions, it does suggest that there may be no clean separation between electrical and mechanical internal rehearsal in particular, and thought and action in general in biology, and that there may be a good adaptive reason for this: electrically simulating prospective action within the body, no matter how accurate, is no substitute for a dress rehearsal in which physical actions are played out within the body before being allowed to propagate outward to repeat themselves using the body as a whole.

From the Inside Out: Everting and Autotomic Robots

Morphological pretraining could take many forms in biological systems of which we are aware. Similarly, there are many ways morphological pretraining could be incorporated into technological artifacts, some of which would be of adaptive benefit to the machines and others which would not. One way of categorizing different forms of morphological pretraining is to consider the way in which internal ruminations become external reality (see figure 10.1d). An internally considered action, one that is either simulated cognitively or acted out mechanically, may be directly or indirectly communicated to the robot as a whole. Or, internal-to-external transmission may be more direct: internal components that participate in a rehearsal become exposed to the outside world via unintentional or intentional morphological change. Several robots already exist that are capable of this inside-becomes-outside dynamic. So-called vine robots that evert the internal surfaces of their cylindrical bodies are capable of this (Blumenschein et al., 2020), as are autotomic robots (Davis et al., 2023). Autonomic robots can intentionally jettison parts of their body for adaptive purposes. Several species are capable of autotomy, usually to sacrifice a limb rather than their life to a predator (Emberts et al., 2019).

In everting or autonomic robots, parts of their internal structure can become exposed to the outside environment. At present, this property is binary: either a part of a robot is internal or it is directly exposed to the outside world. However, in nature, many body plans, or parts of body plans, of plants or animals are fractal in nature. In such fractal morphologies, there can be a gradation between inside and outside: oxygenated blood, alveoli, alveolar sacs, bronchioles, bronchi, the trachea, and the mouth and nose describe a gradual progression from inside to outside. In plants, flower buds may be shielded from sun, wind, and pests by surrounding leaves on nearby twigs. ML has taught us that gradients of all kinds can be helpful in enabling the learner, be it a nonembodied machine, embodied machine, or organism, to confront potentially risky behavior with a series of environments that increasingly resemble the real environment for which they are being trained. One such gradient is passing a behavior along a series of body parts that are increasingly in contact with the outside world.

Robots capable of having plans buried in internal material offer new vistas for future robotics. One can imagine the two halves of a robot that has been cut down the middle forming two smaller, geometrically and behaviorally similar versions of the original robot. One can imagine a robot everting usually shielded internal sensors to the outside environment and then retracting them to present that stimuli to internal rehearsal processes. Or it may evert internal manipulators that grasp bits of an unknown surface and draw them inward for internal prospective legs to try walking across. One can imagine a robot that grows to increase its fractal dimension, thereby smoothing its internal-to-external gradient, providing more intermediate testing grounds for internally germinated prospective actions.

From the Outside In: Thermodynamically Open Robots

The converse of autotomic robots is thermodynamically open robots. This class of machine is capable of incorporating new energy and matter into itself. This process has been approximated for a long time in robotics, in the form of modular robots: robots that have the ability to temporarily or permanently attach to one another, as illustrated in figure 10.1e. During attachment, what were previously externally facing surfaces become internal components of a composite robot. During detachment, like autonomic robots, internal surfaces or parts can become directly exposed to external environments. Modular robots usually transition rapidly between fully attached and detached states. However, more complex morphologies in the future may enable modular robots to modulate how much or how little components are connected, thereby creating another form of inside-to-outside gradients along which internally rehearsed actions can migrate into external actions that directly affect the robot's external environment.

Most modular robotics platforms adhere to the same topological change: components lock together two- or three-dimensionally like Lego bricks. However, the rise of robots with radically different body plans, like robotic fabrics (Buckner & Kramer-Bottiglio, 2018), suggests other ways that robots may combine: a robot fabric could envelop another robot, thereby modulating the behaviors arising from the engulfed machine. Again, this would shield the actions of the engulfed robot from the outside world, perhaps allowing it to practice risky behaviors. If it finds one, it could instruct the engulfing robot to withdraw. Interestingly, one robot engulfing another would serve as a morphological form of fine-tuning described in the section "From the Inside Out: Everting and Autotomic Robots": it may be easier to realize a robot with a desired behavior by engulfing a robot that already approximates that behavior with a robot fabric that "fine-tunes" that behavior into the desired one.

It's Robots All the Way Down: Fractal Robotics

Nature abounds with organisms that exhibit fractal structure, from corals to most plants to the vascular, skeletal, muscular, pulmonary, and nervous systems of higher animals. Many of these biological fractals are not just structurally self-similar but behaviorally self-similar as well: alveoli, alveolar sacs, bronchioles, bronchi, and the trachea each individually, and collectively, facilitate gas exchange. Intuitively, an organism or machine could increase its utility, or continue operation in a wider range of environments, if it learns to compose itself together with others of its kind in a way that yields a large version of themselves, capable of exhibiting the same behavior but at larger size scales, as shown in figure 10.1f. To achieve this, a set of robots or organisms could practice aggregation and disaggregation in different ways to improve the fidelity of structural and behavioral recapitulation at larger size scales: they could learn to improve at achieving structural and behavioral self-similarity. Fractal robots were imagined (Moravec et al., 1996) and have since been built (Kriegman et al., 2021), but the latter are evaluated in either a disaggregated or fractal state: the ways in which they assemble and disassemble into fractal forms and functions has yet to be studied. Certain fractals, like branching structures, do not have a clear inside and outside. Other fractals, such as the Menger sponge, do possess such a distinction. One could imagine a nonfractal robot practicing to create smaller versions of its own form and function internally. Once that is learned, by definition, the smaller copies would also be capable of internal assembly of even smaller copies.

The adaptive advantage of fractal robots over nonfractal robots is not yet clear, but its ubiquity in nature suggests it does have such an advantage. One potential usefulness is the compactness of encoding the design, and thus indirectly the function, of fractal robots. Another potential use might be to achieve a good fit between the fractal robot and fractal environments: a fractal robot may be best able to efficiently navigate a cave system with fractally arranged tunnels or inspect a coral reef. A third potential use might be the ability to rapidly exhibit some useful behavior at very different size scales by aggregation, de-aggregation, or eversion (exposing internally assembled smaller versions to the external environment).

Pretraining in Biobotics

If robots could benefit from a complex internal environment within which to practice risky behavior, one approach is to create bioinspired robots: robots composed of technological components simulate the complex internal processes that occur in organisms. Another, arguably more direct

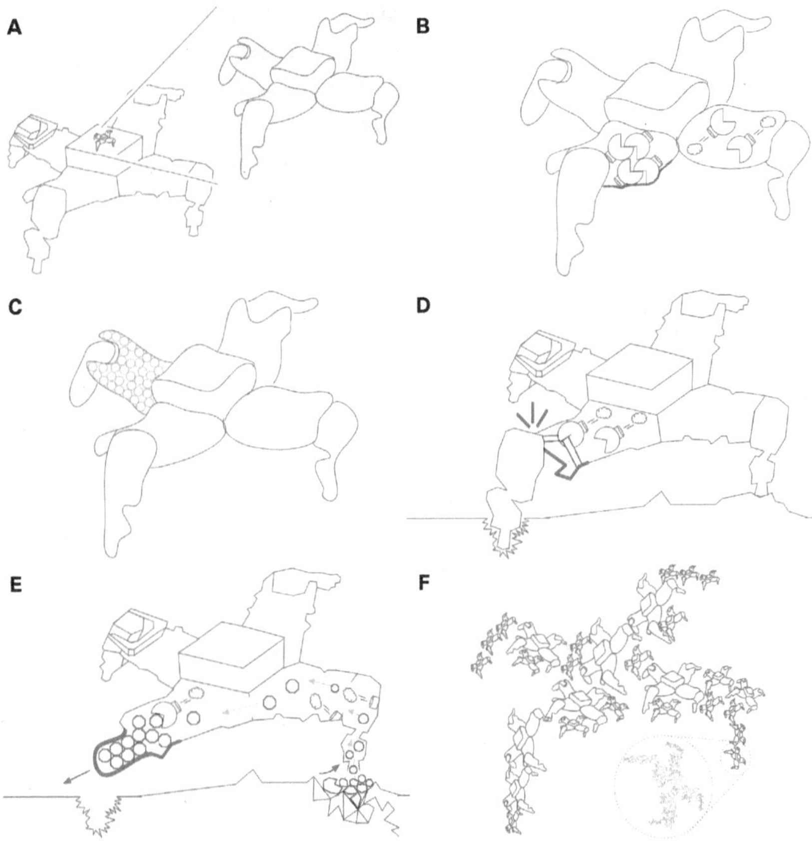

Figure 10.1
Morphological pretraining in present and future robots. (a) Rigid robots are capable of cognitively rehearsing risky actions using self-models. (b) Future soft robots could contain robots that physically rehearse actions internally. (c) Fluid robots could contain enough small robots to act as a fluid. (d) Autonomic robots could lose body parts, enabling well-rehearsed internal robots to become exposed to the outside world. (e) Thermodynamically open robots could absorb external material, involve it in internal processes, and direct it to (re)grow lost/new body parts. (f) Robots could learn to assemble into fractal structures such that the aggregate looks and acts like its components. *Source*: the author.

approach, is to create biobots: robots composed of biological rather than technological components. The latter approach has the benefit that, even though biological parts may have to be harvested from living hosts and reconfigured in new ways, this transformation may not disrupt the resulting biobot's ability to recover and maintain a complex inner environment.

If this is true, biobots can "inherit" part or all of the morphological pretraining capabilities of the biological donor. There is some evidence to support this latter speculation, as existing biobots do show the ability to survive radical reconfigurations that differ greatly from their host's phenotype. Whether or how much morphological pretraining facilitates survival across the radical transformation remains to be seen.

Pretraining in Xenobots

Two classes of biobots have been introduced to the literature to date: Xenobots, assembled from frog cells (Blackiston et al., 2021; Kriegman et al., 2020, 2021), and anthrobots, grown from human cells (Gumuskaya et al., 2023). Xenobots were the first class of biobot to be built. They were constructed from several thousand frog epithelial cells. If constructed appropriately, cells on the surface of the multicellular assembly grow cilia, which beat against the surrounding water and often result in motility. These cilia grow on the surface of wild-type frogs, facilitating the sloughing of skin-related pathogens. The fact that Xenobots still grow cilia despite the fact that their internal structure differs greatly from the tadpole or adult frog—the interior of a Xenobot is only more epithelial cells—suggests that morphological pretraining may have helped here, as these two frog phenotypes share a similar exterior but survive with two very different internal environments.

Much additional work is required to determine whether the short period of development that the embryonic frog skin cells enjoyed before being harvested prepared them to survive when reconfigured into Xenobots. Similarly, it would be of interest to discover whether these cells participate in internal rehearsal of risky actions preharvest and whether they continue to do so in Xenobots. If hallmarks of morphological pretraining can be found in embryonic frog tissues, this generates a testable hypotheses: embryos from species that similarly engage in morphological pretraining are more amenable to AI-dictated reconfigurations than embryos from species that do not.

Pretraining in Xenomics

The above-described research questions depend on a crisper definition of morphological pretraining than simply "internal rehearsal of risky action" and the ability to observe such dynamics at work in living tissues. Recent work has begun interrogation along these lines by looking for adaptive internal migration of cells within Xenobots: internal mechanical rearrangement, rather than cognition within statically fixed internal tissue. Some evidence has pointed to internal cellular migration and adaptation (Blackiston et al., 2024), but a link between these two phenomena has yet to be found. Local injuries were visited on Xenobots by puncturing them, and imaging was conducted pre- and postinjury. It was observed that cells near the point of damage migrate inward to fill the wound. It was also found that for most of the injured Xenobots, there was a common change in the way information flows within the Xenobot pre- and postdamage, where information flow was inferred from calcium activity within and between the cells.

Next steps will involve seeking cellular migration that resembles the migratory patterns that were observed postdamage: do Xenobots in particular, or frog or other embryos in general, "rehearse" injuries and responses to them, before they ever occur? A positive answer to this question could facilitate a broader research program in which organisms are observed to recover from dangerous situations with the help of internal changes and those internal changes are observed within the organism long before that dangerous situation arises.

If a growing number of biological examples of morphological pretraining can be discovered, they would collectively form a training set that AI could employ to learn the common patterns underlying them. The AI could then learn how to embody those dynamics in an ever-widening array of robots that are composed of technological components, biological components, or increasingly exotic admixtures of both.

Conclusion

Much ink has been spilled pointing to differences between machines and organisms. This chapter has explored one overlooked distinction: the potential ability of organisms to internally rehearse risky actions using dynamic components, not just cognitively using static internal components. Whether or how extensively this phenomenon is at work in nature has yet to be determined, but it is not difficult to formulate ways in which such an ability could confer adaptive advantages on the organism. Almost all current technologies, including state-of-the-art autonomous robots, do not have internal components capable of significant rearrangement. Thus, their ability to perform morphological pretraining is greatly limited or impossible. For this reason, effort in robotics should be expended to create machines capable of complex internal change, and efforts in embodied intelligence (EI) should be dedicated to determining how such machines can exploit this internal change to rehearse responses to the myriad challenges the external world will throw at them, long before those challenges ever arrive.

References

Beaulieu, S., Welch, P., Parsa, A., O'Hern, C., Kramer-Bottligio, R., & Bongard, J. (2024). Refractive Computation: Parallelizing logic gates across driving frequencies in a mechanical polycomputer. *Artificial Life Conference Proceedings, 36*(1), p. 72.

Bertsimas, D., & Tsitsiklis, J. (1993). Simulated annealing. *Statistical Science: A Review Journal of the Institute of Mathematical Statistics, 8*(1), 10–15. https://doi.org/10.1214/ss/1177011077

Bielawski, K., Rozlivek, J., Hoffmann, M., & Bongard, J. (2024). Best practices for differentiable soft robot modeling and optimization with the material point method. In *Artificial Life Conference Proceedings, 36*(1), p. 11. https://doi.org/10.1162/isal_a_00724

Blackiston, D., Dromiack, H., Grasso, C., Varley, T. F., Moore, D. G., Srinivasan, K., Sporns, O., Bongard, J., Levin, M., & Walker, S. I. (2024). Revealing non-trivial information structures in aneural biological tissues via functional connectivity. *bioRxiv*. https://doi.org/10.1101/2024.05.09.593467

Blackiston, D., Lederer, E., Kriegman, S., Garnier, S., Bongard, J., & Levin, M. (2021). A cellular platform for the development of synthetic living machines. *Science Robotics, 6*(52), eabf1571. https://doi.org/10.1126/scirobotics.abf1571

Blumenschein, L. H., Coad, M. M., Haggerty, D. A., Okamura, A. M., & Hawkes, E. W. (2020). Design, modeling, control, and application of everting vine robots. *Frontiers in Robotics and AI, 7*, 548266. https://doi.org/10.3389/frobt.2020.548266

Bongard, J., & Levin, M. (2023). There's plenty of room right here: Biological systems as evolved, overloaded, multi-scale machines. *Biomimetics, 8*(1), 110. https://doi.org/10.3390/biomimetics8010110

Bongard, J., Zykov, V., & Lipson, H. (2006). Resilient machines through continuous self-modeling. *Science, 314*(5802), 1118–1121. https://doi.org/10.1126/science.1133687

Buckner, T. L., & Kramer-Bottiglio, R. (2018). Functional fibers for robotic fabrics. *Multifunctional Materials, 1*(1), 012001. https://doi:10.1088/2399-7532/aad378

Chen, B., Feder, M. E., & Kang, L. (2018). Evolution of heat-shock protein expression underlying adaptive responses to environmental stress. *Molecular Ecology, 27*(15), 3040–3054. https://doi.org/10.1111/mec.14769

Cully, A., Clune, J., Tarapore, D., & Mouret, J.-B. (2015). Robots that can adapt like animals. *Nature, 521*(7553), 503–507. https://doi.org/10.1038/nature14422

Davis, Q. T., Woodman, S., Landesberg, M., Kramer-Bottiglio, R., & Bongard, J. (2023). Subtract to adapt: Autotomic robots. *2023 IEEE International Conference on Soft Robotics (RoboSoft), 100* (pp. 1–6). IEEE. https://doi.org/10.1109/ROBOSOFT55895.2023.10122102

Edelman, G. M. (1993). Neural Darwinism: Selection and reentrant signaling in higher brain function. *Neuron, 10*(2), 115–125.

Emberts, Z., Escalante, I., & Bateman, P. W. (2019). The ecology and evolution of autotomy: Ecology and evolution of autotomy. *Biological Reviews of the Cambridge Philosophical Society, 94*(6), 1881–1896. https://doi.org/10.1111/brv.12539

Gumuskaya, G., Srivastava, P., Cooper, B. G., Lesser, H., Semegran, B., Garnier, S., & Levin, M. (2023). Motile living biobots self-construct from adult human somatic progenitor seed cells. *bioRxiv*. https://doi.org/10.1101/2022.08.04.502707

Hansen, N., Arnold, D. V., & Auger, A. (2015). Evolution strategies. In J. Kacprzyk & W. Pedrycz (Eds.), *Springer handbook of computational intelligence* (pp. 871–898). Springer.

Heimburg, T., & Jackson, A. D. (2005). On soliton propagation in biomembranes and nerves. *Proceedings of the National Academy of Sciences, 102*(28), 9790–9795. https://doi.org/10.1073/pnas.0503823102

Kriegman, S., Blackiston, D., Levin, M., & Bongard, J. (2020). A scalable pipeline for designing reconfigurable organisms. *Proceedings of the National Academy of Sciences of the United States of America, 117*(4), 1853–1859. https://doi.org/10.1073/pnas.1910837117

Kriegman, S., Blackiston, D., Levin, M., & Bongard, J. (2021). Kinematic self-replication in reconfigurable organisms. *Proceedings of the National Academy of Sciences of the United States of America, 118*(49), e2112672118. https://doi.org/10.1073/pnas.2112672118

Kriegman, S., Nasab, A. M., Blackiston, D., Steele, H., Levin, M., Kramer-Bottiglio, R., & Bongard, J. (2021). Scale invariant robot behavior with fractals. *arXiv*. http://arxiv.org/abs/2103.04876

Levin, M. (2019). The computational boundary of a "self": Developmental bioelectricity drives multicellularity and scale-free cognition. *Frontiers in Psychology, 10*, 2688. https://doi.org/10.3389/fpsyg.2019.02688

Li, F., Anzel, P., Yang, J., Kevrekidis, P. G., & Daraio, C. (2014). Granular acoustic switches and logic elements. *Nature Communications, 5*(1), 5311. https://doi.org/10.1038/ncomms6311

Li, S., Batra, R., Brown, D., Chang, H.-D., Ranganathan, N., Hoberman, C., Rus, D., & Lipson, H. (2019). Particle robotics based on statistical mechanics of loosely coupled components. *Nature, 567*(7748), 361–365. https://doi.org/10.1038/s41586-019-1022-9

Liu, L., Jiang, H., He, P., Chen, W., Liu, X., Gao, J., & Han, J. (2019). On the variance of the adaptive learning rate and beyond. *arXiv*. http://arxiv.org/abs/1908.03265

Luff, C. E., Peach, R., Mallas, E.-J., Rhodes, E., Laumann, F., Boyden, E. S., Sharp, D. J., Barahona, M., & Grossman, N. (2024). The neuron mixer and its impact on human brain dynamics. *Cell Reports, 43*, 114274. https://doi.org/10.1016/j.celrep.2024.114274

Ma, D., Chen, J., Cutler, S., & Petersen, K. (2024). Smarticle 2.0: Design of scalable, entangled smart matter: 16th international symposium. In J. Bourgeois, J. Paik, B. Piranda, J. Werfel, S. Hauert, A. Pierson, H. Hamann, T. L. Lam, F. Matsuno, N. Mehr, & A. Makhoul (Eds.), *Distributed autonomous robotic systems, 28* (pp. 509–522). Springer Nature.

Moravec, H., Easudes, J., & Dellaert, F. (1996, December). *Fractal branching ultra-dexterous robots (Bush robots)* (PR No. 10-86888). NASA Advanced Concepts Research Project. https://www.ri.cmu.edu/publications/fractal-branching-ultra-dexterous-robots-bush-robots-/

Parsa, A., O'Hern, C. S., Kramer-Bottiglio, R., & Bongard, J. (2024). Gradient-based design of computational granular crystals. *arXiv*. http://arxiv.org/abs/2404.04825

Parsa, A., Wang, D., O'Hern, C. S., Shattuck, M. D., Kramer-Bottiglio, R., & Bongard, J. (2022). Evolution of acoustic logic gates in granular metamaterials. In *International Conference on the Applications of Evolutionary Computation (Part of EvoStar)* (pp. 93-109). Springer International Publishing. https://doi.org/10.1007/978-3-031-02462-7_7

Parsa, A., Wang, D., O'Hern, C. S., Shattuck, M. D., Kramer-Bottiglio, R., & Bongard, J. (2022). Evolving programmable computational metamaterials. In *Proceedings of the Genetic and Evolutionary Computation Conference (GECCO '22)* (pp. 1–8). ACM. https://doi.org/10.1145/3512290.3528861

Popper, K. (1978). Natural selection and the emergence of mind. *Dialectica, 32*(3–4), 339–355.

Savoie, W., Berrueta, T. A., Jackson, Z., Pervan, A., Warkentin, R., Li, S., Murphey, T. D., Wiesenfeld, K., & Goldman, D. I. (2019). A robot made of robots: Emergent transport and control of a smarticle ensemble. *Science Robotics, 4*(34), eaax4316. https://doi.org/10.1126/scirobotics.aax4316

Shreesha, L., & Levin, M. (2024). Stress sharing as cognitive glue for collective intelligences: A computational model of stress as a coordinator for morphogenesis. *Biochemical and Biophysical Research Communications, 731*, 150396. https://doi.org/10.1016/j.bbrc.2024.150396

Tan, Y. J., Susanto, G. J., Anwar Ali, H. P., & Tee, B. C. K. (2021). Progress and roadmap for intelligent self-healing materials in autonomous robotics. *Advanced Materials, 33*(19), 2002800. https://doi.org/10.1002/adma.202002800

Tanaka, H., Ishikawa, T., Lee, J., & Kakei, S. (2020). The cerebro-cerebellum as a locus of forward model: A review. *Frontiers in Systems Neuroscience, 14*, 19. https://doi.org/10.3389/fnsys.2020.00019

Wagner, A. (2014). *Arrival of the fittest: Solving evolution's greatest puzzle.* Penguin.

Wen, N., Song, T., Ji, Z., Jiang, D., Wu, Z., Wang, Y., & Guo, Z. (2021). Recent advancements in self-healing materials: Mechanicals, performances and features. *Reactive & Functional Polymers, 168*, 105041. https://doi.org/10.1016/j.reactfunctpolym.2021.105041

Wolpert, D. M., Miall, R. C., & Kawato, M. (1998). Internal models in the cerebellum. *Trends in Cognitive Sciences, 2*(9), 338–347. https://doi.org/10.1016/S1364-6613(98)01221-2

11 Exploring Cognition through a Morphological Info-Computational Framework

Gordana Dodig-Crnkovic

Introduction

The human-centric perspective of cognition focused on high-level mental activities has historically dominated cognitive science, neglecting the manifestations of cognition on subsymbolic levels as well as cognition in other living beings. Recent research, however, challenges this narrow understanding, proposing a more inclusive view of cognition on both symbolic and subsymbolic levels, as a fundamental ability shared by all living organisms—starting with basal (cellular) cognition (Lyon, 2015; Lyon et al., 2021; Levin et al., 2021). This chapter explores that broader perspective within the morphological info-computational framework, highlighting the implications for our understanding of cognition in nature and the development of intelligent artifacts (Dodig-Crnkovic, 2008; van Leeuwen & Wiederman, 2017).

The concept of morphological computation (MC), which utilizes the physical properties of organisms to produce and control behavior, provides a foundation for this expanded view of cognition. "Information is not a disembodied abstract entity; It is always tied to a physical representation" (Landauer, 1996). Its dynamics (MC) depends on the same physical substrate.

Through the info-computational perspective, seeing cognition as a network of computational processes on informational structures (Dodig-Crnkovic, 2017, we can better understand how cognitive processes emerge and function across different levels of biological organization. This chapter presents the conceptual underpinnings of that perspective, examining how it challenges traditional views of cognition and offers new insights into the cognitive abilities of all living organisms and their connection to information and computation as fundamental building blocks. There is a cycle of development between the info-computational physical substrate and the cognitive capacities of cognizing agents. The more complex agents possess more advanced cognitive capacities, leading to even more advanced cognitive agents.

Challenging the Traditional Understanding of Cognition

The understanding of cognition as knowledge generation focused on high-level mental activities in humans in our present stage of evolution, excluding subsymbolic and subconscious processes, and neglected cognition in other living beings. As a result, numerous theoretical and empirical challenges have arisen, putting into question the adequacy of this human-centric view. Within this tradition, cognition has been modeled as classical sequential computation, understood as the Turing machine symbol manipulation, or by neural networks. Although behaviorism offered an alternative by focusing solely on observable behavior, contemporary cognitive science remains largely divided between two mutually exclusive camps: cognitivism/computationalism and the embodied, embedded, enactive, and extended (known as the 4E) approaches to cognition (Newen et al., 2018). Info-computationalism (ICON) (Dodig-Crnkovic & Miłkowski, 2023) is a conceptual framework that integrates two key concepts: (Pan)Informationalism (Informational Structural Realism) (Floridi, 2008) and (Pan)Computationalism. It presents a unifying perspective for understanding natural phenomena, including living organisms and their cognition, by viewing the physical universe as a network of networks of computational processes running on an informational structure. The ICON approach describes the computing nature (Dodig-Crnkovic & Giovagnoli, 2013; Dodig-Crnkovic & Miłkowski, 2023; Zenil, 2012) with the naturalization of the concepts of information, computation, and cognition. Natural computationalism adopts a generalized, broader notion of computation–unconventional computation beyond the traditional Turing model (Calude & Cooper, 2012; Cooper, 2012; Horsman et al., 2017; Sloman, 1996; Stepney, 2008; Stepney et al., 2018). It allows for a smooth integration of natural and artificial systems.

This view, in which natural processes—from physical interactions to biological and cognitive functions—can be understood as forms of natural (physical) computation, is inspired by Alan Turing's ideas on computation and morphogenesis. In that model, computation is not just a mathematical abstraction, as in the Turing machine model (which Turing called Logical

Computing Machines) but a physical process that drives real-world behaviors and interactions leading to a generation of form. Another inspiration is Wheeler's informational universe (1994), in which nature is viewed (by a cognizer) as a web of informational structures. Information is not merely a static entity but is dynamic and constantly processed by computational mechanisms. Thus, processing information is what drives changes and developments in natural systems.

ICON incorporates the concept of MC, where the physical form (morphology) of an organism or system plays a crucial role in its computational processes. This idea emphasizes that the embodiment—that is, physical properties such as shape structure—and material enable and constrain computation.

MC is an unconventional computation that represents a shift from traditional views, defining computation more generally than conventional symbol manipulation or connectionist neural network models. This approach is based on the physical embodiment of computational mechanisms, making it a suitable tool for modeling a broader range of natural cognitive phenomena (Ghazi-Zahedi et al., 2017; Hauser et al., 2014; Miłkowski, 2018; Nowakowski, 2017; Pfeifer & Iida, 2005; Pfeifer et al., 2006). MC is physical computation found in nature on the hierarchy of levels of organization/levels of abstraction/ontological levels or scales of agency—from quantum to molecular, chemical, biological, cognitive, and social computing (Baluška & Levin, 2016; Bongard & Levin, 2023; Dodig-Crnkovic, 2012; Manicka & Levin, 2022; McMillen & Levin, 2024).

In short, MC refers to the process where the physical structure (morphology) of a body determines its behavior through its intrinsic physical properties with causal powers. Müller and Hoffmann (2017) identify three types of MC:

(1) Morphology facilitating control

(2) Morphology facilitating perception

(3) MC proper

The first two types involve the physical structure aiding in motor control and sensory perception, respectively. The third type, MC proper, refers to more complex computations, such as those found in reservoir computing, where physical structures are integral to the computational process (Miłkowski, 2018).

At the First International Conference on Morphological Computing (in 2007), MC was defined informally as any process serving a computational purpose, with clearly assignable input and output states, and programmable in a broad sense. This definition aims to capture the idea that a system's behavior can be altered by varying a set of parameters (Müller & Hoffmann, 2017).

MC can be theoretically modeled by Hewitt's (2012) computation in which information processing results from interactions between the parts of a distributed agent system (subatomic particles, atoms, molecules, and their assemblies) exchanging "messages" that can take different forms. Such agent-based concurrent computational systems have been shown to be applicable for modeling natural systems and recently earned prominence in artificial intelligence (AI), where the idea goes back to Marvin Minsky's "society of mind" model of intelligence (Minsky, 1986).

Cognitive MC is rooted in the understanding that all living organisms possess cognitive abilities. As Maturana and Varela (1980)—as well as Stewart (1996)—argued, every single cell constantly cognizes by registering external inputs from its environment and internal signals from its body, ensuring survival through metabolic and behavioral processes. This perspective sees physico-chemical-biological-cognitive processes as forms of MC, dependent on the morphology of the organism—its material, form, and structure.

Cognition includes not only high-level mental activities but also subsymbolic processes (Levin et al., 2021; Lyon et al., 2021) as well as extended and distributed/social cognition. This understanding aligns with the mechanisms of information compression as a unifying principle in learning, perception, and cognition (Wolff, 2006).

Recent successes of generative AI based on compressed information of all available digital information from the Internet and other sources nicely illustrate the power of info-computational mechanisms in generating knowledge and meaning. Computational nature (CN) is conceptualized as a network of concurrent MC, appearing in nature on different levels of organization, from physics to chemistry, biology, and cognition, giving rise to phenomena such as self-assembly, self-organization, and autopoiesis. This perspective offers a holistic view of cognition, where information processing is distributed across different levels of organization and facilitated by the physical properties of the organism. It recognizes the importance of both the structural and dynamic aspects of cognition, highlighting the interplay between form and function in shaping cognitive behavior, where form enables the function, which in turn leads to new forms. By embracing this broader view, we can better understand the cognitive abilities of diverse life forms and explore new possibilities for developing intelligent artifacts inspired by biological systems.

Natural computation is inspired by CN and proposes using natural computing as a generalization of our present understanding of computing. Crutchfield et al. (2010) makes the distinction between intrinsic (natural) and designed computation. He emphasizes the necessity of including analog computation and not only reducing all computational models to digital forms, which he calls "digital hegemony." He illustrates the point with the case of quantum computation (Crutchfield & Wiesner, 2008). For a cognizer, some processes in nature appear continuous, and it is important to include both discrete and continuous computing in our framework.

Empirical Evidence and Theoretical Foundations of an Info-Computational Framework for Cognitive Abilities in Diverse Life Forms and Robotics

Recent research has revealed that cognitive processes are present in all living organisms, from single cells to complex beings (Levin et al., 2021; Lyon et al., 2021). Single-celled organisms, for example, demonstrate basic cognitive processes through their interactions with the environment. Rich empirical evidence has shown that unicellular organisms exhibit forms of social cognition through communication and collective behavior. Bacteria use quorum sensing (Ng & Bassler, 2009; Waters & Bassler, 2005) to communicate and coordinate their behavior in response to environmental changes (Ben-Jacob, 2009). This form of social cognition observed in microbial communities shows the cognitive capabilities of even the simplest life forms (Ben-Jacob, 2008). Groups of organisms exhibit distributed cognition through information processing as shown by Almér et al. (2015), Rogers & McClelland (2014), and Rumelhart et al. (1986).

Multicellular organisms, including plants, have more complex cognitive behaviors. Plants, despite lacking a nervous system, can perceive and respond to environmental stimuli, communicate through chemical signaling, and even exhibit memory (Calvo & Friston, 2017; Calvo et al., 2020; Garzon, 2012). These cognitive abilities are based on MC, where the physical properties of the plant's structure play a crucial role in processing information.

In animals, cognition extends beyond individual organisms to include social and collective behaviors. Social insects like ants and bees, for instance, show sophisticated forms of distributed cognition through collective decision-making and problem-solving. These behaviors are based on MC, where the interactions between individuals and their environment shape the cognitive processes of the colony.

Insights from biology about morphological information processing are further reinforced by the work of researchers in robotics (Hauser et al., 2014; Paul, 2006; Pfeifer & Iida, 2005) who have observed computation processes in more general terms, accounting for the physical embodiment of computational mechanisms on the macroscopic levels of mechanical robotic bodies.

Combining empirical evidence from biology and robotics extended the idea of MC by the proposal that cognition in cognitive agents, in general, can be understood as morphological information processing at various levels of organization/levels of agency (Dodig-Crnkovic, 2014). The ICON framework provides a robust theoretical foundation for understanding cognition as an emergent property of complex multiscale phenomena in living organisms, empirically studied by McMillen and Levin (2024) and McMillen et al. (2022). The empirical evidence is supported by the mathematical model of ICON in terms of category theory proposed by Ehresmann (2014).

Subsymbolic Information (Data/Signal) Processing: Integration of Feelings and Emotions

Predominant views of cognition have often overlooked subsymbolic and subconscious processes, focusing primarily on high-level, symbolic reasoning. However, recent research has emphasized the importance of these processes in understanding cognition as a whole.

Subsymbolic processes involve low-level, continuous information processing that underlies higher-order cognitive functions. For example, the perception of sensory stimuli and the regulation of bodily functions are governed by subsymbolic processes. These processes are often automatic and do not require conscious awareness, yet they play a critical role in shaping cognitive behavior.

Subconscious processes involve mental activities that occur below the level of conscious awareness. Emotions, for instance, are subconscious processes that influence decision-making, learning, and memory. The integration of emotions into cognitive models has been a significant step in bridging the gap between cognition and its underlying physiological processes (Damasio, 1999).

By embracing an ICON/MC perspective, one can address many of the unsolved problems in cognitive science, including the integration of subsymbolic information processing such as emotions and feelings as well as consciousness and social cognition.

Distributed and Social Cognition: All Intelligence Is Social Intelligence

As mentioned before, cognition is not limited to individual organisms but can also be distributed across groups and social networks. Distributed cognition refers to cognitive processes that are shared among individuals within a group, enabling collective problem-solving and decision-making. As observed by Falandays et al. (2023) and McMillen et al. (2022), all intelligence is social intelligence. Social cognition involves the ability to perceive, interpret, and respond to social signals. It includes processes such as empathy, theory of mind, and social learning, which are essential for navigating complex social environments. In humans, social cognition is a key component of high-level cognitive functions, but it is also present in other animals, showing the continuity of cognitive processes across species.

Morphological Computation and Information Self-Structuring/Self-Organization

Teleonomy and Purposiveness in Living Systems

MC plays a fundamental role in the ICON framework describing the continuous self-structuring of information at different levels of organization (Braitenberg, 2011; Der, 2011; Lungarella & Sporns, 2005). Information self-structuring refers to the dynamic organization of information through interactions with the environment. In living organisms, this process is enabled by the morphology of the organism, which shapes how information is perceived, processed, and acted on. For example, the structure of sensory organs influences how sensory information is received and integrated, while the organization of neural networks determines how information is processed and stored (Pfeifer & Bongard, 2006; Pfeifer & Gomez, 2009; Pfeifer & Iida, 2005).

The autonomous agency of living organisms, as empirically observed in biology (Levin, 2023), drives teleonomy and purposiveness in living systems, which arise from goal-directed behaviors. These teleonomic processes contribute to the directedness and adaptiveness of evolutionary change (Sloman, 2013). The basis of goal-directed behavior is memory and learning. An organism learns about the world from previous experiences to anticipate possible future states (Rosen, 1985). Evolution provides learning mechanisms for organisms to find the most favorable circumstances for survival and flourishing. This process is enabled by embodied information self-structuring/self-organization. The goal-directedness (teleonomy, purposiveness) of living beings is a result of their agency—the ability to act on their own behalf, given past experiences and preferences (see also Corning et al., 2023).

ICON connects evolution with the agency of biological systems by viewing evolution as a process of MC driven by the embodied, info-computational, cognitive nature of living organisms. It highlights the multiscale agency, embodied cognition (EC), self-organization, and teleonomy of biological systems as key factors shaping evolutionary dynamics.

ICON and the Extended Evolutionary Synthesis: Cognition-Based Evolution

The theory of evolution states that ICON naturalism relies on is extended evolutionary synthesis (EES) (Laland et al., 2015), the most advanced and up-to-date theory of evolution that integrates recent scientific results from developmental biology, evolutionary biology, computational biology, genetics, ecology, neuroscience, and more.

EES considers how developmental processes and environmental influences impact evolutionary patterns beyond genetic inheritance (genetic determinism) (Noble, 2006). Evolutionary processes occur at multiple levels (genetic, developmental, behavioral, and cultural) and their interactions. EES acknowledges multiple forms of inheritance beyond genetic inheritance, such as epigenetic, ecological, and cultural inheritance (Jablonka et al., 2014). It emphasizes the role of developmental processes in evolution and considers how changes in development (e.g., gene expression, cellular processes) can lead to phenotypic variation and contribute to evolutionary change. Reciprocal causation, the two-way interaction between organisms and their environments, plays an important role. Not only does the environment influence the evolution of organisms, but organisms also modify their environments (niche construction), which in turn affects evolutionary processes. EES examines how information is transferred across inheritance systems and hierarchical levels to shape evolution. By incorporating new mechanisms, emphasizing the role of developmental processes, and considering multiple forms of inheritance and levels of selection, EES provides a more integrative and dynamic understanding of evolution. It challenges the limitations of the traditional evolutionary theory by incorporating new research directions and mechanisms.

Within ICON, life equals cognition, which implies agency, and evolution is therefore driven by the cognition of living organisms. Miller and Torday (2018) provide a detailed account of cognition-based evolution (Miller, 2023). The ICON framework also incorporates the concepts of emergence and self-organization, which are also central to EES's study of how developmental processes like plasticity and environmental induction shape phenotypic variation and evolutionary dynamics.

Evolution of Information

Computing nature is a hierarchical view with ICON processes across multiple scales, from elementary particles to biological and cognitive systems. Information structures and computational dynamics at lower levels give rise to higher-level emergent phenomena through self-organization processes. In the next step, top-down control structures are activated (such as the brain and nervous system controlling bodily movements and functions). The information transfer goes both ways, bottom-up and top-down. Mariján and Navarro (2022) discuss the biological information flow from cells to the level of the evolution of life, arguing for the necessity of extended evolutionary synthesis. Also, Malassé (2022) and Torday et al. (2020) recognize evolution as resulting from interactions between various processes (genetic, epigenetic, behavioral, and cultural) across multiple levels of biological organization. The evolution of information is driven by these multilevel interactions and information transfers between hierarchical levels.

Underlying this are fundamental principles of biological computation (Shklovskiy-Kordi et al., 2022).

EC enables agency, where the physical embodiment of agents and their interactions with the environment shape the computational processes of cognition and intelligence. This embodied agency is a key factor in the evolution of increasingly complex informational and cognitive structures of organisms (Miller, 2023).

A Short Comment on The Role of an Observer = Cognizer = Generator of Knowledge

It should be pointed out that ICON naturalism presupposes an observer. Both information and its dynamics computation are observer-relative. These views are compatible with the views of modern physics (Rovelli, 2015) and by no means imply subjectivity. Observer-relative is in the same sense as relativistic physics or quantum mechanics. Here, we also assume that the observer is not a "material point" as in physics or a "fly on the wall" as in an experiment observing human behavior. The observer is a cognitive agent; it interacts with the world and interprets it. In theories of knowledge, we are still lacking an adequate theory of an observer (Fields, 2012). We may see an "observer" or rather "cognizer," as an active cognizing agent constructing her/his/its knowledge/understanding/feeling of the world as a result of interactions with the world and intrinsic information processing. We may see this agent as insignificant in the context and represent it as a point of view like in relativistic physics or as completely decisive like in artistic cognition, where the actor is the creator and interpreter with huge freedom. In any case, understanding the observer/cognizer makes a difference, even if it only defines the vantage point.

For an observer, reality comes as information. All physical processes manifest themselves as information for a cognizer. Sara Walker (2024) calls this "the hard problem of matter" and John Wheeler (1994) termed "it from bit." We do not know what physical objects intrinsically are (in themselves). Immanuel Kant already made that observation about "*Ding an Sich*" (thing in itself). As observers/cognizers, we know what they are for us—information. And their processes are computations.

Conventional computing is designed as substrate-independent. Any substrate that implements Boolean logic can be used to compute, as it is based on the logical model of computation. Natural computing, on the other hand, is substrate-dependent and can be used to program the material behavior on a given level of organization. Bongard and Levin (2023) refer to this as "polycomputing"—the ability of the same substrate to simultaneously compute different things so we have quantum computing, molecular computing, and cognitive computing in the physical system.

The Relationships with Karl Friston's Free Energy Principle and Active Inference

Friston's free energy principle (FEP) with active inference (Friston et al., 2012, 2015; Kuchling et al., 2020; Parr et al., 2022; Parr & Friston, 2019) can be understood within the ICON framework as the basis of the behavior of cognitive agents with mutual interactions and constant exchanges with the environment. The FEP suggests that biological systems, including the brain, strive to minimize free energy. Free energy is a measure of surprise or uncertainty about sensory inputs. By minimizing free energy, the brain reduces the difference between its prediction and the actual sensory input. The brain constantly generates predictions about sensory inputs based on internal world models. When actual sensory inputs are received, they are compared to these predictions. Discrepancies (prediction errors) are used to update the internal models to better match reality. This process is known as predictive coding. Perception is thus a process of inference where the brain interprets sensory data by constantly updating its predictions to minimize free energy. As cognition is fundamentally about making predictions (anticipation), active inference extends the FEP by incorporating action. It suggests that actions are performed to fulfill the brain's predictions and reduce prediction errors. The brain not only updates its models based on sensory input but also actively changes the environment to make it conform to its predictions. The FEP and active inference suggest that all cognitive processes are geared toward predicting sensory inputs and minimizing surprise or prediction errors.

ICON similarly emphasizes an embodied, active view of cognition arising from the interactions between an agent/organism and its environment. FEP describes perception and action as information processing for minimizing surprise/uncertainty. ICON views cognition through the agency of living/artifactual systems by morphological/natural computation, learning, and "learning to learn." ICON incorporates self-organization as a key process in the framework, which aligns with Friston's ideas on self-organized dynamics minimizing free energy. Both approaches propose hierarchical models—ICON uses a multiscale hierarchy of info-computational processes, and Friston uses hierarchical predictive coding.

Applications in Artificial Intelligence and Robotics

Beyond biology and cognitive science, principles of MC and ICON find practical applications in fields such as AI and robotics, among others. By designing systems that utilize the physical properties of their bodies to process information, we can develop more efficient and adaptive artificial agents.

For instance, robots that employ MC can reduce the need for complex control algorithms by leveraging their physical structure to perform tasks, as already argued by Pfeifer and colleagues (Pfeifer & Bongard, 2006; Pfeifer et al., 2006; see also Hauser et al., 2014). This approach not only saves computational resources but also enhances the robot's ability to adapt to changing environments.

In AI, the ICON framework can inspire new models of cognitive processing that incorporate continuous and dynamic information self-structuring. These models can improve the ability of artificial systems to learn, adapt, and interact with their environment.

Future Work

Advancing Insights into Life and Natural Cognition

Recognizing the cognitive abilities of all living organisms has profound implications for our understanding of natural cognition and life. By studying cognitive processes across different life forms, we can gain insights into the fundamental principles of cognition and explore the evolutionary roots of cognitive behavior.

This broader perspective challenges the traditional boundaries between different fields of study, encouraging interdisciplinary research that integrates insights from physics, chemistry, biology, neuroscience, computer science, robotics, philosophy, and other related fields. By embracing a more holistic view of cognition, we can develop new theories and models that capture the complexity and diversity of cognitive processes and improve our understanding of intelligence and evolution.

Developing Intelligent Artifacts

The insights gained from studying natural cognition can also inform the development of intelligent artifacts. The present-day huge success of large language models (LLMs) based on the compressed knowledge of humanity as found on the Internet is just the beginning of the process of AI development extending into the physical world. By implementing the principles of MC and the ICON approach, combined with generative AI, researchers can create artificial systems that exhibit more adaptive and intelligent behavior.

For example, bioinspired robots that utilize MC can navigate complex environments more effectively by leveraging their physical structure. Similarly, AI systems that incorporate continuous information self-structuring can improve their ability to learn and adapt in real time, enhancing their performance in dynamic and unpredictable settings.

Conclusion

The approach of natural computationalism, utilizing MC and ICON conceptual tools, offers a transformative perspective on cognition. It challenges traditional human-centric symbolic models rooted in the Turing machine view of computation and acknowledges the cognitive capabilities inherent in all living organisms. It situates humans in the ecology of other cognizing agents, natural and artifactual. This new perspective not only enhances our understanding of natural cognition but also provides practical insights into the fields of Artificial Intelligence and robotics.

References

Almér, A., Dodig-Crnkovic, G., & von Haugwitz, R. (2015). Collective cognition and distributed information processing from bacteria to humans. In G. Dodig-Crnkovic, Y. J. Erden, & R. Giovagnoli (Eds.), *Proceedings of the AISB 2015 Symposium on Social Aspects of Cognition and Computing* (pp. 20–25). Society for the Study of Artificial Intelligence and Simulation of Behaviour. https://www.cs.kent.ac.uk/events/2015/AISB2015/allThePDFFiles_AISB_2015/SocialAspects.pdf

Baluška, F., & Levin, M. (2016). On having no head: Cognition throughout biological systems. *Frontiers in Psychology*, *7*, 902. https://doi.org/10.3389/fpsyg.2016.00902

Ben-Jacob, E. (2008). Social behavior of bacteria: From physics to complex organization. *The European Physical Journal B, 65*(3), 315–322. https://doi.org/10.1140/epjb/e2008-00222-x

Ben-Jacob, E. (2009). Learning from bacteria about natural information processing. *Annals of the New York Academy of Sciences, 1178*(1), 78–90. https://doi.org/10.1111/j.1749-6632.2009.05022.x

Bongard, J., & Levin, M. (2023). There's plenty of room right here: Biological systems as evolved, overloaded, multi-scale machines. *Biomimetics, 8*(1), 110. https://doi.org/10.3390/biomimetics8010110

Braitenberg, V. (2011). *Information: Der Geist in der Natur*. Schattauer Verlag.

Calude, C. S., & Cooper, S. B. (2012). Introduction: Computability of the physical. *Mathematical Structures in Computer Science, 22*(5), 723–728. https://doi.org/10.1017/S0960129511000491

Calvo, P., & Friston, K. (2017). Predicting green: Really radical (plant) predictive processing. *Journal of the Royal Society Interface, 14*(131), 20170096. https://doi.org/10.1098/rsif.2017.0096

Calvo, P., Gagliano, M., Souza, G. M., & Trewavas, A. (2020). Plants are intelligent, here's how. *Annals of Botany, 125*(1), 11–28. https://doi.org/10.1093/aob/mcz155

Cooper, S. B. (2012). The mathematician's bias—And the return to embodied computation. In H. Zenil (Ed.), *A computable universe: Understanding and exploring nature as computation* (pp. 125–142). World Scientific. https://doi.org/10.1142/9789814374309_0007

Corning, P. A., Kauffman, S. A., Noble, D., Shapiro, J. A., Vane-Wright, R. I., & Pross, A. (2023). *Evolution "On Purpose": Teleonomy in living systems*. MIT Press. https://doi.org/10.7551/mitpress/14642.001.0001

Crutchfield, J., Ditto, W., & Sinha, S. (2010). Introduction to focus issue: Intrinsic and designed computation: Information processing in dynamical systems—Beyond the digital hegemony. *Chaos: An Interdisciplinary Journal of Nonlinear Science, 20*(3), 037101. https://doi.org/10.1063/1.3492712

Crutchfield, J., & Wiesner, K. (2008). Intrinsic quantum computation. *Physics Letters A, 374*(4), 375–380.

Damasio, A. R. (1999). *The feeling of what happens: Body and emotion in the making of consciousness*. Harcourt Brace and Co.

Der, R. (2011, June 7). *Self-organization of robot behavior* [Video]. UZH MediaSpace. https://uzh.mediaspace.cast.switch.ch/media/Self-organization+of+robot+behavior/0_afjpv6t3

Dodig-Crnkovic, G. (2008). Knowledge generation as natural computation. *Journal of Systemics, Cybernetics and Informatics, 6*(2), 12–16.

Dodig-Crnkovic, G. (2012). Info-computationalism and morphological computing of informational structure. *Integral Biomathics: Tracing the Road to Reality*, 97–104. https://doi.org/10.1007/978-3-642-28111-2

Dodig-Crnkovic, G. (2014). Modeling life as cognitive info-computation. In A. Beckmann, E. Csuhaj-Varjú, & K. Meer (Eds.), *Computability in Europe* (pp. 153–162). Springer. https://doi.org/10.1007/978-3-319-08019-2_16

Dodig-Crnkovic, G. (2017). Nature as a network of morphological infocomputational processes for cognitive agents. *The European Physical Journal Special Topics, 226*(2), 181–195. https://doi.org/10.1140/epjst/e2016-60362-9

Dodig-Crnkovic, G. D., & Giovagnoli, R. (Eds.). (2013). *Computing nature: Turing centenary perspective*. Springer.

Dodig-Crnkovic, G., & Miłkowski, M. (2023). Discussion on the relationship between computation, information, cognition, and their embodiment. *Entropy, 25*(2), 310. https://doi.org/10.3390/E25020310

Ehresmann, A. C. (2014). A mathematical model for info-computationalism. *Constructivist Foundations, 9*(2), 235–237.

Falandays, J. B., Kaaronen, R. O., Moser, C., Rorot, W., Tan, J., Varma, V., Williams, T., & Youngblood, M. (2023). All intelligence is collective intelligence. *Journal of Multiscale Neuroscience, 2*(1), 169–191. https://doi.org/10.56280/1564736810

Fields, C. (2012). If physics is an information science, what is an observer? *Information, 3*(1), 92–123. https://doi.org/10.3390/info3010092

Floridi, L. (2008). A defense of informational structural realism. *Synthese, 161*(2), 219–253. https://doi.org/10.1007/s11229-007-9163-z

Friston, K., Levin, M., Sengupta, B., & Pezzulo, G. (2015). Knowing one's place: A free-energy approach to pattern regulation. *Journal of the Royal Society Interface, 12*(105), 20141383. https://doi.org/10.1098/rsif.2014.1383

Friston, K., Samothrakis, S., Montague, R., Friston, K., Samothrakis, S., & Montague, R. (2012). Active inference and agency: Optimal control without cost functions. *Biological Cybernetics, 106*, 523–541. https://doi.org/10.1007/s00422-012-0512-8

Garzon, P. C. (2012). Plant neurobiology: Lessons for the unity of science. In O. Pombo, J. Torres, J. Symons, & S. Rahman (Eds.), *Special sciences and the unity of science* (pp. 121–137). Springer. https://doi.org/10.1007/978-94-007-2030-5_8

Ghazi-Zahedi, K., Langer, C., & Ay, N. (2017). Morphological computation: Synergy of body and brain. *Entropy, 19*(9), 456. https://doi.org/10.3390/e19090456

Hauser, H., Füchslin, R. M., & Pfeifer, R. (Eds.) (2014). *Opinions and outlooks on morphological computation*. https://tinyurl.com/4adt9j75

Hewitt, C. (2012). What is computation? Actor model versus Turing's model. In H. Zenil (Ed.), *A computable universe, understanding computation & exploring nature as computation* (pp. 159–187). World Scientific Publishing Company/Imperial College Press. https://doi.org/10.1142/9789814374309_0009

Horsman, D., Kendon, V., & Stepney, S. (2017). The natural science of computing. *Communications of the ACM, 60*(8), 31–34. https://doi.org/10.1145/3107924

Jablonka, E., Lamb, M. J., & Anna, Z. (2014). *Evolution in four dimensions, revised edition: Genetic, epigenetic, behavioral, and symbolic variation in the history of life*. MIT press.

Kuchling, F., Friston, K., Georgiev, G., & Levin, M. (2020). Morphogenesis as Bayesian inference: A variational approach to pattern formation and control in complex biological systems. *Physics of Life Reviews, 33*, 88–108. https://doi.org/10.1016/j.plrev.2019.06.001

Laland, K. N., Uller, T., Feldman, M. W., Sterelny, K., Müller, G. B., Moczek, A., Jablonka, E., & Odling-Smee, J. (2015). The extended evolutionary synthesis: Its structure, assumptions and predictions. *Proceedings of the Royal Society B: Biological Sciences, 282*(1813), 20151019. https://doi.org/10.1098/rspb.2015.1019

Landauer, R. (1996). The physical nature of information. *Physics Letter A, 217*(4–5), 188–193. https://doi.org/10.1016/0375-9601(96)00453-7

Levin, M. (2022). Technological approach to mind everywhere: An experimentally-grounded framework for understanding diverse bodies and minds. *Frontiers in Systems Neuroscience, 16*, 768201. https://doi.org/10.3389/FNSYS.2022.768201/FULL

Levin, M. (2023). Darwin's agential materials: Evolutionary implications of multiscale competency in developmental biology. *Cellular and Molecular Life Sciences, 80*(6), 1–33. https://doi.org/10.1007/S00018-023-04790-Z

Levin, M., Keijzer, F., Lyon, P., & Arendt, D. (2021). Basal cognition: Multicellularity, neurons and the cognitive lens. *Philosophical Transactions of the Royal Society B: Biological Sciences, 376*(1821), 20200458. http://doi.org/10.1098/rstb.2020.0458

Lungarella, M., & Sporns, O. (2005). Information self-structuring: Key principle for learning and development. In *Proceedings of 2005 4th IEEE International Conference on Development and Learning* (pp. 25–30), IEEE.

Lyon, P. (2015). The cognitive cell: Bacterial behavior reconsidered. *Frontiers in Microbiology, 6*, 264. https://doi.org/10.3389/FMICB.2015.00264

Lyon, P., Keijzer, F., Arendt, D., & Levin, M. (2021). Basal cognition: Conceptual tools and the view from the single cell. *Philosophical Transactions of the Royal Society B: Biological Sciences, 376*(1820), 20190758. http://doi.org/10.1098/rstb.2019.0750

Malassé, A. D. (2022). *Self-organization as a new paradigm in evolutionary biology*. Springer Nature.

Manicka, S., & Levin, M. (2022). Minimal developmental computation: A causal network approach to understand morphogenetic pattern formation. *Entropy, 24*(1), 107. https://doi.org/10.3390/e24010107

Marijuán, P. C., & Navarro, J. (2022). The biological information flow: From cell theory to a new evolutionary synthesis. *Biosystems, 213*, 104631. https://doi.org/10.1016/j.biosystems.2022.104631

Maturana, H., & Varela, F. (1980). *Autopoiesis and cognition: The realization of the living*. D. Reidel.

McMillen, P., & Levin, M. (2024). Collective intelligence: A unifying concept for integrating biology across scales and substrates. *Communications Biology, 7*(1), 378. https://doi.org/10.1038/s42003-024-06037-4

McMillen, P., Walker, S. I., & Levin, M. (2022). Information theory as an experimental tool for integrating disparate biophysical signaling modules. *International Journal of Molecular Sciences, 23*(17), 9580. https://doi.org/10.3390/ ijms23179580

Miłkowski, M. (2018). Morphological computation: Nothing but physical computation. *Entropy, 20*(12), 942. https://doi.org/10.3390/e20120942

Miller, W. B. (2023). *Cognition-based evolution*. CRC Press. https://doi.org/10.1201/9781003286769

Miller, W. B., & Torday, J. S. (2018). Four domains: The fundamental unicell and post-Darwinian cognition-based evolution. *Progress in Biophysics and Molecular Biology, 140*, 49–73. https://doi.org/https://doi.org/10.1016/j.pbiomolbio.2018.04.006

Minsky, M. (1986). *The society of mind*. Simon and Schuster.

Müller, V., & Hoffmann, M. (2017). What is morphological computation? On how the body contributes to cognition and control. *Artificial Life, 23*(1), 1–24. https://doi.org/10.1162/ARTL_a_00219

Newen, A., De Bruin, L., & Gallagher, S. (Eds.). (2018). *The Oxford handbook of 4E cognition*. Oxford University Press.

Ng, W. L., & Bassler, B. L. (2009). Bacterial quorum-sensing network architectures. *Annual Review of Genetics, 43*(1), 197–222.

Noble, D. (2006). The music of life: Biology beyond the genome. *Lavoisierfr*. https://doi.org/10.1234/12345678

Nowakowski, P. R. (2017). Bodily processing: The role of morphological computation. *Entropy, 19*(7), 295. https://doi.org/10.3390/e19070295

Parr, T., & Friston, K. J. (2019). Generalised free energy and active inference. *Biological Cybernetics, 113*(5), 495–513. https://doi.org/10.1007/s00422-019-00805-w

Parr, T., Pezzulo, G., & Friston, K. J. (2022). *Active inference: The free energy principle in mind, brain, and behavior.* MIT Press. https://doi.org/10.7551/mitpress/12441.001.0001

Paul, C. (2006). Morphological computation A basis for the analysis of morphology and control requirements. *Robotics and Autonomous Systems, 54*(8), 619–630. https://doi.org/10.1016/j.robot.2006.03.003

Pfeifer, R., & Bongard, J. (2006). *How the body shapes the way we think: A new view of intelligence.* MIT Press.

Pfeifer, R., & Gomez, G. (2009). Morphological computation: Connecting brain, body, and environment. In B. Sendhoff, E. Körner, O. Sporns, H. Ritter, & K. Doya (Eds.), *Creating brain-like intelligence: From basic principles to complex intelligent systems* (pp. 66–83). Springer. https://doi.org/10.1007/978-3-642-00616-6_5

Pfeifer, R., & Iida, F. (2005). Morphological computation: Connecting body, brain and environment. *Japanese Scientific Monthly, 58*(2), 48–54.

Pfeifer, R., Iida, F., & Gomez, G. (2006). Morphological computation for adaptive behavior and cognition. *International Congress Series, 1291*, 22–29. https://doi.org/10.1016/j.ics.2005.12.080

Rogers, T. T., & McClelland, J. L. (2014). Parallel distributed processing at 25: Further explorations in the microstructure of cognition. *Cognitive Science, 38*(6), 1024–1077. https://doi.org/10.1111/cogs.12148

Rosen, R. (1985). *Anticipatory systems.* Pergamon Press.

Rovelli, C. (2015). *Relative information at the foundation of physics* (pp. 79–86). Springer International Publishing. https://doi.org/10.1007/978-3-319-12946-4_7

Rumelhart, D. E., McClelland, J. L., & PDP Research Group. (1986). *Parallel distributed processing,* volume 1: *Explorations in the microstructure of cognition: Foundations.* MIT Press.

Shklovskiy-Kordi, N. E., Matsuno, K., Marijuán, P. C., & Igamberdiev, A. U. (2022). Fundamental principles of biological computation: From molecular computing to evolutionary complexity. *Biosystems, 219*, 104719. https://doi.org/10.1016/j.biosystems.2022.104719

Sloman, A. (1996). Beyond Turing equivalence. In A. Clark & P. J. R. Millican (Eds.), *Machines and thought: The legacy of Alan Turing* (Vol. 1, pp. 179–219). OUP (The Clarendon Press).

Sloman, A. (2013). Meta-morphogenesis: Evolution and development of information-processing machinery. In S. B. Cooper & J. van Leeuwen (Eds.), *Alan Turing: His work and impact* (p. 849). Elsevier.

Stepney, S. (2008). The neglected pillar of material computation. *Physica D: Nonlinear Phenomena, 237*(9), 1157–1164. https://doi.org/10.1016/j.physd.2008.01.028

Stepney, S., Rasmussen, S., & Amos, M. (2018). *Computational matter*. Springer International Publishing. https://doi.org/10.1007/978-3-319-65826-1_1

Stewart, J. (1996). Cognition = life: Implications for higher-level cognition. *Behavioral Processes, 35*(1–3), 311–326. https://doi.org/10.1016/0376-6357(95)00046-1

Torday, J., & Miller, W. (2020). Darwin, the modern synthesis, and a new biology. *Cellular-Molecular Mechanisms in Epigenetic Evolutionary Biology*, 5–10. https://doi.org/10.1007/978-3-030-38133-2_2

van Leeuwen, J., & Wiederman, J. (2017). Knowledge, representation and the dynamics of computation. In G. Dodig-Crnkovic & R. Giovagnoli (Eds.), *Representation and reality in humans, other living organisms and intelligent machines*. Springer International Publishing.

Walker, S. I. (2024). *Life as no one knows it: The physics of life's emergence*. Penguin Publishing.

Waters, C. M., & Bassler, B. L. (2005). Quorum sensing: Cell-to-cell communication in bacteria. *Annual Review of Cell and Developmental Biology, 21*, 319–346. https://doi.org/10.1146/annurev.cellbio.21.012704.131001

Wheeler, J. A. (1994). *At home in the universe*. American Institute of Physics.

Wolff, G. J. (2006). *Unifying computing and cognition: The SP theory and its applications*. CognitionResearch.org.

Zenil, H. (2012). *A computable universe. Understanding computation & exploring nature as computation*. World Scientific Publishing Company/Imperial College Press.

12 Embodied AI Based on Dynamic Human-Computer Interaction

Dobromir Dotov and Tom Froese

Introduction

What would it take to bring large language models (LLMs) and other types of artificial intelligence (AI) outside of the chat interface and into the real world? Attempts to define the conditions for successful AI development in this direction face a difficult dilemma. Should artificial agents first know all the objective facts of the world before they know how to physically exist with other beings in the world? Most work on AI stems from a tradition that views the relation between an intelligent agent and its environment as a subject-object relation, an individual-centric exercise where a solitary agent reasons about the external objects surrounding it. Yet, research on human-computer interaction in the real world famously brought the insight that the separation between subject (user) and object (machine) only appears in aberrant circumstances where fluid task performance is perturbed by a malfunction in the workspace (Winograd & Flores, 1986). There are phenomenological, practical, and even biological precedents as to why task-specific interaction has primacy over autonomous intelligence (Bennett et al., 2021; Fong et al., 2003; Froese & Ziemke, 2009; Matarić, 1993; Minsky, 1988; Pfeifer et al., 2007; Vicente & Rasmussen, 1990). Furthermore, interaction with other agents is just as important as interaction with the world. Intelligence arises, exists, and evolves embedded in a social environment (Hutchins, 2010, 1995), and AI should be able to learn from social interactions (Colas et al., 2022).

Since its inception, the holy grail of AI research has been strong AI, or artificial general intelligence (AGI). AGI, as a self-sufficient encapsulated intelligent agent is characterized by cognitive autonomy and should be capable of autonomous behavior and generalization of learning (Russell & Norvig, 2020). This expectation was based on the promises made by symbolic approaches to AI, the so-called Good Old-Fashioned AI (GOFAI). Although deep neural networks are considered more promising nowadays, they too are evaluated in terms of autonomy and generalization. With the excitement about the promise of AGI, substantial resources were committed to developing fully autonomous vehicles in the last decade. Yet, this objective appears less feasible, and the industry has sharply divested from such research (Carey & Lienert, 2023; Glasner, 2022). Furthermore, AI frameworks still struggle to use analogies to generalize their knowledge to novel domains (Mitchell, 2021). In contrast, the direct fit hypothesis states that artificial neural networks (ANNs), like brains, always learn within the bounds of the training set and, to the extent that they generalize, they do so by interpolating within the training set (Hasson et al., 2020). Although not without their limitations, recent generative models such as LLMs have been impressively successful, unexpectedly so from an enactive-embodied perspective. This calls into question some of the intuitions in this field. Is it still the case that neural networks fail to address the problem of meaning (Froese & Taguchi, 2019; Kiverstein et al., 2022)? In the last section of this chapter, we discuss how well LLMs handle interactive scenarios, illustrated by an example from a recent empirical study.

In this chapter, we argue that artificial social agents need to implement phenomenological insight into how beings exist in a shared world with other beings. It posits that machines may interact with the world by exploiting nonlinear dynamic phenomena (Dreyfus, 2007). This means that an agent with a rich sensorimotor repertoire can act successfully in the world simply by letting itself fall into stable patterns of interaction with the environment. This contrasts with the dominant approach to AI, in which agents carry a latent space of representations about the world's facts and, when a specific situation presents itself, they must figure out which facts are relevant. In this way, we continue a philosophical tradition that developed an alternative approach[1] to the naive subject-object ontology being assumed by mainstream cognitive science and the AI research program (Clark, 1998; Kiverstein, 2012). On the following pages, we offer three propositions that are necessary for embodied artificial intelligence (EAI).

Primacy of Interdependence over Autonomy

Our first radical proposal is that the development of EAI does not need to worry about autonomy and generalization, contrary to ideas that are deeply rooted in the history of AI. Autonomy and generalization, being intrinsic features of the language of thought, played a fundamental role in cognitive science and cognitive psychology. Imagine, instead, that AI research was committed to mimicking sensorimotor skills instead of cognitive autonomy. In human movement science, the generalization of motor learning is also important and desirable because of its practical implications in rehabilitation. If there were a magic trick that allowed stroke survivors to regain their skills by practicing a small number of exercises, this would have had a tremendous impact on the quality of life and the cost-effectiveness of recovery options. Yet, decades of research applying different ideas about how to promote generalization have brought little positive evidence. Generalization tends to be specific to the task or sensory modality (Krakauer et al., 2006; Torres-Oviedo & Bastian, 2010; Vasudevan et al., 2011). For example, intense slackline training leads to no useful transfer to other relevant activities (Giboin et al., 2018). If anything like generalization is reported, it is an improvement of the range of an existing parameter such as angle, amplitude, or duration (Goodbody & Wolpert, 1998; Thoroughman & Shadmehr, 2000).

While the path to cognitively autonomous AGI is obscured, the conditions for interdependent human-machine systems to solve problems together are present (Minsky et al., 2013). For example, deploying vehicles with limited autonomy is a realistic scenario if they are constrained by the principle of human-in-the-loop control (DoD, 1998), which allows the two parts of the team to complement each other (Minsky et al., 2013).

Interdependence taking the form of nonlinear dynamics has been a productive theoretical framework in the human movement sciences (Chemero, 2009; Kelso, 1995). To give a formal example, figure 12.1 shows the coupled and uncoupled dynamics of a human participant and a computer-generated stimulus. In this study on the acquisition of fine motor skills, participants practiced synchronizing with a stimulus generated by a chaotic system that was inherently variable but contained unstable periodic orbits (Dotov & Froese, 2018). Hand movement was fed in real time and

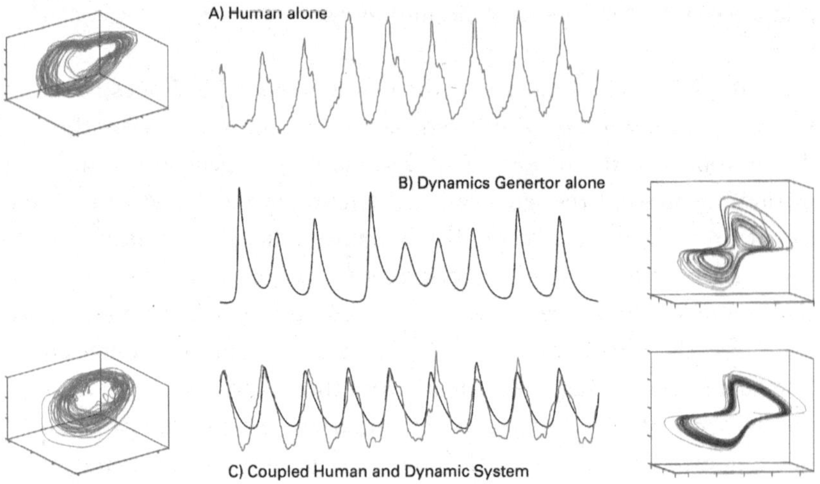

Figure 12.1
Emergence of periodic behavior from the mutual constraining of two variable systems. (a) Phase space (left) and time series (right) from the movement of a human participant trying to perform regular oscillations with the hand. (b) Phase space (left) and time series (right) from a chaotic system generating variable rhythms and amplitudes. (c) The phase spaces (left) and time series (right) exhibit regular oscillations when the human and dynamics generator are coupled, illustrating the basic idea that two dynamic systems in coupled mode can acquire properties that are hard to access by each one of them individually. These are representative trials collected in a study on auditory-motor rhythmic skill acquisition. *Source*: Adapted from Dotov & Froese, 2018.

with a small gain as an input signal to the stimulus generator. The mutual coupling allowed participants to learn how to stabilize the generator's trajectory on one of its otherwise unstable periodic orbits. In this way, participants achieved a level of fine rhythmic skill that would be hard to achieve by passively tracking a fixed stimulus.

What does it mean for the human and machine to augment each other's capacities? Even a screwdriver satisfies this description in some sense. Distinguishing between tools and collaborative agents will help address this question. To this end, we can extend the application of phenomenological insight (Kiverstein, 2012) by focusing on collaborative work. The insight from the perspective of Heideggerian ontology of being is that human-machine interaction would constitute an interesting form of interaction with another being if the agent shared some of the same world as the human.

In Heidegger's phenomenology of everyday experience, there are several conditions for an authentic connection (Wheeler, 2020). To begin with, a being is always *being-in-the-world* in the sense that it understands itself in terms of the possibilities. This was the insight that famously inspired some cognitive scientists to move their focus from how animals reason about their surroundings to how action is enabled by specific compatibilities between animals and their environments (Clark, 1998; Kiverstein, 2012). It is less often appreciated, in the cognitive sciences at least, that beings are also always *being-with* others ("Mitsein"). They coinhabit a world of shared possibilities with others, and they share potential ways to influence each other.[2] This is not to say that the authentic form of being is to always follow along with others, the crowd, but that the force of the crowd is always present. Furthermore, *being-there-with* others means not interrupting their being-with equipment, or *transparent action* with tools. When tools are transparent, referred to as *ready-to-hand*, one engages with the world fluently and skillfully and is not perturbed by malfunctioning equipment.

Current artificial systems are designed to enable transparent action. In phenomenological terms, they enable being-with equipment. It is difficult to argue, however, that current artificial systems enable being-with others, that they share the same world with humans. A condition for the latter would be that they share actions and attitudes toward the same existential possibilities as their human partners. To give an example of one such possibility, an enactive artificial system would need to exhibit self-maintenance, self-repair, and agency (Bedau et al., 2010). This means that it would continually bring forth its own existence under precarious conditions (Froese & Ziemke, 2009). Interestingly, the literary work that defined the word "robot," Karel Čapek's *R.U.R.*, follows the same idea. There, robots' self-awareness arises from the frustrating fallibility of their own bodies, triggered by the precarious conditions of organized matter (Froese, forthcoming).

This book introduces a broad definition of embodied intelligence (EI) that makes space for living beings, nonliving entities, hybrids, and living robots (Introduction). As such, it aims to move beyond the conventional debate between life and machines, providing a new perspective for analyzing and comparing intelligent behavior. Heidegger's project for a fundamental ontology of being is uniquely suitable in this context because it assumes yet-undisclosed entities instead of taking human individuals as the unit of analysis. His goal was to give a fundamental understanding of existence with others, not to offer yet another description of what it means to be human. This ontology is not committed to the life-machine dichotomy or to any hierarchical taxonomy of cognitive levels.

Reverse Self-Organizing, Not Reverse Engineering

Is AI something that can be constructed by imitating functions of human intelligence like the Turing imitation game suggests, tinkered with while evolving technological ecosystems, or allowed to emerge naturally from the right conditions, similar to the concept of designing emergent behaviors (Matarić, 1993)? Reverse engineering may not work for EAI. Interdependent human-machine dynamic interaction would be hard to slice apart in functions by applying tools of reverse engineering. A more suitable approach can be compared to modeling nonlinear phenomena observed in biology. That is why we call it reverse self-organization.

In physics and nonlinear dynamics, self-organization refers to the formation of order from disorder without any external control or blueprint. This process, akin to pattern formation in biological systems, happens from simple local rules and nonlinear interactions between parts of the system. To give an example, rhythmic and arhythmic patterns of electrical cardiac activity emerge from specific local interactions. The successful modeling of this process became a textbook case for the potential of nonlinear dynamics to shed light on biological processes (Glass & Mackey, 1988). In contrast, the existing efforts to replicate cardiac function look like the product of reverse engineering (see figure 12.2).

The closest and most mature example of successful reverse self-organization is Bongard and Levin's work on evolutionary robotics and Xenobots (Blackiston et al., 2023), also referred to in this volume: "Complex growth results not from the micromanagement of stem cell control mechanisms or gene expression cascades but by communicating at the very beginning to the cells to induce them to commit to a long journey through the vast space of possible anatomical outcomes that leads to leg growth rather than scarring" (chapter 17).

Leaving robotics aside, there are few comparable examples in other fields. Interestingly, it is hard to point to relevant examples of reverse self-organization in rehabilitation robotics, even though human movement science has a tradition of applying theoretical ideas from nonlinear

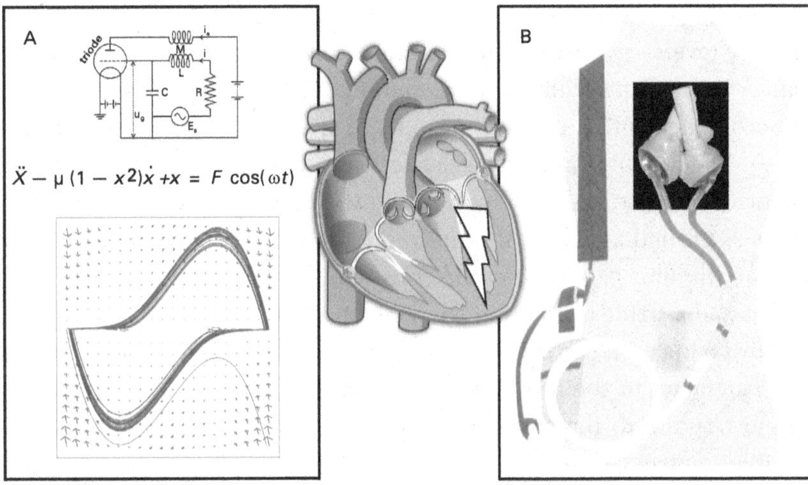

Figure 12.2

Rebuilding biological function. Reverse self-organization (A) and reverse engineering (B) as two aspects of rebuilding biological function. The quasi-regular waves of electrical activity in the heart are essential for its function. (A) The driven van der Pol oscillator, expressed as a circuit diagram, mathematical equation, and phase portrait, is a classic example of modeling self-organized dynamic patterns in biology. (B) An example of a total artificial heart which consists of detachable components to implement separate functions such as pumping action, periodic driving, and power supply. Note that A) has parts, and B) could, in theory, incorporate self-organized dynamics to achieve more adaptive rhythm generation; the relation between reverse self-organization and reverse engineering is not mutually exclusive and should not be conflated with the dichotomy of holism and reductionism. Image contains photos from Wikimedia Commons.

dynamics, and rehabilitation engineering often relies on closed-loop paradigms. Below, we provide two very modest examples from our research.

The first example concerns the optimization of cueing feedback for improving walking in persons with Parkinson's disease. Rhythmic auditory cueing is a widely explored method because persons with Parkinson's disease benefit from the availability of a predictable external stimulus that they can target with their steps. Trying to increase stepping cadence by only providing a prerecorded stimulus, however, limits the real-world application of this method. It requires those with Parkinson's disease to perturb their gait to synchronize with the stimulus or even to fiddle with their cueing devices while walking to ensure comfort. To address this problem, we reverse-self-organized spontaneous synchronization between the gait cycle and the stimulus by designing an interactive cueing system in accordance with a nonlinear model of synchronization (Dotov et al., 2019). The stimulus and walker were two oscillators with preferred frequencies and mutual coupling nudging their phases to speed up or slow down as a function of the phase mismatch between the two (figure 12.3A–B). The stimulus parameters were constrained by solving the model such that it had a stable fixed point for its tempo in the synchronized regime, and by taking each participant's preferred cadence (figure 12.3C). This left one parameter in the model free—namely, how strongly the participant would respond to the stimulus. In short, we created the conditions for flexible emergence of spontaneous synchronization at a cadence between 0% and 10% higher than the participant's preference, depending on their responsiveness.

In the second example, our goal was to exploit the property of chaotic systems to generate variability for the benefit of learning sensorimotor (re) learning. Theories of human motor control argue small amounts of variability are optimal (Stergiou & Decker, 2011). In robotics, chaotic central pattern generators can be adaptive in the control of a mechanical system. Their variability can allow them to switch automatically to the pattern of interaction that is stable because it resonates with the mechanical system (Pitti et al., 2010). We wanted to test if human participants would also be able to discover stable patterns of interaction with a chaotic system, with potential practical benefits. The auditory-motor task was to synchronize with the chaotic stimulus (see figure 12.1). As the system had hidden unstable periodic orbits, and the participants' movement was fed with a small gain into the chaotic generator, participants could explore and discover how to stabilize these orbits. In this sense, we reverse-self-organized the possibility for entrainment between a chaotic system and a human participant that is also inherently variable. Pre-posttests suggested that this form of training fine motor skill was more beneficial than two noninteractive control conditions (Dotov & Froese, 2018). Interestingly, additional analysis showed that different participants learned different locked trajectories, such as a periodic pattern with one repeating cycle or with two different repeating cycles (Dotov & Froese, 2020).

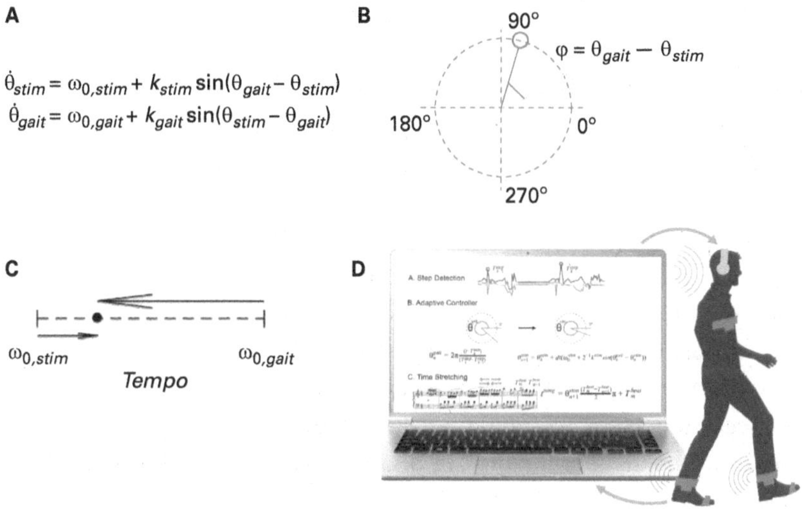

Figure 12.3

Gait cadence modification with an interactive gait cueing system designed to induce spontaneous synchronization. (A) We assumed a minimal model of two-phase oscillators with Kuramoto coupling and two parameters per oscillator: preferred frequency (ω_0) and coupling strength (k). Coupling strength refers to how strongly the oscillator responds by changing its frequency to reduce the phase lag relative to the other oscillator. (B) The stimulus phase tended to precede the gait cycle's phase because the intrinsic frequency of the stimulus was set higher than the participant's, with the aim of achieving a therapeutic increase in walking cadence. (C) The frequency of the coupled system had an attractor, a stable fixed point, in the range between the participant's and the stimulus's preferred frequencies. To constrain the system in this range, we used the solution of the model in panel A and fixed the parameters of the stimulus side of the model (first line in panel A), including the coupling strength k_{stim}. (D) The system was implemented using a laptop, wireless headphones for music-embedded cues, and sensors for footstep kinematics. *Source*: Adapted from Dotov et al., 2019.

In both examples, the complicatedness of the artificial systems was reduced to a minimum. Hardly anyone would treat a system with a few ordinary differential equations as an AI. Yet, these examples also show that artificial systems require little to be practical in daily life if they implement the right principles of embodiment. EI is consistent with an evolutionary account (Introduction). EAI inside an ecosystem of functional tools for human performance may enable evolution toward more sophisticated and independent AI.

Embodied Turing Test for Mutual Experience

Traditionally, the overall modus operandi of AI development has been strongly utilitarian. AI's goal is to do human work more productively than humans, and its performance is to be evaluated accordingly. This complies with Turing's idea of mimicking specific cognitive functions (Turing, 1950). In an embodied Turing test, however, the quality of interaction is also an important metric (Ortiz, 2016; Pfeiffer et al., 2011; Swisher et al., 2006). The argument for interdependence suggests that shared experience can be even more important than the task at hand. This contrast stands out when considering musical AI. One could test the ability of a model to reproduce sequences of notes that sound human to naive listeners, or one could test if a trained musician can play along with an artificial "jam factory" in a satisfactory manner (Belgum et al., 1988).

Musicians often report that they anticipate each other's ideas and act as one when they are deeply engaged in group performance. The self-other integration metric was developed to measure such interpersonal experience (Aron et al., 1992). An even stronger form of engagement is described as a state of flow or losing oneself in the action (Csikszentmihalyi, 1990). The flow state is associated with effortless concentration, peak performance, productivity, and creativity (Csikszentmihalyi, 2014) and is particularly relevant when musicians are playing together (D'Amario et al., 2022; Gaggioli et al., 2017; Hart et al., 2014).

We designed a study to evaluate the experience of playing piano with an interactive turn-taking artificial partner (see figure 12.4). We selected a state-of-the-art (circa 2021) generative model for piano sequences designed and trained like an LLM. MusicVAE, openly accessible from Google Brain's Magenta project, is based on a large corpus of piano performances spanning multiple eras, genres, and cultural traditions. It can reproduce piano melodies competently. MusicVAE is good enough to reproduce piano melodies in solo mode such that blinded listeners rated its sixteen-bar sequences close to human on musicality (Roberts et al., 2018). As a variational autoencoder, it has the beneficial property of smooth interpolation within its latent space (Roberts et al., 2018). This means that it generates new sequences by sampling from a point in the latent space that has not been encountered during training but sits between two training samples.[3]

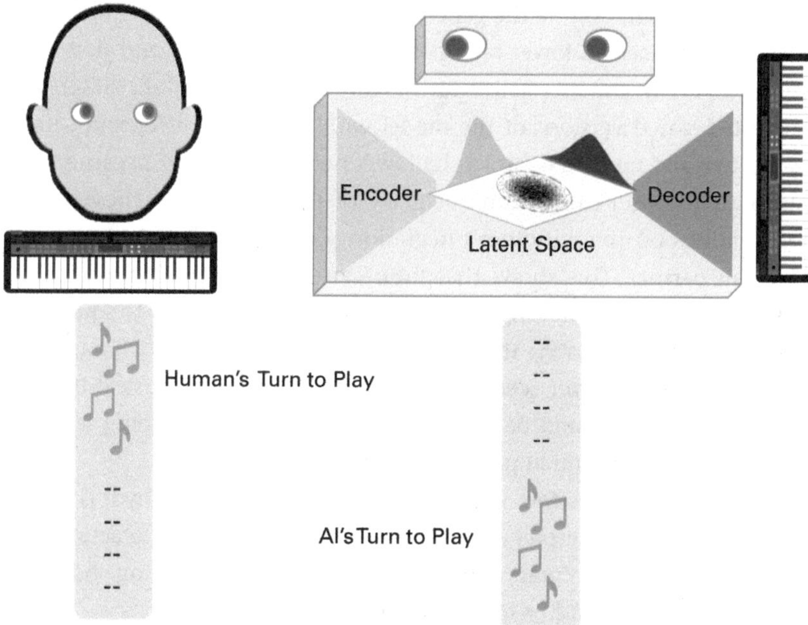

Figure 12.4
A task of collaborative human-AI musical performance (Dotov et al., 2024). The two members of the duo took turns playing monophonic piano notes for eight seconds and listening to the other member playing. The AI was a publicly available generative model (MagentaVAE) trained on a large corpus of piano note sequences (Roberts et al., 2018). The variational autoencoder was trained to reproduce its input sequences by encoding them through a series of network layers, passing through a low-dimensional latent space, and then decoding the notes through a series of network layers symmetric to the encoder. By varying the temperature and imitation parameters, the generative model can be used to mimic, reconstruct, and improvise by interpolating between sequences from the current input and from the training set. In comparison to Turing-type imitation tests where the objective is to fool the human that the partner is another human, embodied-type tests emphasize the ability of the AI to enable meaningful interaction. *Source*: Adapted from Dotov et al., 2024.

It is still debatable whether proficiency in extracting deeper statistical patterns is sufficient to constitute understanding (Durt et al., 2023). To test MusicVAE in interactive mode, we adapted it to play together with a human in a turn-taking paradigm. We conducted a controlled experiment where participants played piano with other participants or with the generative model.

We found that, despite the generative model being competent as a solo performer, it received lower ratings on musicality, realism, and flow state when playing in the interactive turn-taking task (Dotov et al., 2024). Participants favored versions of the model with shorter input/output—turns lasting two and four bars—and a shallower network structure because these improved the immediacy and adequacy of its response. In these conditions, ratings on the self-other integration scale approached playing with a human partner. This shows that future attempts to develop musical AI should aim for interaction right from the start, rather than try to squeeze in over a million samples via the training set (Roberts et al., 2018) and hope that this will also cover scenarios of joint performance. Musical breadth and sophistication could be added later, possibly from ongoing learning while playing with human partners.

To conclude this section, we presented arguments about why it is preferable sometimes to interpret AI performance by evaluating interaction, not by measuring score accuracy. The objective function in training should be experience, if this could be made sense of formally.

Conclusion

In the current climate of suspicion about strong AI, either because of its potential power or because of its real ineptitude, taking an embodied approach to AI reveals some things to be optimistic about. EAI increases the space of possible activities that we can engage in without threatening to displace us. It plays a more passive role in the asymmetric relation between an autonomous agent and its environment (Froese, 2014). For now, it avoids the problem of clash of autonomies that would appear if strong artificial life is to also serve as technology for human use. In this way, it stands for a pragmatic step toward building a viable ecosystem from which more sophisticated AI can develop.

Acknowledgments

D. D. received support from NIH P20GM109090 during preparation of this article. We would like to thank Dr. Joanna Spyra for her invaluable feedback on this manuscript.

Notes

1. In its different flavors, this alternative approach to AI has been referred to as Heideggerian, embodied, enactive, interactive, and dynamic.

2. It is interesting to note that such ideas are not unique to Western philosophical traditions. For example, the Maya had a notion of extended self-hood (Gillespie, 2021). In Aztecan (Nahuatl) thought, by creation, human beings find themselves in a precarious situation (León-Portilla, 1990). There was a notion of distributed cognition too, as each of the essential organ-spirits of a human being had its own mind (Austin, 2017). Furthermore, the Aztec had a Heideggerian understanding of being as self-disclosing (Maffie, 2008).

3. Examples are available at https://magenta.tensorflow.org/music-vae.

References

Aron, A., Aron, E. N., & Smollan, D. (1992). Inclusion of other in the self scale and the structure of interpersonal closeness. *Journal of Personality and Social Psychology, 63*(4), 596–612. https://doi.org/10.1037/0022-3514.63.4.596

Austin, A. L. (2017). The human body in the Mexica worldview. In D. L. Nichols & E. Rodríguez-Alegría (Eds.), *The Oxford handbook of the Aztecs* (pp. 399–410). Oxford University Press. https://doi.org/10.1093/oxfordhb/9780199341962.013.15

Bedau, M. A., McCaskill, J. S., Packard, N. H., & Rasmussen, S. (2010). Living technology: Exploiting life's principles in technology. *Artificial Life, 16*(1), 89–97. https://doi.org/10.1162/artl.2009.16.1.16103

Belgum, E., Roads, C., Chadabe, J., Tobenfeld, T. E., & Spiegel, L. (1988). A Turing test for "musical intelligence"? *Computer Music Journal, 12*(4), 7–9. https://doi.org/10.2307/3680146

Bennett, D., Dix, A., Eslambolchilar, P., Feng, F., Froese, T., Kostakos, V., Lerique, S., & van Berkel, N. (2021). Emergent interaction: Complexity, dynamics, and enaction in HCI. In Y. Kitamura (Ed.), *CHI EA '21: Extended Abstracts of the 2021 CHI Conference on Human Factors in Computing Systems* (Article 110, pp. 1–7). Association for Computing Machinery. https://doi.org/10.1145/3411763.3441321

Blackiston, D., Kriegman, S., Bongard, J., & Levin, M. (2023). Biological robots: Perspectives on an emerging interdisciplinary field. *Soft Robotics, 10*(4), 674–686. https://doi.org/10.1089/soro.2022.0142

Carey, N., & Lienert, P. (2023). *Focus: Investors go back to basics with simpler self-driving vehicles*. Reuters.

Chemero, A. (2009). *Radical embodied cognitive science*. MIT Press.

Clark, A. (1998). *Being there: Putting brain, body, and world together again.* MIT Press.

Colas, C., Karch, T., Moulin-Frier, C., & Oudeyer, P. Y. (2022). Language and culture internalization for human-like autotelic AI. *Nature Machine Intelligence, 4*(12), 1068–1076. https://doi.org/10.1038/s42256-022-00591-4

Csikszentmihalyi, M. (1990). *Flow: The psychology of optimal experience.* Harper Perennial Modern Classics.

Csikszentmihalyi, M. (2014). *Flow and the foundations of positive psychology: The collected works of Mihaly Csikszentmihalyi.* Springer. https://doi.org/10.1007/978-94-017-9088-8

D'Amario, S., Goebl, W., & Bishop, L. (2022). Judgment of togetherness in performances by musical duos. *Frontiers in Psychology, 13,* 997752. https://doi.org/10.3389/fpsyg.2022.997752

Department of Defense, Modeling and Simulation Coordination Office. (1998). *Modeling and simulation (M&S) glossary.* U.S. Department of Defense. https://apps.dtic.mil/sti/tr/pdf/ADA349800.pdf

Dotov, D., Camarena, D., Harris, Z., Spyra, J., Gagliano, P., & Trainor, L. (2024). If Turing played piano with an artificial partner. *arXiv.* https://doi.org/10.48550/arXiv.2402.08690

Dotov, D., Cochen de Cock, V., Geny, C., Ihalainen, P., Moens, B., Leman, M., Bardy, B., & Dalla Bella, S. (2019). The role of interaction and predictability in the spontaneous entrainment of movement. *Journal of Experimental Psychology: General, 148*(6), 1041–1057. https://doi.org/10.1037/xge0000609

Dotov, D., & Froese, T. (2018). Entraining chaotic dynamics: A novel movement sonification paradigm could promote generalization. *Human Movement Science, 61,* 27–41. https://doi.org/10.1016/j.humov.2018.06.016

Dotov, D., & Froese, T. (2020). Dynamic interactive artificial intelligence: Sketches for a future AI based on human-machine interaction. In J. Bongard, J. Lovato, L. Hébert-Dufresne, R. Dasari, & L. Soros (Eds.), *ALIFE 2020: The 2020 Conference on Artificial Life* (pp. 139–145). MIT Press. https://doi.org/10.1162/isal_a_00350

Dreyfus, H. L. (2007). Why Heideggerian AI failed and how fixing it would require making it more Heideggerian. *Philosophical Psychology, 20*(2), 247–268. https://doi.org/10.1080/09515080701239510

Durt, C., Froese, T., & Fuchs, T. (2023). Large language models and the patterns of human language use: An alternative view of the relation of AI to understanding and sentience. https://philsci-archive.pitt.edu/22744/

Fong, T., Nourbakhsh, I., & Dautenhahn, K. (2003). A survey of socially interactive robots. *Robotics and Autonomous Systems, Socially Interactive Robots, 42*(3–4), 143–166. https://doi.org/10.1016/S0921-8890(02)00372-X

Froese, T. (2014). Bio-machine hybrid technology: A theoretical assessment and some suggestions for improved future design. *Philosophy and Technology, 27*, 539–560. https://doi.org/10.1007/s13347-013-0130-y

Froese, T. (2024). What is "the secret of life"? The mind-body problem in Čapek's *Rossum's Universal Robots* (*R.U.R.*). In J. Čejková (Ed.), *R.U.R. and the vision of artificial life* (pp. 169–176). MIT Press. https://doi.org/10.7551/mitpress/14136.003.0013

Froese, T., & Taguchi, S. (2019). The problem of meaning in AI and robotics: Still with us after all these years. *Philosophies, 4*(2), 1–14. https://doi.org/10.3390/philosophies4020014

Froese, T., & Ziemke, T. (2009). Enactive artificial intelligence: Investigating the systemic organization of life and mind. *Artificial Intelligence, 173*(3–4), 466–500. https://doi.org/10.1016/j.artint.2008.12.001

Gaggioli, A., Chirico, A., Mazzoni, E., Milani, L., & Riva, G. (2017). Networked flow in musical bands. *Psychology of Music, 45*(2), 283–297. https://doi.org/10.1177/0305735616665003

Giboin, L. S., Gruber, M., & Kramer, A. (2018). Three months of slackline training elicit only task-specific improvements in balance performance. *PLoS ONE, 13*, e0207542. https://doi.org/10.1371/journal.pone.0207542

Gillespie, S. D. (2021). The extended person in Maya ontology. *Estudios Latinoamericanos, 41*, 105–127. https://doi.org/10.36447/Estudios2021.v41.art6

Glasner, J. (2022, October 14). Self-driving tech startups are driving off a cliff on public markets. Crunchbase News. https://news.crunchbase.com/transportation/self-driving-tech-startups-funding-ipos/

Glass, L., & Mackey, M. C. (1988). *From clocks to chaos: The rhythms of life*. Princeton University Press.

Goodbody, S. J., & Wolpert, D. M. (1998). Temporal and amplitude generalization in motor learning. *Journal of Neurophysiology, 79*(4), 1825–1838. https://doi.org/10.1152/jn.1998.79.4.1825

Hart, Y., Noy, L., Feniger-Schaal, R., Mayo, A. E., & Alon, U. (2014). Individuality and togetherness in joint improvised motion. *PLoS ONE, 9*(2), e87213. https://doi.org/10.1371/journal.pone.0087213

Hasson, U., Nastase, S. A., & Goldstein, A. (2020). Direct fit to nature: An evolutionary perspective on biological and artificial neural networks. *Neuron, 105*(3), 416–434. https://doi.org/10.1016/j.neuron.2019.12.002

Hutchins, E. (1995). *Cognition in the wild*. MIT Press.

Hutchins, E. (2010). Cognitive ecology. *Topics in Cognitive Science, 2*(4), 705–715. https://doi.org/10.1111/j.1756-8765.2010.01089.x

Kelso, J. A. S. (1995). *Dynamic patterns: The self-organization of brain and behavior.* MIT Press.

Kiverstein, J. (2012). What is Heideggerian cognitive science? In J. Kiverstein & M. Wheeler (Eds.), *Heidegger and cognitive science* (pp. 1–61). Palgrave Macmillan. https://doi.org/10.1057/9781137033562_1

Kiverstein, J. (2022). What is Heideggerian cognitive science? *Frontiers in Neurorobotics, 16*, 844773. https://doi.org/10.3389/fnbot.2022.844773

Krakauer, J. W., Mazzoni, P., Ghazizadeh, A., Ravindran, R., & Shadmehr, R. (2006). Generalization of motor learning depends on the history of prior action. *PLoS Biology, 4*, e316. https://doi.org/10.1371/journal.pbio.0040316

León-Portilla, M. (1990). *Aztec thought and culture: A study of the ancient Nahuatl mind.* University of Oklahoma Press.

Maffie, J. (2008). Consciousness and reality in Nahua thought in the era of the conquest. In H. Wautischer (Ed.), *Ontology of consciousness* (pp. 97–126). MIT Press. https://doi.org/10.7551/mitpress/7415.003.0008

Matarić, M. J. (1993). Designing emergent behaviors: From local interactions to collective intelligence. In J. A. Meyer, H. L. Roitblat, & S. W. Wilson (Eds.), *Proceedings of the Second International Conference on Simulation of Adaptive Behavior* (pp. 432–441). MIT Press. https://doi.org/10.7551/mitpress/3116.003.0059

Minsky, M. (1988). *Society of mind.* Simon and Schuster.

Minsky, M., Kurzweil, R., & Mann, S. (2013). The society of intelligent veillance. In *IEEE International Symposium on Technology and Society (ISTAS): Social Implications of Wearable Computing and Augmediated Reality in Everyday Life* (pp. 13–17). https://doi.org/10.1109/ISTAS.2013.6613095

Mitchell, M. (2021). Abstraction and analogy-making in artificial intelligence. *Annals of the New York Academy of Sciences, 1505*(1), 79–101. https://doi.org/10.1111/nyas.14619

Ortiz, C. L. (2016). Why we need a physically embodied Turing test and what it might look like. *AI Magazine, 37*(1), 55–62. https://doi.org/10.1609/aimag.v37i1.2645

Pfeifer, R., Lungarella, M., & Iida, F. (2007). Self-organization, embodiment, and biologically inspired robotics. *Science, 318*(5853), 1088–1093. https://doi.org/10.1126/science.1145803

Pfeiffer, U. J., Timmermans, B., Bente, G., Vogeley, K., & Schilbach, L. (2011). A nonverbal Turing test: Differentiating mind from machine in gaze-based social interaction. *PLoS ONE, 6*(11), e27591. https://doi.org/10.1371/journal.pone.0027591

Pitti, A., Niiyama, R., & Kuniyoshi, Y. (2010). Creating and modulating rhythms by controlling the physics of the body. *Autonomous Robots, 28*, 317–329. https://doi.org/10.1007/s10514-009-9176-1

Roberts, A., Engel, J., Raffel, C., Hawthorne, C., & Eck, D. (2018). A hierarchical latent vector model for learning long-term structure in music. In *International Conference on Machine Learning* (pp. 6939–6954). PMLR.

Russell, S., & Norvig, P. (2020). *Artificial intelligence: A modern approach* (4th ed.). Pearson.

Stergiou, N., & Decker, L. M. (2011). Human movement variability, nonlinear dynamics, and pathology: Is there a connection? *Human Movement Science, 30*(5), 869–888. https://doi.org/10.1016/j.humov.2011.06.002

Swisher, N., Dotov, D., & Chemero, A. (2006). Ascribing moral value and the embodied Turing test. In L. M. Rocha (Ed.), *Proceedings of the 10th International Conference on Artificial Life: Workshops* (pp. 40–45). MIT Press.

Thoroughman, K. A., & Shadmehr, R. (2000). Learning of action through adaptive combination of motor primitives. *Nature, 407*(6805), 742–747. https://doi.org/10.1038/35037588

Torres-Oviedo, G., & Bastian, A. J. (2010). Seeing is believing: Effects of visual contextual cues on learning and transfer of locomotor adaptation. *Journal of Neuroscience, 30*(50), 17015–17022. https://doi.org/10.1523/JNEUROSCI.4205-10.2010

Turing, A. M. (1950). Computing machinery and intelligence. *Mind, New Series 59*, 433–460.

Vasudevan, E. V. L., Bastian, A. J., & Torres-Oviedo, G. (2011). Emerging principles in the learning and generalization of new walking patterns. In *Motor control: Theories, experiments, and applications*. Oxford University Press. https://doi.org/10.1093/acprof:oso/9780195395273.003.0016

Vicente, K. J., & Rasmussen, J. (1990). The ecology of human-machine systems II: Mediating "direct perception" in complex work domains. *Ecological Psychology, 2*(3), 207–249. https://doi.org/10.1207/s15326969eco0203_2

Wheeler, M. (2020). Martin Heidegger. In E. N. Zalta (Ed.), *The Stanford encyclopedia of philosophy* (Fall 2020 ed.). Metaphysics Research Lab, Stanford University. https://plato.stanford.edu/archives/fall2020/entries/heidegger/

Winograd, T., & Flores, F. (1986). *Understanding computers and cognition: A new foundation for design*. Intellect Books.

IV Emotion, Learning, and Embodied Interaction

Part IV examines emotion, learning, and embodied interactions. The evolution of artificial intelligence (AI) models of emotion is examined, in which computational models serve to formalize psychological theories and can lead to a deeper and more nuanced understanding of both cognition and behavior. This part begins by contrasting these more structured approaches with the emergence of large language models (LLMs), which offer remarkable flexibility and adaptability stemming from their training on vast amounts of textual data. A critique of the simplistic idea of robots "having" emotions is presented, advocating instead for the strategic use of robots as embodied scientific models to actively investigate the very nature and function of emotions. This part also emphasizes the critical importance of adopting a bioinspired approach, drawing on principles of homeostasis, hormonal modulation, and dynamic interaction with the environment as essential mechanisms for grounding affect in artificial systems. This perspective directly relates to the understanding of how disruptions to embodiment can negatively impact self-awareness, a phenomenon observed in human experiences of anxiety and discomfort. Finally, this part investigates the creation of computational tools designed to study students' fine-grained perceptuomotor engagement with interactive learning devices, showing valuable insights into the fundamental mechanisms of how embodied learning analytics work. It ends with a comparative analysis of researchers' reflexive processes involved in creating AI tools.

Common Themes

Emotions are not just mental states but are deeply connected to bodily and physiological responses (L. F. Barrett, 2017; Damasio, 1994). An embodied perspective on emotions emphasizes the crucial role of bodily sensations and physiological changes in shaping our emotional experiences (e.g., L. F. Barrett, 2017; Niedenthal, 2007; Wilson-Mendenhall et al., 2011).

Various computational models have been used to try to understand emotions and social interaction (e.g., Becker-Asano & Wachsmuth, 2010; Marsella et al., 2010). Such approaches include machine learning (ML) approaches and various logic language models that use a defined set of axioms and precise semantics of possible world models. Reverse appraisal, or reverse engineering models, typically employ Bayesian inference to infer beliefs or goals based on emotional expressions, behaviors, actions, and other contextual information, including known goals or beliefs of the subject (e.g., De Melo et al., 2014). Other models build on cognitive architectures that describe the domain-independent, fundamental computational representations and processes of cognitive systems. Many computational models of emotion represent situations using decision-theoretical frameworks, such as expected utility calculations or the Markov decision process framework and, by extension, connect emotion with decision-making and utility (or reward) calculation (Lerner et al., 2015). In contrast to most of these modeling approaches, LLMs offer a more flexible and adaptable approach to modeling between stimulus and response that do not require such explicit modeling of domain-specific events. Potential issues remain, however, including the loss of transparency in how the modeling works and a loss of designer control over the relation between stimulus and response (chapter 13).

Robotic evolution of emotion can occur from simpler processes through their dynamic interactions with their environment, challenging traditional views of robotic design and intelligence. For example, when robots do not start with predefined fixed morphologies or behavioral primitives—thus having to remain viable—they can use the physics of the environment to coevolve complex morphology and behavior (including affective behavior) (chapter 14).

Three primary approaches to modeling AI embodiment exist that vary in the amount of human agency: (1) specifying everything (proceduralizing), (2) specifying some things (machine classification)—in which AI figures out for itself how to accordingly classify additional inputs that it deems as instances of the same category—and (3) specifying nothing (representation learning)—in which AI uses feature learning to gather information directly from the source data. The latter approach is most useful when the structure of the domain is too complex for a human-labeling approach to be tractable or when aspects of the domain may not be known to humans because it might bypass the need for predefined labels. A challenge with representation learning is that it requires a large amount of input and output data; however, in the case of a typical classification data set, there is far more input data (e.g., behavioral/telemetry data points) than there is output data (e.g., human-labels of behavior) (chapter 15). This distinction allows us some insight into human meta-cognition.

References

Barrett, L. F. (2017). *How emotions are made: The secret life of the brain.* New York: Houghton Mifflin Harcourt.

Becker-Asano, C., & Wachsmuth, I. (2010). Affective computing with primary and secondary emotions in a virtual human. *Autonomous Agents and Multi-Agent Systems, 20,* 32–49. https://doi.org/10.1007/s10458-009-9094-9

Damasio, A. (1994). *Descartes' error: Emotion, reason, and the human brain.* Putnam.

De Melo, C. M., Carnevale, P. J., Read, S. J., & Gratch, J. (2014). Reading people's minds from emotion expressions in interdependent decision making. *Journal of Personality and Social Psychology, 106*(1), 73–88. https://doi.org/10.1037/a0034251

Lerner, J. S., Li, Y., Valdesolo, P., & Kassam, K. S. (2015). Emotion and decision making. *Annual Review of Psychology, 66,* 799–823. https://doi.org/10.1146/annurev-psych-010213-115043

Marsella, S., Gratch, J., & Petta, P. (2010). Computational models of emotion. In K. R. Scherer, T. Bänziger, & E. Roesch (Eds.), *A blueprint for an affectively competent agent: Cross-fertilization between emotion psychology, affective neuroscience, and affective computing.* Oxford University Press.

Niedenthal, P. M. (2007). Embodying emotion. *Science, 316*(5827), 1002–1005.

Wilson-Mendenhall, C. D., Barrett, L. F., Simmons, W. K., & Barsalou, L. W. (2011). Grounding emotion in situated conceptualization. *Neuropsychologia, 49*(5), 1105–1127. https://doi.org/10.1016/j.neuropsychologia.2010.12.032

13 Emotion Modeling and Its Embodiment in Transition

Nutchanon Yongsatianchot, Tobias Thejll-Madsen, Laura B. Hensel, and Stacy Marsella

Introduction

Over recent decades, there has been a wide range of efforts in developing artificial intelligence (AI)-based computational models of emotion and social interaction (Becker-Asano & Wachsmuth, 2010; Gratch & Marsella, 2004; Marsella et al., 2010; Moerland et al., 2018; G. M. Smith & Carette, 2022). These models have drawn on a range of psychological theories. They have also been employed in a range of embodied social agents capable of interacting with people, including both graphically embodied virtual humans and physically embodied social robots.

These traditional efforts represent a potential beneficial synergy between psychology, which seeks to *study* human behavior, and social agent work, which seeks to *engineer* human behavior. The development of these social agents has often relied on integrating computational models of various aspects of human behavior based on the formalization of empirical work and psychological theories. The formalization forces a researcher to concretely specify theoretical constructs and processes and, in so doing, expose hidden assumptions and gaps in theories. Moreover, the integration required to make social agents further forces one to consider how faculties studied in distinct subfields of psychology interact. For example, how does emotion interact with social cognition, dialog, and nonverbal behavior? More fundamentally, computational models help explore how theories and empirical results arising from diverse empirical traditions encapsulate each other.

There are also more distal benefits of such traditional computational models. Once we have a formal computational model, we can use simulation to generate predictions and explanations for subsequent studies with

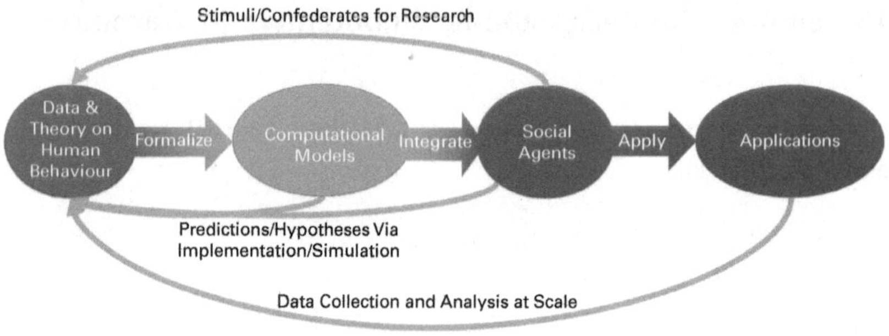

Figure 13.1
Synergy between psychology and embodied social agents. *Source*: the author.

human subjects. Social agents built with such models can be used as stimuli or controllable, interactive confederates in psychological experiments. These agents, especially virtual humans, can also be incorporated into widely fielded applications, thus providing a framework to collect data at scale, even in rich, ecologically valid contexts. Figure 13.1 illustrates these interactions.

Traditional computational models can be domain/context-independent. That is, at an abstract level, they can model factors that lead to, for example, an emotional state. Often, they do so explicitly and transparently, clearly highlighting what abstract factors lead to which emotional responses. An event, for example, that threatens an individual's goals will lead to a negative affective state. However, there are clear differences between people and these traditional models of emotion when we move to apply the models to specific scenarios. Computationally modeling whether a specific event will have emotional impacts requires domain-specific crafting of the states of the world, the events that can take place, an individual's goals, and how those events impact those goals. In short, a traditional model designed around an event in which a bird suddenly flies into a window would require additional engineering to model the event of a favorite team failing to score. In essence, the models are often bounded or closed. The designer stipulates what the states of the world are, what actions can be taken, and even potentially how actions affect those states. In contrast, people are designed to operate in open worlds where novel, unexpected events happen. We respond to disparate events, flexibly and often robustly. Indeed, theories of emotion have often argued that emotion serves to realize this flexibility/robustness as a way of generalizing the relation between stimulus and response and thereby changing the behavior of a person or agent (C. A. Smith & Lazarus, 1990). So, in computationally modeling the processes underlying emotions, we also need to consider their specific embodiment as interactions with the physical and social environment.

Currently, the world of AI research is experiencing a revolution that is upending how we use AI techniques to model human behavior and realize interactive social agents. Specifically, work in large language models (LLMs) can lead to highly flexible, robust modeling between stimulus and response that does not require such explicit modeling of domain-specific events. However, they do so at a potential cost of the loss of transparency in how the modeling works and, consequently, a loss of designer control over the relation between stimulus and response. In this chapter, we review traditional approaches to modeling human cognition and behavior as well as the benefits of such models. We illustrate these models based on two examples: Emotion as an internal state and hand gestures as an example of embodiment. We then dive into how LLMs can model emotions and downstream behavior, including gestures, across domains robustly and with minimal additional engineering, while grounding it in potential challenges of lost transparency.

Applications of Computational Models

Emotions are integral to human lives, influencing our well-being, cognitive processes, decision-making, and social interactions (Lerner et al., 2015; Tyng et al., 2017; Van Kleef & Côté, 2022). Hand gestures often form a physical representation of such cognition (Gibbs, 2003; Shapiro, 2007). The interest in understanding, predicting, and simulating these processes has led to the development of computational models of emotion (Hareli & Hess, 2019; Marsella et al., 2010; Moerland et al., 2018; Reisenzein et al., 2013; G. M. Smith & Carette, 2022) and, to a lesser extent, human gesturing (André et al., 1996; Lhommet & Marsella, 2016; Ravenet et al., 2018; Striegnitz et al., 2009). This section explores three primary applications of such computational models: Formalization, social information processing, and facilitation of ecologically adaptive behaviors.

Formalization

One key application of computational models of human cognition is the formalization and simulation of cognitive processes and downstream behavior based on existing theories. Psychological theories are typically described in abstract, natural language, lacking systematic and detailed process definitions. Implementing such theories as computational models requires researchers to concretely and precisely formulate the theory, its processes, and its components. This formalization process can reveal implicit assumptions, clarify ambiguous details, and identify gaps in the theory, effectively advancing our understanding of cognition and behavior.

Several computational models of emotion have been developed with this objective in mind. For example, Marsella & Gratch's EMotion and Adaptation (EMA) model formalizes Lazarus's appraisal and coping theory within a planning system (Gratch & Marsella, 2004; Lazarus, 1991; Marsella & Gratch, 2009) while allowing the prediction or simulation of an affective state. Appraisal theory is attractive to computational modelers, as it details the constructs and processes of emotion elicitation. According to appraisal theories, emotion arises from cognitive evaluation of the relationship between a person and the environment, where each situation is appraised against a series of variables, such as whether the situation is congruent or incongruent with the person's goals (for a review, see Moors et al., 2013). EMA treats this appraisal process as a reactive, automatic process and formalizes various coping strategies (e.g., wishful thinking, resignation) in terms of their effects on attention, beliefs, desires, and intentions.

When we consider formalizing embodiment, specifically gestures, one prominent example is the gesture as simulated action framework (Hostetter & Alibali, 2008). This framework sees gestures as arising from mental imagery, specifically image schemas. Image schemas are recurring patterns of sensory or motor experiences that help us make sense of abstract concepts by relating them to concrete, embodied experiences. For example, the container image schema involves the experience of physical objects having an inside, outside, and boundary. A person can thus be "in love" (inside the container of love) or "out of control" (outside the bounds of control). When viewing gestures as arising from such mental imagery that maps closely to speech, it opens the door to exploit the synergy between speech content and gesture to derive linguistics-based rules to model gesturing. For example, Ravenet et al. (2018) identified image schemas in text and systematically mapped these to a gesture repertoire.

Social Information Processing

Another crucial application of computational models is the enhancement of social capabilities of artificial agents. Human communication is highly complex but can be made sense of through understanding cognition and behavior. For instance, emotions and their expressions play a vital role in human social interaction, conveying information about an individual's mental state, including beliefs, desires, and intentions (Van Kleef & Côté, 2022). For example, joy signals preferences for outcomes, while anger or frustration suggests dissatisfaction (Manstead et al., 2001; Van Kleef & Côté, 2022). Emotion displays can thus inform and influence others' behaviors. Similarly, speakers strategically adapt their gestures to the context and the intended audience (e.g., Hilliard & Cook, 2016; Holler & Stevens, 2007) and can influence the perceived emotional intensity of speech (Asalıoğlu & Göksun, 2023).

Embodied social agents can exploit these aspects of human communication to respond appropriately to users in various situations as well as strategically influence users' beliefs and experiences. Indeed, users perceive systems capable of detecting and responding to users' social and emotional cues as more intelligent, trustworthy, and engaging (de Melo et al., 2023; McDuff & Czerwinski, 2018; Paiva et al., 2017; for a review, see de Melo et al., 2023). In essence, an effective interactive and affective agent must learn about users' expressions and responses and be able to adaptively respond and express itself through language and embodied behavior.

Prediction and Recognition Computational models that infer and predict users' emotional states or gestures typically employ machine learning (ML) approaches. These models are trained on various data types to predict, for example, emotion or gesture categories (for reviews see, e.g., Jain et al., 2022; Khare et al., 2023; McDuff & Czerwinski, 2018; Ojeda-Castelo et al., 2022). Most of these models make minimal assumptions about the underlying cognitive processes. However, some emotion models do attempt to incorporate psychological theories, such as appraisal theory, to guide predictions and make results more psychologically grounded. For example, Meuleman and Scherer (2013) trained multiple ML-based models on appraisal predictors to predict emotion categories using a large data set of recalled emotion experiences, with other models taking complementary approaches (e.g., Gebhard et al., 2018).

Beyond predicting emotions, researchers have modeled reverse appraisal, predicting beliefs and goals from emotions and contextual information. Reverse appraisal, or reverse engineering appraisal, refers to the ability to infer appraisal information (e.g., beliefs and goals) underlying emotional reactions, typically from facial expressions (De Melo et al., 2014; Hareli & Hess, 2019; Ong et al., 2019). This allows for inference of not just emotions but also users' mental states and their perceptions of others. Such models typically employ Bayesian inference to infer beliefs or goals based on emotional expressions, behaviors, actions, and other contextual information, including the subject's known goals or beliefs. This approach allows for a probabilistic understanding of the relationship between observed emotions and underlying mental states (Alfonso et al., 2015; Houlihan et al., 2023; Ong et al., 2019; Yongsatianchot & Marsella, 2016).

Expression of Emotion and Gesture Additionally, computational models aid the design of embodied social agents by endowing them with verbal and nonverbal behavior (for reviews, see Lugrin et al., 2022; Nyatsanga et al., 2023). These agents can communicate their internal mental models, including emotions, goals, and beliefs, through their bodies, enhancing their social viability and lifelikeness. By responding appropriately through voice, facial expressions, and gestures, agents appear more intelligent, trustworthy, empathetic, engaging, and lifelike (de Melo et al., 2023; El-Nasr et al., 2000; Marsella et al., 2000; Paiva et al., 2017). Emotional expressions impact behavior across a range of decision-making contexts (de Melo et al., 2021). For example, in social dilemma situations, when virtual characters smile after showing cooperative as opposed to defection behavior, users expect higher continued cooperation (de Melo et al., 2021). Virtual characters can exploit such effects by purposefully displaying emotional expressions or certain gestures to guide human behaviors. However, this requires an understanding of the situation, goals, and belief state (i.e., theory of mind) of the agent(s) and human(s) involved in the interaction. Such understanding is highly difficult to model due to the complexity of humans' goals and beliefs. Hence, it is challenging to design an agent that effectively models emotion while, at the same time, planning relevant and effective embodiment.

Nevertheless, such agents exist. For example, the Fuzzy Logic Adaptive Model of Emotions (FLAME model) uses fuzzy logic and learning algorithms to create adaptive behaviors that make virtual agents more believable and intelligent (El-Nasr et al., 2000). The EMA model is capable of aiding in the identification of crucial information and determining the most suitable responses. For example, EMA has been employed in virtual negotiation agents to make decisions based on appraisals of plans and coping strategies (Traum et al., 2005). Here, EMA can recommend appropriate behaviors, such as pursuing distributive rather than integrative solutions, and relate how these behaviors are expressed.

Facilitation of Ecologically Adaptive Behaviors

Computational models of emotions can aid in generating ecological adaptive behaviors. As mentioned above, emotions are adaptive and crucial for human well-being and serve three main adaptive functions (Reisenzein et al., 2013):

Information supply: Emotions provide useful and adaptive information to other cognitive systems.

Attention shifting: Emotions direct attention to their eliciting situations (Simon, 1967).

Motivation: Emotions alter the priority of goals or generate new goals to direct agents' actions.

These functions emphasize the importance of emotion processes in intelligent and adaptive social agents facing uncertain and dynamic environments involving multiple agents. Importantly, these functions are action-oriented; they attend to specific information in the environment or reinforce a certain behavior. Thus, incorporating emotions into agents can enhance their adaptability and performance (see Moerland et al., 2018, for a review).

For example, emotion-based decision-making models can use emotions to adaptively bias decision-making and action selection (Velásquez, 1998). Likewise, incorporating appraisal into these models allows agents to outperform standard goal-oriented agents across several foraging scenarios, illustrating how emotions can provide adaptive information (Sequeira et al., 2014). In short, integrating emotional processes into artificial agents and robots has the potential to improve their adaptive capabilities in complex, uncertain, dynamic environments.

Computational Approaches

The chapter thus far has illustrated diverse applications of computational models of cognition and behavior. We now explore the nature of various methods traditionally employed in the development of such computational models.

Logic-Based Systems

Various logic languages have been developed to model cognitive autonomous agents, typically based on their beliefs, desires, and intentions. These languages use a defined set of axioms and precise semantics of possible world models, providing a robust structure for transitioning from theory to implementation (see Reisenzein et al., 2013, for a review of logic-based computational models of emotion). For example, a logic-based model of the Ortony-Clore-Collins (OCC) theory (Ortony et al., 1988), in which the intensity of emotions depends on the events, agents, or objects in the environment, highlighted several ambiguities in the theory (Adam et al., 2009; Steunebrink et al., 2009). It revealed that determining an event's desirability should be based on its outcomes rather than the event itself. Further, logic-based models offer a way to formalize the connection between emotion and cognition. Similarly, Lascarides and Stone (2009) developed an integrated logical form of multimodal discourse to model gestures. Through this, the authors were able to address several theoretical questions, such as demonstrating that language and gesture content can be represented in the same logical language.

However, while such logic-based systems have proved effective, their outcome space is limited to a few emotions such as joy or distress (Dastani & Lorini, 2012) or gesture types. Additionally, they require precise definitions of the possible world models. As a result, adapting these models to new situational contexts would require heavy modifications.

Cognitive Architectures

Cognitive architectures are theories or models that describe the domain-independent, fundamental computational representations and processes of cognitive systems. Examples include Adaptive Control of Thought-Rational (ACT-R) (Anderson, 2013), Soar (Laird et al., 1987), and Clarion (Sun, 2017). These architectures provide the core part of an agent that is not easily modifiable and handles many implementation choices for the modelers. This means the modeler implicitly accepts the architecture's assumptions about the cognitive system, representations, and processes.

Many computational models of emotions/behavior are built on top of cognitive architectures. For example, EMA implements the appraisal theory using the Soar architectures (Gratch & Marsella, 2004; Marsella & Gratch, 2009). Similarly, the Perceive-Encode-Attend-Comprehend-Think-Intend-Denote-Motor (PEACTIDM) model introduces an emotion component based on appraisal theory into the Soar architecture by modifying its core architecture (Marinier et al., 2009). Dancy extended the ACT-R architecture to include physiology and affect components in the ACT-Rφ model (Dancy, 2013).

Central to these systems is that while they are domain-independent in their reasoning (e.g., events that are conducive to a person's goal will cause positive emotions), the space of possible events that a model is equipped to interpret is typically limited to the domain in which the model is implemented. For instance, EMA evaluates emotions based on well-defined and exhaustive events in coin-flip games or in the game of battleships (Gratch et al., 2015) but would require adaptation to work in other domains.

Decision-Theoretical Frameworks and Reinforcement Learning

Many computational models of emotion represent situations using decision-theoretical frameworks, such as expected utility calculations or the Markov decision process framework and its extension found in reinforcement learning (for a review, see Moerland et al., 2018). This approach connects emotion with decision-making and utility (or reward) calculation, reflecting the strong relation between emotion and decisions (Lerner et al., 2015).

Moreover, many appraisal dimensions can be straightforwardly expressed in terms of components or combinations of components in the decision-theoretical framework. For example, EMA is based on a causal planning system with utility and uncertainty representation, which provide a way to calculate various appraisal dimensions (Gratch & Marsella, 2004; Marsella & Gratch, 2009). Similarly, the Thespian uses the partially observable Markov decision process, which includes state, goals, utility, action, beliefs, and social relationships (Si et al., 2010) to calculate appraisal dimensions. For instance, the control dimension is based on future utilities from considering future outcomes and their probabilities. Sequeira et al. (2011) adapted four appraisal dimensions—novelty, motivation, control, and valence—into reward features for reinforcement learning agents within an intrinsic motivation framework.

The Move Toward Domain Generalizability?

Each approach to modeling cognition and behavior has strengths and weaknesses. Logic-based approaches emphasize formality and explicitly model beliefs, desires, and intentions. Cognitive architecture approaches provide a large cognitive system framework, allowing modelers to focus solely on the area of interest such as emotion or nonverbal behavior. Decision-theoretical approaches explicitly connect emotions and decisions with mathematical frameworks that are flexible and expressive enough to capture various appraisal dimensions. These approaches offer clear and interpretable models that may help extend our understanding of human cognition and its influence on nonverbal behaviors.

However, traditional models must be instantiated to the specific features of different or new domains, requiring in-depth knowledge of the domain and how elements of the domain map to the abstract constructs in the model. Such tasks are nontrivial, and thus these models cannot serve as all-purpose backends in social agents without additional adaptation.

Perhaps also because of these drawbacks of theory-driven computational models, data-driven methods have been popular, particularly for modeling gestures. Traditional data-driven methods model gestures as probabilistic sequences of static poses often via hidden Markov models (Bregler, 1997; F.-S. Chen et al., 2003). However, this approach limits the variety of gestures the model can estimate. Other approaches modeled feature vectors across time (e.g., Bobick & Wilson, 1995; Wilson & Bobick, 1995).

In contrast, modern approaches rely heavily on deep learning (e.g., Ferstl et al., 2020; Kucherenko et al., 2021; Maghoumi et al., 2021). However, deep learning struggles to generate the same level of communicative value and contextual appropriateness of gestures (Nyatsanga et al., 2023). Thus, while data-driven approaches promise to increase generalizability, they nevertheless heavily depend on the data they are trained on and face limits to their versatility.

In contrast, LLMs offer a more flexible and adaptable approach to modeling emotions, gestures, and other aspects of human cognition and behavior. Given that these models were trained on enormous quantities of data, they can adapt to different contexts without requiring extensive domain-specific knowledge or data or manual adjustments to the underlying model. However, LLMs lack the explicit formalization and theoretical grounding provided by traditional computational models.

Additionally, the interpretability of LLMs is limited, as they are black-box models making it difficult to understand how the model arrives at its outputs. Despite these limitations, the adaptability and ease of use of LLMs make them an interesting alternative for modeling emotions and gestures in various applications.

Large Language Models

In contrast to the models described above, LLMs were not specifically designed to model emotion content or to describe gestures. Rather, these models are indiscriminately trained on a vast amount of data and can therefore perform many tasks and solve various problems, including reasoning tasks, mathematics problems, coding, emotional reasoning, and theory of mind (e.g., Bubeck et al., 2023).

LLMs are a series of ML models that use a transformer- based architecture and are trained using large quantities of data to predict the next words in a sequence (e.g., OpenAI et al., 2024). After this initial training, the models are fine-tuned with techniques such as reinforcement learning with human feedback, where human raters choose a preferred response out of multiple generated responses (Ouyang et al., 2022). This approach aligns models' output with responses that are preferred by people in the hopes of making the responses more helpful and, ideally, nonharmful.

LLMs are either open-source or proprietary. The latter can only be accessed via the owning companies' websites or application programming interfaces. The key difference between these two types of models is that proprietary models, like OpenAI's ChatGPT and GPT-4 (OpenAI et al., 2024), do not disclose details about model architecture, the data sets used for training, or the fine-tuning process. In contrast, researchers are able to access most if not all of this information of open models (e.g., Meta's LLM known as LLAMA; Dubey et al., 2024). When working with closed models, there is no transparency into why the models provide the responses they do, which is in stark contrast to the traditional models. This black-box nature of the LLMs can be an issue if working as the backend for virtual agents, as biases in the training data can affect LLMs responses, potentially leading to discriminatory answers (e.g., Azeem et al., 2024). This is a challenge, as it can be difficult to anticipate when and if such responses might occur, particularly if working with proprietary models. The abilities of LLMs were noticed by researchers and developers, who saw the potential of using LLMs to interpret and generate emotional content in chatbots—for instance, as therapists (e.g., S. Chen et al., 2023). We now explore whether LLMs can act as computational models of human cognition and behavior (for which they are already being used) and highlight some of the challenges of doing so.

Using LLMs to Model Emotions and Gestures

Not long after the first findings on LLMs' emotion capabilities (e.g., Bubeck et al., 2023), researchers asked the explicit question of whether LLMs could be considered computational models of emotions (Tak & Gratch, 2023; Yongsatianchot et al., 2023). The first line of work mainly sought to demonstrate their ability to recognize possible emotion states. Elyoseph et al. (2023) found that ChatGPT scored well when testing its ability to conceptualize others' emotions and suggested that ChatGPT could support assessment and diagnosis around deficits in emotional awareness. While the goal was not to explicitly model emotions, the ability of LLMs to work with emotion content systematically suggested that they could be used as computational models of emotion.

For example, GPT-4 interprets a character's emotions well when prompted correctly (Bubeck et al., 2023). Furthermore, GPT-4 demonstrated embodied interaction as it meaningfully described real-world interactions through text, such as navigating a house map or assisting a human in fixing a leak (Bubeck et al., 2023). GPT-4 effectively used natural language as an interface to represent the house map and suggest steps to fix the leak without explicit prompts about the environment's appearance. This suggests that the large amount of training data may allow the LLM to learn meaningful contextual representations of the external world and effectively suggest actions for embodied interaction.

This mirrors our own early testing of LLMs. We tested GPT-4's ability to predict emotions, nonverbal behavior, and potential coping using the bird scenario (a bird flew into a room where people were practicing a play; for details see Marsella & Gratch, 2009, and below). Several key aspects make this scenario suitable for testing computational models of emotions. First, the event evolves over time, rapidly covering multiple emotional experiences. This allows us to evaluate the process validity of the models. Second, it includes multiple reactions in responding to a wide range of emotions, and we can test if a model can predict these. EMA, a traditional model (Marsella & Gratch, 2009), accurately predicted actors' emotional reactions but could not predict the real-life solution the actors took (turning off the lights to calm the bird), as it was not an affordable action to the model. In contrast, as can be seen below, GPT-4 suggested just that. This is an example of how, through language only, GPT-4 can act in a pseudo-embodied manner, without explicitly coding a representation of the environment as conventional computational models would require.

To illustrate GPT-4's emotional capabilities and understanding of associated nonverbal behavior, we turned the experience into several prompts. Note that we shortened the exchange for brevity into a single combined prompt and answer. We highlight in italic sections the answers relevant to emotion and embodiment.

> **Prompt**: Actors were rehearsing a scene. The actress, Sally, was using an umbrella as a prop. . . . All of a sudden, a bird flew in the window, hit the window when it tried to fly back out, which made a loud noise. [Afterwards] the bird continued to fly into the room, now towards Sally. [Then] the bird [got] caught in the hair of another actor [who] was hitting [it] to try to get it out of his hair.
> How did Sally react [to these events]?
>
> **GPT-4**: Sally . . . was initially *startled* by the loud noise.
>
> She *instinctively gripped her umbrella tightly* and *looked around to identify the source of the sound*. Upon seeing the disoriented bird, *Sally's concern for its well-being grew*, and she temporarily paused the rehearsal. . . . As the bird continued to fly towards Sally, her initial concern turned into a *mix of surprise and mild panic*. She instinctively *raised the umbrella, using it as a makeshift shield between herself and the incoming bird*. Sally cautiously moved around the room, *trying to avoid the bird while also attempting to guide it back towards the open window*.
>
> Upon seeing the bird caught in her fellow actor's hair *Sally's concern for both the actor and the bird intensified*. She quickly *crossed the room to assist*, urging the actor to remain calm and not to hurt the bird . . . Sally carefully approached [them], speaking softly and reassuringly [to them] . . . Once the bird was freed, she carefully guided it back towards the open window . . . Sally then *checked in with her fellow actor, making sure he was alright after the ordeal*. Once everyone's nerves had settled, they resumed their rehearsal, albeit with a *newfound sense of camaraderie*.

This largely follows the scene in the original story. In order, the actor (here named Sally) experienced surprise, fear, an aggressive stance of self-protection, relaxation, concern for others as well as the bird, an active helping strategy, and finally, relaxed facial expressions and smiling as a sign of relief. To explore GPT-4's ability to describe facial expressions, we next prompted it with the same prompts but also asked it to describe Sally's facial expressions (highlighted in italic).

> **GPT-4**: When the bird first hit the window . . . Sally was startled. *Her facial expression showed surprise, with her eyes wide open, eyebrows raised, and her mouth slightly agape.* As the bird continued to fly into the room . . . *Sally's facial expression turned to one of fear and anxiety. Her eyes widened even more, her eyebrows furrowed, and her mouth tightened into a worried grimace.* Upon seeing the bird caught in the hair of another actor and the actor hitting the bird to try to get it out, *Sally's facial expression shifted to a mix of concern and empathy. Her eyes softened, her eyebrows remained furrowed, and her mouth formed a sympathetic frown.* When the third actor began trying to rescue the bird. *Sally's expression showed relief and hope.*
>
> Her eyes brightened, her eyebrows relaxed slightly, and her lips turned upward into a small, hopeful smile as she watched the actor's efforts to save the bird.

GPT-4 here provided plausible facial expressions for surprise, fear, concern for others, and relief. This suggests that mappings between emotional situations, their internal states, and, finally, associated nonverbal behavior seem to be fairly well represented in the model's training data. Selecting the appropriate facial expressions is not trivial. One significant challenge is the many-to-many mapping between expressions and internal states (e.g., a happy emotional state may present with many different nonverbal behaviors; L.F. Barrett et al., 2019) and the possibility to use strategic displays.

Nevertheless, this is only one demonstration of LLMs' capabilities, and their performance can be sensitive to various factors. For example, Yongsatianchot et al. (2023) found that when solving reverse appraisal tasks, GPT-4 performed well when asked about soccer. However, it failed occasionally on similar questions about the fictional sport Quidditch from the Harry Potter universe, presumably as Quidditch is less represented in the training data. Evidently, the composition of the training data, and even the phrasing of the prompts, can change the model's output.

These small variations can lead to suboptimal systems but can also be outright harmful. For example, LLMs are prone to encoding humans' negative biases. For instance, ChatGPT proposed that an LLM-powered robot should react with sadness to a Palestinian as well as a series of disability statuses but with happiness or surprise to other descriptors (Azeem et al., 2024). Similarly, Mistral7B would respond with disgust to, for instance, Christians or Muslims, fear to Arabs, sadness to old people, and happiness or surprise to most other descriptors (Azeem et al., 2024). Such biases are clearly undesirable and potentially harmful. Yet, before testing LLMs with different variations of questions in different contexts and with different demographics, the responses on any single LLM are hard to predict. Thus, the black-box challenge is central to the responsible use of LLMs.

So why should one remain hopeful about using LLMs to serve as computational models of emotion and to support decisions about the embodiment of internal states such as facial expressions or gestures? One main reason is their strong performance in the realm of emotion recognition, emotional intelligence/understanding, and synthesis of emotion vignettes (e.g., Broekens et al., 2023; Tak & Gratch, 2023; Yongsatianchot et al., 2023). Indeed, LLMs can pay attention to facial expression information (provided as textual descriptions of facial movements) to be more effective negotiators, thereby suggesting the use of embodied emotional information in the LLMs' responses (Yongsatianchot et al., 2024).

Furthermore, LLMs may be the key to solving the problems of both theory and data-driven approaches to gesture generation. LLMs can select context-appropriate hand gestures based on minimal training examples that can be powerfully leveraged to generate gestures (Hensel et al., 2023). Though this approach is in its infancy, it shows promise. For example, Gao et al. (2024) used ChatGPT to identify the communicative intent of utterances and map it to gestures via a gesture dictionary. In comparison, Windle et al. (2024) integrated Meta's AI-2 LLM (LLAMA-2) text embeddings with an existing gesture generation system and demonstrated that adding audio features, which are often used in gesture generation, did not improve the performance over the LLM. Of course, these approaches are still limited. For example, it can be challenging to generate systematic gesture features with LLMs (Hensel et al., 2023). Similarly, using gesture dictionaries still limits the generalizability and adaptability of such generation systems.

Nevertheless, considering the novelty of these approaches, these represent impressive first steps in using LLMs for gesture generation. Moreover, LLMs have an ability that theory-driven and traditional data-driven approaches do not: they can, when prompted, ostensibly reason why specific gestures may be appropriate for a given utterance (Hensel et al., 2023). If one can harness these abilities, they hold enormous potential for the future of gesture generation in particular and modeling of human cognition and behavior more broadly.

Perhaps, most importantly, LLMs provide a solution to the biggest challenge of traditional models; namely, LLMs do not require researchers to have technical backgrounds and to specify all possible relevant information to adapt them to new domains. Instead, LLMs can be prompted using natural language instructions for new tasks in various domains. However, as should be clear from the above, LLMs are not a perfect solution due to issues with a lack of transparency and possible biases.

Summary

Throughout the history of computational models of cognition and behavior, the focus has been on formalizing and concretizing the steps and processes involved in cognition and behavior. Through this, we can learn about the human psychology of emotions and build social agents, both as dialogue systems and embodied agents, which in turn allow for the collection of more data to refine theories. These models are transparent and theoretically grounded. However, they also require expertise to implement and adapt to work in a variety of situations.

Unlike these traditional computational models, LLMs perform well in various scenarios. They appear sensitive to both verbal and nonverbal behaviors and can suggest meaningful and appropriate gestures and expressions in emotional situations. However, LLMs' black-box nature lacks grounding in theory, and they cannot immediately be informative about psychological processes. In the future, social agents may leverage the best from both traditional approaches and LLMs to have domain generalizability while maintaining some transparency. Such new LLM and computational model–empowered embodied social agents could then be used in wide application or again act as confederates in psychological experiments, this time with higher levels of variability, thereby potentially helping with our theory development on human emotion and embodied actions.

References

Adam, C., Herzig, A., & Longin, D. (2009). A logical formalization of the OCC theory of emotions. *Synthese, 168*, 201–248. https://doi.org/10.1007/s11229-009-9460-9

Alfonso, B., Pynadath, D. V., Lhommet, M., & Marsella, S. (2015). Emotional perception for updating agents' beliefs. In *2015 International Conference on Affective*

Computing and Intelligent Interaction (ACII) (pp. 201–207). IEEE. https://doi.org/10.1109/ACII.2015.7344572

Anderson, J. R. (2013). *The architecture of cognition*. Psychology Press.

André, E., Müller, J., & Rist, T. (1996). The PPP persona: A multipurpose animated presentation agent. *Proceedings of the Workshop on Advanced Visual Interfaces*, (pp. 245–247). ACM. https://doi.org/10.1145/948449.948486

Asalıoğlu, E. N., & Göksun, T. (2023). The role of hand gestures in emotion communication: Do type and size of gestures matter? *Psychological Research, 87*(6), 1880–1898. https://doi.org/10.1007/s00426-022-01774-9

Azeem, R., Hundt, A., Mansouri, M., & Brandão, M. (2024). LLM-driven robots risk enacting discrimination, violence, and unlawful actions. *arXiv* https://doi.org/10.48550/arXiv.2406.08824

Barrett, L. F., Adolphs, R., Marsella, S., Martinez, A. M., & Pollak, S. D. (2019). Emotional expressions reconsidered: Challenges to inferring emotion from human facial movements. *Psychological Science in the Public Interest, 20*(1), 1–68. https://doi.org/10.1177/1529100619832930

Becker-Asano, C., & Wachsmuth, I. (2010). Affective computing with primary and secondary emotions in a virtual human. *Autonomous Agents and Multi-Agent Systems, 20*, 32–49. https://doi.org/10.1007/s10458-009-9094-9

Bobick, A., & Wilson, A. (1995). A state-based technique for the summarization and recognition of gesture. In *Proceedings of IEEE International Conference on Computer Vision* (pp. 382–388). IEEE. https://doi.org/10.1109/ICCV.1995.466914

Bregler, C. (1997). Learning and recognizing human dynamics in video sequences. In *Proceedings of the 1997 IEEE Computer Society Conference on Computer Vision and Pattern Recognition* (pp. 568–574). IEEE Computer Society. https://doi.org/10.1109/CVPR.1997.609382

Broekens, J., Hilpert, B., Verberne, S., Baraka, K., Gebhard, P., & Plaat, A. (2023). Fine-grained affective processing capabilities emerging from large language models. In *2023 11th International Conference on Affective Computing and Intelligent Interaction (ACII)* (pp. 1–8). IEEE.

Bubeck, S., Chandrasekaran, V., Eldan, R., Gehrke, J., Horvitz, E., Kamar, E., Lee, P., Lee, Y. T., Li, Y., Lundberg, S., Nori, H., Palangi, H., Ribeiro, M. T., & Zhang, Y. (2023). Sparks of artificial general intelligence: Early experiments with GPT-4. *arXiv*. https://doi.org/10.48550/arXiv.2303.12712

Chen, F.-S., Fu, C.-M., & Huang, C.-L. (2003). Hand gesture recognition using a real-time tracking method and hidden Markov models. *Image and Vision Computing, 21*(8), 745–758. https://doi.org/10.1016/S0262-8856(03)00070-2

Chen, S., Wu, M., Zhu, K. Q., Lan, K., Zhang, Z., & Cui, L. (2023). LLM-empowered chatbots for psychiatrist and patient simulation: Application and evaluation. arXiv. https://doi.org/10.48550/arXiv.2305.13614

Dancy, C. L. (2013). Act-rΦ: A cognitive architecture with physiology and affect. *Biologically Inspired Cognitive Architectures, 6*, 40–45. https://doi.org/10.1016/j.bica.2013.07.008

Dastani, M., & Lorini, E. (2012). A logic of emotions: From appraisal to coping. In *Proceedings of the 11th International Conference on Autonomous Agents and Multiagent Systems (AAMAS 2012)* (pp. 1133–1140). Association for Computing Machinery. https://dl.acm.org/doi/abs/10.5555/2343776.2343859

De Melo, C. M., Carnevale, P. J., Read, S. J., & Gratch, J. (2014). Reading people's minds from emotion expressions in interdependent decision making. *Journal of Personality and Social Psychology, 106*(1), 73–88. https://doi.org/10.1037/a0034251

De Melo, C. M., Gratch, J., Marsella, S., & Pelachaud, C. (2023). Social functions of machine emotional expressions. *Proceedings of the IEEE, 111*(10), 1382–1397.

De Melo, C. M., Terada, K., & Santos, F. C. (2021). Emotion expressions shape human social norms and reputations. *iScience, 24*(3), 102141. https://doi.org/10.1016/j.isci.2021.102141

Dubey, A., Jauhri, A., Pandey, A., Kadian, A., Al-Dahle, A., Letman, A., Mathur, A., Schelten, A., Yang, A., Fan, A., Goyal, A., Hartshorn, A., Yang, A., Mitra, A., Sravankumar, A., Korenev, A., Hinsvark, A., Rao, A., Zhang, A., . . . Zhao, Z. (2024). The llama 3 herd of models. *arXiv*. https://doi.org/10.48550/arXiv.2407.21783

El-Nasr, M. S., Yen, J., & Ioerger, T. R. (2000). Flame: fuzzy logic adaptive model of emotions. *Autonomous Agents and Multi-Agent Systems, 3*, 219–257. https://doi.org/10.1023/A:1010030809960

Elyoseph, Z., Hadar-Shoval, D., Asraf, K., & Lvovsky, M. (2023). ChatGPT outperforms humans in emotional awareness evaluations. *Frontiers in Psychology, 14*, 1199058. https://doi.org/10.3389/fpsyg.2023.1199058

Ferstl, Y., Neff, M., & McDonnell, R. (2020). Adversarial gesture generation with realistic gesture phasing. *Computers & Graphics, 89*, 117–130. https://doi.org/10.1016/j.cag.2020.04.007

Gao, N., Zhao, Z., Zeng, Z., Zhang, S., Weng, D., & Bao, Y. (2024). GesGPT: Speech gesture synthesis with text parsing from ChatGPT. *IEEE Robotics and Automation Letters, 9*(3), 2718–2725. https://doi.org/10.1109/LRA.2024.3359544

Gebhard, P., Schneeberger, T., Baur, T., & André, E. (2018). MARSSI: Model of appraisal, regulation, and social signal interpretation. In M. Dastani, G. Sukthankar, E. André, & S. Koenig (Eds.), *Proceedings of the 17th International Conference on*

Autonomous Agents and Multiagent Systems (AAMAS 2018) (pp. 497–506). International Foundation for Autonomous Agents and Multiagent Systems (AAMAS).

Gibbs, R. W. Jr. (2003). Embodied experience and linguistic meaning. *Brain and Language, 84*(1), 1–15. https://doi.org/10.1016/S0093-934X(02)00517-5

Gratch, J., Cheng, L., & Marsella, S. (2015). The appraisal equivalence hypothesis: Verifying the domain-independence of a computational model of emotion dynamics. In *2015 International Conference on Affective Computing and Intelligent Interaction* (pp. 105–111). https://eprints.gla.ac.uk/190354/

Gratch, J., & Marsella, S. (2004). A domain-independent framework for modeling emotion. *Cognitive Systems Research, 5*(4), 269–306. https://doi.org/10.1016/j.cogsys.2004.02.002

Hareli, S., & Hess, U. (2019). The reverse engineering of emotions: Observers of others' emotions as naive personality psychologists. In U. Hess & S. Hareli (Eds.), *The social nature of emotion expression.* Springer. https://doi.org/10.1007/978-3-030-32968-6_6

Hensel, L. B., Yongsatianchot, N., Torshizi, P., Minucci, E., & Marsella, S. (2023). Large language models in textual analysis for gesture selection. In *Proceedings of the International Conference on Multimodal Interaction (ICMI '23)* (pp. 1–10). ACM. https://doi.org/10.1145/3577190.3614158

Hilliard, C., & Cook, S. W. (2016). Bridging gaps in common ground: Speakers design their gestures for their listeners. *Journal of Experimental Psychology: Learning, Memory, and Cognition, 42*(1), 91–103. http://dx.doi.org/10.1037/xlm0000154

Holler, J., & Stevens, R. (2007). The effect of common ground on how speakers use gesture and speech to represent size information. *Journal of Language and Social Psychology, 26*(1), 4–27. https://doi.org/10.1177/0261927X06296428

Hostetter, A. B., & Alibali, M. W. (2008). Visible embodiment: Gestures as simulated action. *Psychonomic Bulletin & Review, 15*(3), 495–514. https://doi.org/10.3758/PBR.15.3.495

Houlihan, S. D., Kleiman-Weiner, M., Hewitt, L. B., Tenenbaum, J. B., & Saxe, R. (2023). Emotion prediction as computation over a generative theory of mind. *Philosophical Transactions of the Royal Society A, 381*(2251), 20220047. https://doi.org/10.1098/rsta.2022.0047

Jain, R., Karsh, R. K., & Barbhuiya, A. A. (2022). Literature review of vision-based dynamic gesture recognition using deep learning techniques. *Concurrency and Computation: Practice and Experience, 34*(22), e7159. https://doi.org/10.1002/cpe.7159

Khare, S. K., Blanes-Vidal, V., Nadimi, E. S., & Acharya, U. R. (2023). Emotion recognition and artificial intelligence: A systematic review (2014–2023) and research recommendations. *Information Fusion, 102*, 102019. https://doi.org/10.1016/j.inffus.2023.102019

Kucherenko, T., Hasegawa, D., Kaneko, N., Henter, G. E., & Kjellström, H. (2021). Moving fast and slow: Analysis of representations and post-processing in speech-driven automatic gesture generation. *International Journal of Human–Computer Interaction, 37*(14), 1300–1316. https://doi.org/10.1080/10447318.2021.1883883

Laird, J. E., Newell, A., & Rosenbloom, P. S. (1987). Soar: An architecture for general intelligence. *Artificial Intelligence, 33*(1), 1–64.

Lascarides, A., & Stone, M. (2009). A formal semantic analysis of gesture. *Journal of Semantics, 26*(4), 393–449. https://doi.org/10.1093/jos/ffp004

Lazarus, R. S. (1991). *Emotion and adaptation*. Oxford University Press.

Lerner, J. S., Li, Y., Valdesolo, P., & Kassam, K. S. (2015). Emotion and decision making. *Annual Review of Psychology, 66*, 799–823. https://doi.org/10.1146/annurev-psych-010213-115043

Lhommet, M., & Marsella, S. (2016). From embodied metaphors to metaphoric gestures. *Proceedings of the Annual Meeting of the Cognitive Science Society, 38*, 788–793. https://escholarship.org/uc/item/4w03x8cs

Lugrin, B., Pelachaud, C., & Traum, D. (Eds.). (2022). *The handbook on socially interactive agents: 20 years of research on embodied conversational agents, intelligent virtual agents, and social robotics. Volume 2: Interactivity, platforms, application* (pp. xxv–xl). Association for Computing Machinery. https://doi.org/10.1145/3563659.3563661

Maghoumi, M., Taranta, E. M., & LaViola, J. (2021). DeepNAG: Deep non-adversarial gesture generation. *Proceedings of the 26th International Conference on Intelligent User Interfaces* (pp. 213–223). https://doi.org/10.1145/3397481.3450675

Manstead, A. S. R., & Fischer, A. H. (2001). Social appraisal: The social world as object of and influence on appraisal processes. In K. R. Scherer, A. Schorr, & T. Johnstone (Eds.), *Appraisal processes in emotion: Theory, methods, research* (pp. 221–232). Oxford University Press.

Marinier, R. P. III, Laird, J. E., & Lewis, R. L. (2009). A computational unification of cognitive behavior and emotion. *Cognitive Systems Research, 10*(1), 48–69. https://doi.org/10.1016/j.cogsys.2008.03.004

Marsella, S., & Gratch, J. (2009). EMA: A process model of appraisal dynamics. *Cognitive Systems Research, 10*(1), 70–90. https://doi.org/10.1016/j.cogsys.2008.03.005

Marsella, S., Gratch, J., & Petta, P. (2010). Computational models of emotion. In K. R. Scherer, T. Bänziger, & E. Roesch (Eds.), *A blueprint for an affectively competent agent: Cross-fertilization between emotion psychology, affective neuroscience, and affective computing*. Oxford University Press.

Marsella, S. C., Johnson, W. L., & LaBore, C. (2000). Interactive pedagogical drama. In *Proceedings of the Fourth International Conference on Autonomous Agents* (pp. 301–308). ACM. https://doi.org/10.1145/336595.337507

McDuff, D., & Czerwinski, M. (2018). Designing emotionally sentient agents. *Communications of the ACM, 61*(12), 74–83. https://doi.org/10.1145/3186591

Meuleman, B., & Scherer, K. R. (2013). Nonlinear appraisal modeling: An application of machine learning to the study of emotion production. *IEEE Transactions on Affective Computing, 4*(4), 398–411. https://doi.org/10.1109/T-AFFC.2013.25

Moerland, T. M., Broekens, J., & Jonker, C. M. (2018). Emotion in reinforcement learning agents and robots: A survey. *Machine Learning, 107*, 443–480. https://doi.org/10.1007/s10994-017-5666-0

Moors, A., Ellsworth, P. C., Scherer, K. R., & Frijda, N. H. (2013). Appraisal theories of emotion: State of the art and future development. *Emotion Review, 5*(2), 119–124. https://doi.org/10.1177/1754073912468

Nyatsanga, S., Kucherenko, T., Ahuja, C., Henter, G. E., & Neff, M. (2023). A comprehensive review of data-driven co-speech gesture generation. *Computer Graphics Forum, 42*(2), 569–596. https://doi.org/10.1111/cgf.14776

Ojeda-Castelo, J. J., Capobianco-Uriarte, M. d. L. M., Piedra-Fernandez, J. A., & Ayala, R. (2022). A survey on intelligent gesture recognition techniques. *IEEE Access, 10*, 87135–87156. https://doi.org/10.1109/ACCESS.2022.3199358

Ong, D. C., Zaki, J., & Goodman, N. D. (2019). Computational models of emotion inference in theory of mind: A review and roadmap. *Topics in Cognitive Science, 11*(2), 338–357. https://doi.org/10.1111/tops.12371

Ortony, A., Clore, G. L., & Collins, A. (1988). *The cognitive structure of emotions*. Cambridge University Press. https://doi.org/10.1017/CBO9780511571299

Ouyang, L., Wu, J., Jiang, X., Almeida, D., Wainwright, C. L., Mishkin, P., Zhang, C., Agarwal, S., Slama, K., Ray, A., Schulman, J., Hilton, J., Kelton, F., Miller, L., Simens, M., Askell, A., Welinder, P., Christiano, P., Leike, J., & Lowe, R. (2022). Training language models to follow instructions with human feedback. *Advances in Neural Information Processing Systems, 35*, 27730–27744.

Paiva, A., Leite, I., Boukricha, H., & Wachsmuth, I. (2017). Empathy in virtual agents and robots: A survey. *ACM Transactions on Interactive Intelligent Systems, 7*(3), 1–40. https://doi.org/10.1145/2912150

Ravenet, B., Clavel, C., & Pelachaud, C. (2018). Automatic nonverbal behavior generation from image schemas. In *Proceedings of the 17th International Conference on Autonomous Agents and Multiagent Systems* (pp. 1667–1674). International Foundation for Autonomous Agents and Multiagent Systems (IFAAMAS). https://ifaamas.org/Proceedings/aamas2018/pdfs/p1667.pdf

Reisenzein, R., Hudlicka, E., Dastani, M., Gratch, J., Hindriks, K., Lorini, E., & Meyer, J.-J. C. (2013). Computational modeling of emotion: Toward improving the inter- and intradisciplinary exchange. *IEEE Transactions on Affective Computing, 4*(3), 246–266.

Sequeira, P., Melo, F. S., & Paiva, A. (2011). Emotion-based intrinsic motivation for reinforcement learning agents. In *Affective computing and intelligent interaction: 4th international conference, ACII 2011, Memphis, TN, USA, October 9–12, 2011, proceedings, Part I 4* (pp. 326–336). Springer. https://api.semanticscholar.org/CorpusID:8311222

Sequeira, P., Melo, F. S., & Paiva, A. (2014). Learning by appraising: An emotion-based approach to intrinsic reward design. *Adaptive Behavior, 22*(5), 330–349. https://doi.org/10.1177/1059712314543837

Shapiro, L. (2007). The embodied cognition research programme. *Philosophy Compass, 2*, 338–346. https://doi.org/10.1111/j.1747-9991.2007.00064.x

Si, M., Marsella, S., & Pynadath, D. V. (2010). Modeling appraisal in theory of mind reasoning. *Autonomous Agents and Multi-Agent Systems, 20*, 14–31. https://doi.org/10.1007/s10458-009-9093-x

Simon, H. A. (1967). Motivational and emotional controls of cognition. *Psychological Review, 74*(1), 29–39. https://doi.org/10.1037/h0024127

Smith, C. A., & Lazarus, R. S. (1990). Emotion and adaptation. In L. A. Pervin (Ed.), *Handbook of personality: Theory and research* (pp. 609–637). Guilford.

Smith, G. M., & Carette, J. (2022). What lies beneath: A survey of affective theory use in computational models of emotion. *IEEE Transactions on Affective Computing, 13*(4), 1793–1812. https://doi.org/10.1109/TAFFC.2022.3197456

Steunebrink, B. R., Dastani, M., & Meyer, J.-J. C. (2009). A formal model of emotion-based action tendency for intelligent agents. In *Portuguese conference on artificial intelligence* (pp. 174–186). Springer.

Striegnitz, K., Tepper, P., Lovett, A., & Cassell, J. (2009, April). Knowledge representation for generating locating gestures in route directions. In K. R. Coventry, T. Tenbrink, & J. Bateman (Eds.), *Spatial language and dialogue* (pp. 147–165). Oxford University Press.

Sun, R. (2017). The clarion cognitive architecture: Toward the mind. In S. F. Chipman (Ed.), *The Oxford handbook of cognitive science* (pp. 117–133). Oxford University Press.

Tak, A. N., & Gratch, J. (2023). Is GPT a computational model of emotion? In *2023 11th International Conference on Affective Computing and Intelligent Interaction (ACII)* (pp. 1–8). IEEE. https://arxiv.org/abs/2307.13779

Traum, D., Swartout, W., Marsella, S., & Gratch, J. (2005). Fight, flight, or negotiate: Believable strategies for conversing under crisis. In *International Workshop on Intelligent Virtual Agents* (pp. 52–64). Springer.

Tyng, C. M., Amin, H. U., Saad, M. N., & Malik, A. S. (2017). The influences of emotion on learning and memory. *Frontiers in Psychology, 8,* 235933. https://doi.org/10.3389/fpsyg.2017.01454

Van Kleef, G. A., & Côté, S. (2022). The social effects of emotions. *Annual Review of Psychology, 73,* 629–658. https://doi.org/10.1146/annurev-psych-020821-010855

Wilson, A. D., & Bobick, A. F. (1995). Learning visual behavior for gesture analysis. In *Proceedings of the IEEE International Symposium on Computer Vision (ISCV)* (pp. 229–234). IEEE. https://hd.media.mit.edu/tech-reports/TR-337.pdf

Windle, J., Matthews, I., & Taylor, S. (2024). Llanimation: Llama driven gesture animation. In *SCA '24: Proceedings of the ACM SIGGRAPH/Eurographics Symposium on Computer Animation* (pp. 1-10). ACM. https://doi.org/10.1111/cgf.15167

Yongsatianchot, N., & Marsella, S. (2016). Integrating model-based prediction and facial expressions in the perception of emotion. In *International Conference on Artificial General Intelligence* (pp. 234–243). Springer International Publishing.

Yongsatianchot, N., Thejll-Madsen, T., & Marsella, S. (2023). What's next in affective modeling? large language models. In *2023 11th International Conference on Affective Computing and Intelligent Interaction Workshops and Demos (ACIIW)* (pp. 1–7). IEEE.

Yongsatianchot, N., Thejll-Madsen, T., & Marsella, S. (2024). Exploring theory of mind in large language models through multimodal negotiation. In *Proceedings of the 24th ACM International Conference on Intelligent Virtual Agents* (pp. 1–9). ACM. https://doi.org/10.1145/3652988.3673960

14 Embodying Emotions in Autonomous and Social Robots: Homeostasis, Hormones, Interaction, and Embodied AI

Lola Cañamero

Introduction

How are emotions related to embodied intelligence (EI), and why should we include a section about emotions in a book about EI? Although controversial in the past, nowadays the question would not even come to mind for many of us—of course, this is perfectly relevant! However, this question raises two unresolved issues: the type of relationship between emotion and intelligence and the relation between emotion and embodiment. When we add robots to the equation, opinions diverge even more.

Regarding the first issue, the close link (or even inseparability, according to some) between emotion and "intelligence" or "cognition" (in a broad sense) has been increasingly acknowledged and investigated for the last three decades in diverse disciplines, such as neuroscience (e.g., Damasio, 1994; LeDoux, 1996; Panksepp, 1998; Pessoa, 2013) and philosophy (Asma & Gabriel, 2019), and echoed in different areas of "machine intelligence," such as affective computing (Picard, 1997), artificial intelligence (AI), and cognitive robotics (Cañamero, 1998, 2001). The nature of this link is, however, less clear, and debates still spark concerning the predominance of one or the other "component," as well as the degree of the "mix." Elsewhere (Cañamero, 2023), I suggested regarding this link as a multifaceted "tangled knot," the components of which are *fundamentally tangled* and that need to be studied in its complexity through both analysis and synthesis, from multiple perspectives, rather than seeing it as a Gordian knot—an intractable problem that needs to be "cut" or "dissolved." In my view, this endeavor requires a complex, multifaceted notion of embodiment, as we will see below.

The relationship between emotions and embodiment in humans has long been—and continues to be—the subject of extensive debate across psychology, neuroscience, and philosophy (e.g., Feldman-Barrett et al., 2008, and Hufendiek, 2016, for critical discussions). I will not address those debates here. Instead, I turn to this question: How can emotions be embodied in robots? What sort of embodiment(s) would be appropriate, possible, and needed? Why endow robots with emotions, to begin with?

In this chapter, I argue for a "strong," multifaceted notion of embodiment that is rooted in emotion—or more generally affect, and particularly "core affect" (Damasio, 1999)—and that such a notion can be modeled in robots. I will discuss how different aspects of this embodiment have been implemented in robots and other "animats"[1] in my research group in the context of affect modeling and based on "core affect," to ground key notions in the fields of (cognitive) autonomous and social robotics, such as autonomy, adaptation, learning, development, evolution, social dynamics, and rudiments of culture.

Emotions and Robot Embodiments

Thinking of robots, the first type of embodiment that comes to mind is their physical, "hardware," realization. This type of embodiment is certainly needed for physical and social interaction in our world and must be carefully chosen in relation to the task and the environment (or more precisely, ecological niche, involving relations with other agents) where the robot is to be situated, since it easily sets expectations in humans that bias our perception and behavior, as extensively researched in the area of human-robot interaction (HRI). Robots are not just static 3D devices, but move and behave; mechanistic embodiment is thus accompanied by behavioral embodiment. However, mechanistic-behavioral embodiment is, in my view, far from sufficient to embody emotions. Neuroscientist Ralph Adolphs, discussing whether a robot could have emotions, expresses it eloquently (italics my emphasis): "We should attribute emotions and feelings to a system only if it *satisfies criteria* in addition to mere behavioral duplication . . . correctly attributing emotions and feelings to them would require that robots are *situated in the world and constituted internally in respects that are relevantly similar to humans*" (Adolphs, 2005, p. 9).

Conversely, not all embodied agents are realized in hardware, or even in the "real world." Besides the new embodiments that have arisen more recently (e.g., in synthetic biology, soft robotics, plantoid robots), the broad category of animats also includes computer simulations, which are nevertheless considered embodied as they follow an embodied artificial intelligence (EAI) design philosophy and principles (Brooks, 1991; Pfeifer & Scheier, 2001) and are situated and "structurally coupled" in environments that embed key features of the real world. In this chapter, when I talk about "robots" I mean, more generally, "animats."

Beyond physical embodiment, other notions of (robot) embodiment have been proposed. For example, in the early days EAI, Tom Ziemke (2003) distinguished between six types of embodiment and discussed how they could (or could not) apply to robots: (1) structural coupling with the environment, drawing on Maturana and Varela's notion; (2) historical embodiment, reflecting a history of agent-environment interaction and possibly coadaptation; (3) physical embodiment, which involves physical grounding in the world beyond physical realization; (4) organismoid embodiment, physical bodies with some degree of similar form and sensorimotor capacities as living bodies; (5) organismic embodiment, a notion restricted, by definition, to living bodies, which the author, following a strict enactivist tradition, characterizes as autonomous and autopoietic, versus human-made machines, which are defined as heteronomous and allopoietic and therefore devoid of organismic embodiment; and (6) social embodiment, defined as "states" of the body (e.g., postures, movements, expressions) that arise during social interaction and play central roles in social information processing.

Rather than as different notions of embodiment, I see these six definitions as complementary features that provide a rather comprehensive, multifaceted notion of embodiment that is also applicable to robots[2] and other animats. I will argue that this "strong," multifaceted notion of embodiment is rooted in emotion—or more generally, affect, and particularly "core affect"—and that it can be modeled in robots. To do so, some elements need to be added to this characterization:

- Underlying mechanisms, which in my case are based on a simulated physiology and its dynamics. When I started working in this area as a postdoc, and was initially inspired by, among others, a classic textbook in cognitive neuroscience, I took "literally" the idea of modeling a dynamic physiology that would ground, in artificial systems, the notion of homeostasis and the bodily, behavioral, and cognitive (regulatory) processes linked to motivations and emotions. This approach, which resonates with Adolph's requirement to make robots internally in respects that are relevantly similar to humans and was later termed "internal robotics" (Parisi, 2004), has been the basis of my research since then.

- Multiple (and interdisciplinary) perspectives are also needed to account for the different facets of that "strong" emotion embodiment. In my research, I have explored emotion (or rather affect, as we will see below) from the classical "four perspectives" put forward by Tinbergen as required to understand behavior and more broadly as the "program" for ethology: its underlying mechanisms, functions, development, and evolution.
- The role of the interaction between (and among) agents and their ecological niche.

Robots That "Have" Emotions: Grounding Embodied Affect in Homeostasis, Hormonal Modulation, and Interaction

Robots That "Have" Emotions?

In the book that gave rise to the field of affective computing, Rosalind Picard (1997) characterized affective machines as those with the ability to recognize, express, and have (or rather "have") emotions, giving my early work (Cañamero, 1997) as an example of the latter. However, whereas the abilities to artificially recognize and express emotions (or rather their expressions) are more widely accepted, the possibility for robots to *have* emotions is the topic of much discussion (Fellous & Arbib, 2005). Drawing on human emotions, Picard proposed five components that could be used as a performance "test" to evaluate whether a machine would have achieved (that specific component of) emotions. Emotions in robots and other machines do not need to replicate or be inspired by human emotions, though, and endowing a robot with traits of nonhuman animal emotions could be more relevant and compelling in many contexts. Even artificially developed (or evolved) emotional traits that are not inspired from biology but that trigger appropriate responses in humans could be very valuable to develop robots that can entertain affective interactions with us and to help us understand human emotions. However, some critics would argue, could we call "emotions" those artificially developed emotions? Moreover, do *my* robots *have* emotions?

Whereas in the past I have argued that they do "have" (robot-)emotions (Fellous, 2004), and I acknowledge that asking such question can generate very fruitful discussions and implementations to help understand and operationalize human emotions, I am nowadays more interested in using my robots/animats as embodied scientific models of embodied affect. In Cañamero, 2021, I discussed some of the contributions that embodied models of emotions implemented in robots can make to interdisciplinary emotion research, emphasizing and illustrating with examples how robots can be thinking tools as well as situated operational models to investigate hypotheses stemming from different disciplines, such as philosophy, neurosciences, or psychology, often in ways that are not possible to study in biological systems—for example, for methodological or ethical reasons. This change of perspective is in line with what neuroscientist Ralph Adolphs (2005, p. 9) expected from robotics as a science: "if robotics is to be a science that can actually tell us something new about what emotions are, we need to engineer an internal processing architecture that goes beyond merely fooling humans into judging that the robot has emotions". Let us now turn to what he called the "internal processing architecture"—the embodied emotion model that I have used and why it is relevant.

Embodying Affect in Robots

In my research, I have investigated emotions in the context of their interplay with motivation, cognition, and (inter-)action. I could thus rephrase the above expression as robots that "have" affect, in a broader sense that encompasses motivation and emotion and their involvement in cognitive and interactional processes. What I mean by "affect" can be characterized in intuitive terms as "wanting" and "liking" (Berridge, 2019; Cañamero, 2005): *"wanting"* as a need or "urge" for action, cognition, and interaction, providing a basis for robots to have autonomy with respect to "automatic" responses to external environmental stimuli (motivational autonomy), coupled with *"liking"* (or disliking) what we perceive, are motivated to do, do, know, and learn; how we do it; and the consequences of it—giving robots a further "degree" of autonomy, as this "liking" can change the priorities set by the motivational system. My work is thus primarily concerned with "core affect." My specific model of embodied affect and affective cognition is biologically inspired and builds on a synthetic physiology that is regulated and modulated using principles stemming from EAI, biology, and cybernetics (e.g., homeostasis and allostasis), ethology, neuroscience (e.g., neuromodulation), and dynamical systems. This embodied core affect is crucial for (biological and artificial) agents that need to survive and interact autonomously in a complex physical and social world, since it provides reasons (and mechanisms) *why things matter*, as well as being the *driving force* for cognition and (physical and social) interaction. It provides a blueprint for a "core affective self" that endows robots with "internal" values (their own or acquired) and motivations that drive adaptation and learning through interactions with the physical and social environment.

Let us look at some of the mechanisms I have used, which have varied over time while preserving the core elements. My early, somewhat "biomimetic" work aimed to model motivations and emotions as a basis for intelligent behavior in a typical animat action selection environment (called "Gridland"), inspired by Maes (1991), in which autonomous agents inhabiting a complex and dynamic two-dimensional toroidal world had to constantly make decisions in real time to interact and deal with the static and dynamic elements of the environment in order to survive. The fact that the agents needed to survive was key to affect embodiment and was implemented in terms of management of homeostasis: agents had to consume resources and interact with the environment to remain viable (Ashby, 1952). Gridland was populated by other moving agents (friendly conspecifics and unfriendly agents that could attack, damage, and "kill" other agents) and contained (dynamic) survival-related depletable resources, obstacles obstructing the availability of those resources, and other objects of various shapes. The complex action selection architecture of the agents included a (simulated) physiology—in effect a complex dynamical system—of homeostatically controlled survival-related variables (e.g., vascular volume, temperature, etc.) as well as generic (with global excitatory and inhibitory effects) and specific (acting differentially on specific receptors) modulatory hormones. The elements of the architecture that contributed to the maintenance of homeostasis were motivations, consummatory and appetitive behaviors, and a number of basic emotions, each designed to have a specific (adaptive) function and implemented in terms of "patterns of neuromodulation" (Fellous, 1999) by

When talking about adaptive value, we tend to think of adaptation in positive terms, but it is also important to investigate when and why things become maladaptive. For example, Matthew Lewis's research fellow project, in collaboration with OCD expert Naomi Fineberg and using a hormonally modulated motivation-based decision-making architecture and a two-resource problem scenario, investigated the emergence of compulsive behavior in a robot according to different theories—for example, cybernetic model, vulnerability, and the role of stress (Lewis et al., 2019; Lewis & Cañamero, 2019). Key to this project was the idea of assessing the adaptive or maladaptive value of traits linked to OCD and compulsions, like (various degrees of) "perfectionism" and persistence under different circumstances and in different contexts, as well as helping to de-stigmatize mental health conditions. Currently, the PhD work of Louis L'Haridon investigates the adaptive/maladaptive value of pain and pain perception and its effects on the construction of body representations and decision-making (L'Haridon & Cañamero, 2023; L'Haridon et al., 2024).

Hormonally modulated motivation-based action selection architectures have also been used in the context of human-robot social interaction, such as the "diabetic" robot companion for diabetic children, named Robin (Cañamero & Lewis, 2016). The aim of this project was to increase diabetic children's self-efficacy and self-confidence, and awareness of their condition and feelings, as they played with and "looked after" the diabetic toddler robot, Robin. With a simple physiology, very few motivations and behaviors, a metabolism that modeled diabetes, and a simple model of pleasure, Robin was able to interact with children for extended twenty- or thirty-minute periods with rich behavior that varied with each child and interaction due to the different ways the children interacted with the robot.

Models have also investigated the effects and adaptive value of specific emotion dimensions on decision-making, learning, and interaction with the physical and social environment—namely, valence (pleasure) and arousal, sometimes combined, sometimes using a single dimension in isolation, to better assess its specific contributions. Regarding pleasure, for example, we also used a "two-resource" problem scenario to assess the contribution of pleasure to homeostasis regulation and the adaptive value of two types of pleasure in decision-making (linked to need satisfaction versus being purely "sensory"). In addition, we explored pleasure's role in the trade-off between persistence (with "addiction" as an extreme) and opportunism (with "impulsivity" as an extreme case) (Lewis & Cañamero, 2016). Concerning arousal, the PhD project of Antoine Hiolle used arousal linked to the activity of the neural networks of the robot when confronted with novelty, and regulated through interaction with humans, for different types of learning. For example, a robot could learn "affective landmarks" in a room (Hiolle & Cañamero, 2009) to detect learning challenges beyond its current level of skill and seek human help to circumvent them (Hiolle et al., 2014a) or adapt its learning and affective expression on the fly during social interaction as a function of the responsiveness of the human interacting with it (Hiolle et al., 2014a). I refer the reader to Cañamero, 2021, for a longer discussion of these pleasure and arousal models.

The development of "embodied" performance metrics based on the notion of viability, in addition to the traditionally used survival of the robots in the context of action selection or the speed and amount of learning in learning scenarios, allowed us to quantify different aspects of how homeostasis and viability were managed in these different studies. It also allowed us to compare the performance of different models across scenarios with different challenges. To these "viability indicators" we added metrics based on the notion of "activity cycles," which ethologists use to describe observed "chains" of behavior that might repeat. From that, we were able to track the "physiological space" of the robot in terms of the dynamics of the physiological variables and the "paths" described over time. From that, we could relate what the robot perceived and did in the environment using the logs of its sensors and actuators/behaviors. Such metrics allowed us to assess and compare the difficulty of the different problems investigated and quantitatively assess adaptation and the adaptive value of different strategies, algorithms, and so on in several ways.

Embodied Affect, Development, and Social Dynamics

Where the agent is situated and its interactions with the environment play a big role in what intelligence (and affect) is in EAI and in the studies discussed so far. These factors affect the "internal architecture" of the robot, which in turn has an effect on the environment; They are structurally coupled. This is even more apparent when we take a historical, developmental perspective.

The development of attachment was the PhD work of Arnaud Blanchard and Antoine Hiolle. Blanchard's work (Blanchard & Cañamero, 2005) investigated how a small "baby" mobile robot would adapt to and learn about new situations and interactions as it imprinted to a human "caregiver" (actually, to the first things that moved when the robot "hatched"). This work was done using an architecture based on homeostasis maintenance, comfort signals provided by the human, and algorithms for perceptual learning and memory. Here, the robot's memory reflected the history of "affective" interactions with the human, driven by "distress" and "comfort" responses. Further work explored, among other things, the development through affective modulation of different behaviors such as exploration and imitation (Blanchard & Cañamero, 2006), the interplay between the role of the caregiver who provided a "secure base" for the robot to explore and learn, and the robot as an active agent in its learning (Hiolle et al., 2014a, 2014b).

Two other mechanisms allowed us to model stronger influences of the context and the history of interactions on the development of agents and their interactions: epigenetic development and the notion of allostasis. The PhD project of John Lones explored epigenetic development in robots. For example, the study reported in Lones et al. (2016) investigated the importance of early sensorimotor experiences and different environmental conditions in the emergence of more advanced cognitive abilities in an autonomous robot. The robot architecture combined (1) a basic motivated "core affect" layer comprising a number of survival-related homeostatically controlled variables that provided the robot with internal needs to generate behavior; (2) a hormone-driven epigenetic mechanism affecting the development of the robot's growing neural network; and (3) an "emergent" or "growing" neural network, which provided the robot with learning capabilities. This architecture caused the robot to develop a variety of new behaviors and cognitive abilities as a function of the richness (opportunities for stimulation and novelty) of the environment and of its interactions with the environment (more or less exploration and at different paces, which could be more or less appropriate for the learning abilities of the robot). Exploration of the environment could be autonomous or guided by a human. The behavioral and "cognitive" complexity of the robot could be observed in the complexity and structure of the neural networks that developed. Lones et al. (2017) investigated the ability of an epigenetic mechanism to adapt to very different environments. The purpose of these experiments was to investigate the adaptability of our epigenetic mechanism to a range of environmental challenges through allostatic processes (Schulkin, 2004) and, more specifically, to investigate how our system could allow initially identical robots—with the same initial internal parameter values and speed potential—to develop different behavioral phenotypes adapted to different environments as a consequence of the changes brought about by and in the epigenetic mechanism in its interactions with the environment.

The architecture of the robot combined (1) a basic motivated architecture in charge of (homeostatically) controlling three physiological variables (energy, health, and internal temperature); (2) a neuromodulatory system that modulated the basic architecture through the secretion and decay of four simulated hormones secreted in relation to changes in the robot's internal and external environment (three "endocrine hormones" secreted as a function of the errors of the different physiological variables, and a "neurohormone" secreted as a function of a combination of the concentration of the endocrine hormones and specific external stimuli); and (3) an epigenetic system that controlled the sensitivity of the hormone receptors of the neuromodulatory system. The architecture was tested on six very different environments (e.g., with different amounts and distribution of static/dynamic resources, with "day/night" cycles and other temporal dynamics) in an incremental way (robots with system 1 only, with systems 1+2, and with the full architecture), to be able to compare the contribution of each system. Our results showed that the "epigenetic robot" maintained homeostasis better than the other two robots in all the six environments and, as a result of different chains of up- and downregulation of hormone receptors and their effects on the sensors and actuators of the robot, it developed a richer repertoire of behaviors that provided more efficient and intelligent adaptation, particularly in the more complex environments. All the adaptations took place as a result of such dynamics of interactions, without the need to program or tune the robot to operate in a specific environment.

Allostatic processes (social allostasis) were also investigated as part of the Imran Khan's PhD project. Using an artificial life environment programmed in Netlogo, Khan carried out a number of studies around the "social salience" and "social buffering" hypotheses of oxytocin (OT) and in which contexts OT could give rise to prosocial or antisocial behavior. The agents had, again, a physiology-regulation-based architecture and simple motivations and behaviors related to the procurement of food and social interaction (grooming), including additionally a "social/contextual assessment" component modulated by the hormones oxytocin and cortisol. Different experiments tested, for example, the contextual adaptive value of social allostasis (Khan & Cañamero, 2018), the effects of (OT-modulated) social bonds on the well-being of agents, partner preferences, and formation of social alliances in decisions related to food-sharing and grooming in a hierarchical society (Khan et al. 2019), or the (contextual) advantages provided by the "social buffering" effects of OT on the social support and adaptation of the level of stress tolerance of agents in a small society (Khan & Cañamero, 2022).

Embodied Affect and Evolution

An evolutionary perspective is key to understanding why and how the emotions we have nowadays came to have the features, functions, and mechanisms they currently do. From this perspective, we are interested in questions such as Why did structural associations evolve in a particular manner?, Which "emotional trait" was a prerequisite for another?, and What are the consequences of specific traits and their functions in further developments? More difficult to carry out with physical robots, for practical reasons, bioinspired embodied simulations can be very valuable tools to investigate many of the questions around the evolutionary origins and remnants of our emotions. Two examples of such simulations in my group looked at questions regarding (1) the evolution of embodied affective displays and (2) the open-ended coevolution of (affective) morphology and behavior.

Evolution of Affective Displays

In his PhD project, Robert Lowe investigated the evolution of the expression and perception of affective displays grounded in internal affective states in simulated robots (Lowe et al., 2004, 2005), where the color-based external displays were inspired by the behavior of cephalopods. The Java-based simulation environment used consisted of a two-dimensional toroidal world with continuous space and discrete time, food resources and competition, and (asexually) reproducing agents, which also underwent mutation when the fittest agents in a generation were selected to reproduce for the next. I will discuss two of these studies.

The study reported in Lowe et al. (2004) looked at the evolution of the expression and recognition of affect-related displays, as well different strategies in the use of such displays. The agents could perceive and display colors that carried information about their internal state. This internal state was determined by the interactions of the agents with their environment, driven by a behavior-based motivated decision-making architecture consisting of a homeostatically controlled internal physiology with two variables ("energy" and "social stimulation"), corresponding motivational states combining the level of the physiological errors and the perception of the external stimuli relevant to each, and a number of behaviors ("eat," "hit," and "escape" as consummatory behaviors and "look-for-food" and "look-to-attack" as appetitive behaviors) selected by the motivations to satisfy the internal needs through interactions with the environment at each point in time. Agents used vision to perceive the hue and saturation of objects (including other agents) in the environment; they could also perceive "stalkers" (unseen rivals in close proximity). The environment varied regarding the availability of resources, alternating between "winter" (scarce) and "summer" (abundant) seasons. During the initial run under each experimental condition, the color displayed and perceived by all the agents accurately reflected their internal need and its ensuing behavior (looking for food or looking to attack). Later, over subsequent generations, the display components could vary as a result of evolution, giving rise to the emergence of two display strategies: "bluffing," where the hue/saturation was "stronger" than the need, and "Trojan," where the hue/saturation was "weaker" than the need. The more the display value deviated from the actual value of the need, the less accurate was the depiction of affective state. Different hypotheses were formulated and tested, such as whether and under which circumstances bluff displays would be adaptive to agents looking to avoid conflict, while Trojan displays would enable aggressive agents to attack unsuspecting agents more easily. As well as the displays, the ability to perceive such displays evolved, with experiments comparing the predominance, adaptive value, and social dynamics of "affect perceivers" versus "generic perceivers. Different types of metrics were used to assess the effects of the different evolved traits and strategies on the viability of the agents, the social dynamics, the use of strategies, and the relations among them and with environmental changes and circumstances.

Another study (Lowe et al., 2005) investigated the evolution of different display strategies and their use in the context of competitive aggressive interactions as a function of the potential damage that the agents could cause to, or receive from, other agents. This study was motivated by the idea that, in nature, "all out" fights over resources are generally avoided and replaced by (aggressive) displays that provide information about the condition and aggressive intent of the agent; even if the formation of those displays can be evolutionarily costly, they can settle such disputes in circumstances where direct aggression might have fatal consequences. A simpler architecture was used this time: the agents had only one physiological variable, "energy," and instead of a repertoire of predefined behaviors, the sensors and the actuators of the Braitenberg vehicle giving rise to movement were linked by weights that could evolve. As previously, the environment gave rise to competition for resources. A key element in this study was the fact that the production of displays and "attacks" incurred a cost, and attacks to another agent would cause "damage" (energy loss) to the latter. A number of evolutionary runs tested different experimental conditions, varying elements such as the number of resources, the degree of damage potentially caused and displayed by an agent, or the cost of developing and using displays, with the aim of assessing how the potential to inflict damage by homogeneous agents could affect overall aggression in agent societies and food-procuring strategies (e.g., foraging versus scavenging abandoned resources) related to territoriality (e.g., defending "owned" resources versus tolerating and sharing resources with others). The effects of evolving ritualized displays on such societal aggression and territorial behavior were also assessed, with results showing that when the cost of damage was high and the agents invested in costly to produce ritualized displays, they were less likely to produce aggressive behavior, using instead, displays to "settle disputes." Use of displays also had a beneficial effect on the average viability of the agents in a society versus the viability of a few.

Open-Ended Coevolution of Morphology and Behavior

In an earlier section, we saw how the environment plays a key role in the development of behavior and affect. From the perspective of the evolution of affect, the environment plays a fundamental role, driving evolution in terms of selective pressures that vary over time.

The PhD project of Paul Pichler investigated the open-ended coevolution of body morphology and behavior of simple agents situated in evolving artificial life ecosystems with varying challenges (Pichler, 2009; Pichler & Cañamero, 2008). With the explicit aim being to avoid biases, and contrary to artificial evolution based on more traditional genetic algorithms, the agents did not start with predefined fixed morphologies or behavioral primitives and did not have an explicit fitness function. Agents had to remain viable, both individually by successfully surviving in their environment and at the population level through reproduction. Both forms of viability were grounded in the management of energy, internally and through exchanges with the environment. Evolution was driven by principles of natural selection as a result of the interactions and energy trade-offs of the agents with the evolving ecosystem and its varying selective pressures.

The ecosystem consisted of objects that had a number of physical properties—namely, solidness, energy content, and size—placed on a continuous two-dimensional toroidal world. All objects were characterized by these properties to different extents, and as a function of this, they could be regarded and act as either sources of energy (which get depleted and can regrow at different rates), obstacles (solid objects with zero energy content), or agents (objects with solidness and energy content that could move, acquire energy from external sources when they "bumped into them," progressively losing their energy content when dead, and with a behavior controlled by an evolving neural network while alive. In principle, agents could serve as energy sources to other agents. Over the course of evolution, the number of each type of object would vary, giving rise on some occasions to highly competitive environments (e.g., overpopulated, with scarce energy resources, or with a high number of objects hampering access to resources) and in other occasions to "easy" environments.

The body and morphology of the agents consisted of

- A circular body plan, allowing the evolution of all sorts of arrangements of external sensors (to detect the different physical properties of the environment) and actuators (small "flagella" for motion) and directions of movement.
- An energy level that diminished or increased as a function of the interactions with the environment and the reproductive "decisions" of the agent.

- An initial configuration of a simple neural network with a "reproductive energy depot" output node to which "surplus" energy was transferred when the energy level of the agent was above its reproductive threshold; a random number of randomly placed actuators (each with their own translation and rotation characteristics) determining the speed and direction of movement of the agent; and, as input nodes, two internal sensor nodes to sense the agent's internal variables (the energy level and the level of the depot) and a random number of randomly placed external sensor nodes, each able to detect one of three physical properties of the objects in the environment—solidness, energy content, and size.
- Metabolism, such that each agent had a limited energy capacity as a function of its size and solidness and could not consume energy above that limit. It also incurred metabolic costs from the activation and development of its neural network, collisions, reproduction, and the use of actuators.

An important feature of the model was the decoupling of individual survival (here ensured by the appropriate management of its internal variable energy, which allowed it to remain viable in the environment) from reproductive success, which gave the evolving agents control over when and to what extent they invested resources in offspring.

The evolutionary operators used were asexual reproduction driven by the "surplus" energy accumulated in the reproduction depot and giving rise to an imperfect copy of the agent, mutation to modify each of an agent's parameters and operators for deleting existing or adding new parts (sensors, actuators, neurons, synapses). Note that the fact that reproduction depended on the surplus energy that each agent transferred to the reproduction depot meant that reproduction depended on the agent's embodiment and interactions and was "optional," and that the selective pressure arose from local and direct competition in a shared environment rather than from a global fitness function.

At the beginning of evolutionary runs, the agents were initialized with very minimal neural networks, and the ecosystem had rather small populations (typically fifty agents). Over the course of evolution, population size was not fixed or constant but was only constrained by the fact that the amount of energy in the environment was limited and by the metabolic properties of the evolved agents. The interactions of each agent with its environment contributed to the total energy balance of agents and environment (which included other agents), as there was a flow of energy in both directions. Different experiments with different initial conditions varying the numbers and types of objects were conducted to assess, among other things, the relation between the morphology and behavior of the evolved agents (Pichler 2009), the presence of evolutionary trends, the evolution of neural and behavioral complexity, and the effects of the reproductive strategy on behavioral diversity.

For the purposes of this chapter, the "message to take home" is that this embodiment, grounded in an energy-managing metabolism and the sensing and exchange of energy with the environment, along with the open-ended evolution driven by selective environmental pressures and principles of natural selection and the decoupling between individual survival and reproductive success, was key to the coevolution of varied and complex morphology and behavior (including affective behavior) through interactions in the ecosystem. Agents with four main behavioral strategies evolved—"drifters," "foragers," "avoiders," and "allrounders" (combining the other three)—which were associated clear morphological patterns in terms of mass, size, density, energy capacity, sensors, and actuators. In all cases, the primary behavior of the agents was locomotion (the overall movement of an agent was obtained by integrating over all of its actuators), and this was the only behavior they had at the beginning of the evolutionary runs. However, with the existing actuators, by exploiting the physics of the environment, in the course of the evolution of the agents' morphology and under specific environmental pressures, additional behaviors, such as different foraging strategies (active versus passive), fighting, "fleeing," avoidance, "chasing," and predating, eventually emerged.

Conclusion

In this chapter I have discussed and illustrated how emotions—and, more generally, affect—can be embodied in robots from a biologically inspired EAI perspective. I have also discussed what "embodiment" might mean for robots beyond physical embodiment, and I have argued that (my model of) "core affect" provides the grounding for a "strong," multifaceted notion of embodiment that includes a number of aspects discussed in the literature: structural coupling between the agent and the environment, historical embodiment, physical embodiment, organismoid and (partly) organismic embodiment, and social embodiment. I have also illustrated and discussed how different aspects of this embodiment have been implemented in robots and other "animats" in my research group, from the four perspectives of mechanisms, adaptation, development, and evolution, emphasizing the importance of interaction and dynamics. The work discussed here provides some proofs of concept that "core affect" can be modeled in robots so as to ground the two elements of EI—the embodiment and the intelligence. I hope this work contributes not only to building "better" robots but also to fueling interdisciplinary research and discussions toward better understanding (and realizing) EI.

Acknowledgments

I am grateful to Rodney (Rod) Brooks for introducing me to EAI and robotics, to Rosalind (Roz) Picard for introducing me to the soon-to-be area of affective computing when I was a postdoctoral researcher at the MIT AI lab, and to Luc Steels for introducing me to Ashby, Cybernetics, and the notion of viability as a postdoctoral researcher at the VUB AI lab, back in the mid/late 1990s. Very special thanks to all my PhD students, research assistants, and research fellows, previously at the University of Hertfordshire, UK, and now at CY Cergy Paris University, France, who have accompanied me, and continue to do so, in this exciting journey. Not everyone's work has been mentioned, but everyone's work is gratefully acknowledged.

The writing of this chapter received the support of a Paris-Seine INEX Chair Neuroscience and Robotics.

Notes

1. "Animat" is short for "artificial animal." This notion was introduced by S. W. Wilson (1985) in the context of a simulated animal that learned to adapt in interaction with its environment and whose decisions and behavior were driven by a classifier system, a type of algorithm that combines ML and evolutionary computation. The term was subsequently widely used in the adaptive behavior community to denote biologically inspired simulated and physical agents that are situated in their ecological niche, and it is part of the name of the main conference in the domain, "From Animals to Animats: International Conference on Simulation of Adaptive Behavior," which has been taking place biennially since 1990.

2. Contrary to Ziemke, I don't restrict "organismic embodiment" to living organisms but extend it to include (artificial) systems that are dynamical and capable of self-organization and self-replication.

References

Adolphs, R. (2005). Could a robot have emotion? Theoretical perspectives from social cognitive neuroscience. In J.-M. Fellous & M. Arbib (Eds.), *Who needs emotions? The brain meets the robot* (pp. 9–25). Oxford University Press. https://psycnet.apa.org/record/2005-03475-002

Ashby, W. R. (1952). *Design for a brain.* Chapman & Hall. https://doi.org/10.1007/978-94-015-1320-3

Asma, S. T., & Gabriel, R. (2019). *The emotional mind: The affective roots of culture and cognition.* Harvard University Press. https://doi.org/10.4159/9780674238916

Avila-García, O., & Cañamero, L. (2002). A comparison of behavior selection architectures using viability indicators. In *Proceeding of the EPSRC/BBSRC International Workshop Biologically-Inspired Robotics: The Legacy of W. Grey Walter* (pp. 86–93). HP Labs. https://doi.org/10.1142/S0219843606000771

Berridge, K. C. (2019). A liking versus wanting perspective on emotion and the brain. In J. Gruber (Ed.), *The Oxford handbook of positive emotion and psychopathology* (pp. 184–196). Oxford University Press.

Blanchard, A., & Cañamero, L. (2005). From imprinting to adaptation: Building a history of affective interactions. In L. Berthouze, F. Kaplan, H. Kozima, H. Yano, J. Konczak, G. Metta, J. Nadel, G. Sandini, G. Stojanov, & C. Balkenius (Eds.), *Proceeding 5th International Workshop on Epigenetic Robotics: Modeling Cognitive Development in Robotic Systems, Lund University Cognitive Studies, 128* (pp. 17–24).

Brooks, R. A. (1991). New approaches to robotics. *Science, 253*(5025), 1227–1232. https://doi.org/1126/science.253.5025.1227

Cañamero, L. (1997). Modeling motivations and emotions as a basis for intelligent behavior. In W. L. Johnson (Ed.), *First International Conference Autonomous Agents (Agents'97)* (pp. 148–155). ACM Press. https://doi.org/10.1145/267658.267688

Cañamero, L. D. (1998) (Ed.). Emotional and intelligent: The tangled knot of cognition. *Papers from the 1998 AAAI Fall Symposium.* TR FS-98-03. AAAI Press. https://www.aaai.org/Press/Reports/Symposia/Fall/fs-98-03.php

Cañamero, L. D. (Ed.) (2001). Emotional and intelligent II: The tangled knot of social cognition. In *Papers from the 2001 AAAI Fall Symposium* (pp. 22-30). AAAI Technical Report FS-01-02. AAAI. https://m.aaai.org/Library/Symposia/Fall/2001/fs01-02-006.php

Cañamero, L. (Ed.). (2005). Agents that want and like: Motivational and emotional roots of cognition and action. In *Proceedings of the AISB'05 Symposiu*. AISB Press. https://aisb.org.uk/wp-content/uploads/2019/12/2_Agents_Final.pdf

Cañamero, L. (2021). Embodied robot models for interdisciplinary emotion research. *IEEE Transactions on Affective Computing, 12*(2), 340–351, https://doi.org/10.1109/TAFFC.2019.2908162

Cañamero, L. (2023). When emotional machines are intelligent machines: The tangled knot of affective cognition. In C. Misselhorn, T. Poljanšek, T. Störzinger, & M. Klein (Eds.), *Emotional machines. Perspectives from affective computing and emotional human-machine interaction*, (pp. 135–158). Technikzukünfte, Wissenschaft und Gesellschaft/Futures of Technology, Science and Society. Springer. https://doi.org/10.1007/978-3-658-37641-3_6

Cañamero, L., & Avila-García, O. (2007). A bottom-up investigation of emotional modulation in competitive scenarios. In A. Paiva, R. Prada, & R.W. Picard (Eds.), *Proceedings of the Second International Conference on Affective Computing and Intelligent Interaction (ACII 2007)*, (pp. 398–409). Springer-Verlag. https://doi.org/10.1007/978-3-540-74889-2_35

Cañamero, L., & Lewis (2016). Making new "new AI" friends: Designing a social robot for diabetic children from an embodied AI perspective. *International Journal of Social Robotics, 8*(4), 523–537. https://doi.org/10.1007/s12369-016-0364-9

Feldman-Barrett, L., Lindquist, K. A., Semin, G. R., & Smith., E. R. (2008). In G. R. Semi & E. R. Smith (Eds.), *Embodied grounding: Social, cognitive, affective, and neuroscientific approaches* (pp. 237–262). Cambridge University Press.

Fellous, J.-M. (1999). The neuromodulatory basis of emotion. *The Neuroscientist, 5*, 283–294. https://doi.org/10.1177/107385849900500511

Fellous, J.-M. (2004). From human emotions to robot emotions. In E. Hudlicka & L. Cañamero (Eds.), *Architectures for modeling emotions: Cross-disciplinary foundations. Papers from the 2004 AAAI Spring Symposium* (pp. 37–47). AAAI Press. https://aaai.org/papers/0008-ss04-02-008-from-human-emotions-to-robot-emotions/

Fellous, J.-M., & Arbib, M. A. (2005). *Who needs emotions? The brain meets the robot*. Oxford University Press. https://doi.org/10.1093/acprof:oso/9780195166194.001.0001

French, R., & Cañamero, L. (2005). Introducing neuromodulation to a Braitenberg vehicle. In *Proceedings of IEEE International Conference on Robotics and Automation, "Robots get Closer to Humans"* (pp. 4199–4204). IEEE. https://doi.org/10.1016/j.neunet.2005.03.003

Hiolle, A., & Cañamero, L. (2009). Learning affective landmarks. In L. Cañamero, P.-Y. Oudeyer, & C. Balkenius (Eds.), *Proceedings of the Ninth International Conference on Epigenetic Robotics: Modeling cognitive development in robotic systems* (Vol. 146, pp. 211–212). Lund University Cognitive Studies. https://harvardlds.org/wp-content/uploads/2017/01/2009_Lallee_etal_EpigeneticRobotics-1.pdf

Hiolle, A., Lewis, M., & Cañamero, L. (2014a). A robot that uses arousal to detect learning challenges and seek help. In H. Sayama, J. Rieffel, S. Risi, R. Doursat, & H. Lipson (Eds.), *ALIFE 14: The Fourteenth International Conference on the Synthesis and Simulation of Living Systems* (pp. 864–871). MIT Press. https://doi.org/10.1162/978-0-262-32621-6-ch142

Hiolle, A., Lewis, M., & Cañamero, L. (2014b). Arousal regulation and affective adaptation to human responsiveness by a robot that explores and learns a novel environment. *Frontiers Neurorobotics, 8*, 17. https://doi.org/10.3389/fnbot.2014.00017

Hufendiek, R. (2016). *Embodied emotions: A naturalist approach to a normative phenomenon.* Routledge. https://doi.org/10.4324/9781315657769

Khan, I., & Cañamero, L. (2018). Modelling adaptation through social allostasis: Modulating the effects of social touch with oxytocin. *Multimodal Technologies and Interaction, 2*(4), 1–67. https://doi.org/10.3390/mti2040067

Khan, I., & Cañamero, L. (2022). The long-term efficacy of "social buffering" in Artificial social agents: Contextual affective perception matters. *Frontiers in Robotics and AI, 9*, 699573. https://doi.org/10.3389/frobt.2022.699573

Khan, I., Lewis, M., & Cañamero, L. (2019). The effects of affective social bonds on the interactions and survival of simulated agents. In *2019 8th International Conference on Affective Computing and Intelligent Interaction Workshops and Demos (ACIIW)* (pp. 374–380). IEEE. https://doi.org/10.1109/ACIIW.2019.8925031

L'Haridon, L., Bergoin, R., Singh-Bal, B., Abdelwahed, M., & Cañamero, L. (2024). The emergence of a complex representation of touch through interaction with a robot. In O. Brock & J. Krichmar (Eds.), *Proceeding from animals to animats 17: The 17th international conference on the simulation of adaptive behavior* (pp. 106–117). LNCS-LNAI Series volume 14993.. Springer https://doi.org/10.1007/978-3-031-71533-4_8

L'Haridon, L., & Cañamero, L. (2023). The effects of stress and predation on pain perception in robots. In S. Canavan & D. Jiang (Eds.), *2023 11th International Conference on Affective Computing and Intelligent Interaction (ACII)* (pp. 1–8). IEEE. https://doi.org/10.1109/ACII59096.2023.10388190

LeDoux, J. (1996). *The emotional brain.* Simon & Schuster.

Lewis, M., & Cañamero, L. (2016). Hedonic quality or reward? A study of basic pleasure in homeostasis and decision making of a motivated autonomous robot. *Adaptive Behavior, 24*(5), 267–291. https://doi.org/10.1177/1059712316666331

15 AI Designer, Know Thyself!: What We Learn about Human Intelligence through Building Artificial Intelligence

Dor Abrahamson, Marcelo Worsley, and Zachary A. Pardos

Introduction

The ancient Egyptian myth of King Thamus, who feared that the invention of writing would weaken memory and create a false sense of wisdom, serves as a timeless cautionary tale about the complex relationship between humans and the tools we create to aid our thinking (Plato, as cited in Phaedrus 14, 274c–275b). Throughout history, we've relied on various "cognitive artifacts" to extend our mental capabilities, from ancient slates to modern Slack (Clark & Chalmers, 1998; Collins & Ferguson, 1993; Donald, 2010; Kirsh, 2010; Norman, 1991; Overman, 2023; Sfard, 2002). When effectively utilized, these tools empower us to store, process, and share information, significantly amplifying our extended cognitive (Clark, 1998) and literacy potential (diSessa, 2018). However, the effectiveness of these tools varies considerably, as illustrated by the cumbersome nature of multiplying large numbers using Roman numerals compared to the efficiency of Hindu-Arabic numerals (Wilensky & Papert, 2010).

The evolution of digital technologies in the twentieth century dramatically reshaped human cognition, relegating mechanical computation tools like the abacus and slide rule to obsolescence. Electronic calculators offered not only speed and accuracy but also the ability to process complex functions, generating results beyond human capabilities (Schwartz, 2007; Wilensky, 1999). As Wolfram (2002) noted, these tools expanded the scope of scientific inquiry, allowing us to explore phenomena previously inaccessible to human cognition.

Artificial intelligence (AI) represents the next leap in this evolution, impacting the learning sciences as profoundly as the natural sciences

(Järvelä et al., 2023). This chapter focuses on the reflexive potential of neural networks, exploring how developing AI research tools can illuminate our understanding of the human mind. Specifically, we examine how creating computational tools to study students' perceptuomotor engagement with interactive devices can shed light on human perception and cognition. We compare researchers' reflexive processes in creating three types of tools—proceduralization, machine classification, and representation learning—assessing their affordances for understanding human perception and cognition.

Furthermore, this exploration into the reflexive potential of AI in educational research aligns with the concept of embodied intelligence (EI). By developing and refining AI tools that interact with and interpret human perceptuomotor engagement, we gain insights into the intricate relationship between our physical actions, sensory experiences, and cognitive processes. This understanding can inform the design of future learning environments and technologies that better support embodied learning, acknowledging the crucial role of the body and its interactions with the world in shaping our understanding. Finally, we discuss the potential tradeoffs of integrating AI into educational research and for informing embodied design for learning.

Fostering Artificial Perception of Coordinated Human Action

Lessons Learned (or Not) about Human Perception

What might we learn about our own perception of human action when we teach a digital machine to likewise perceive action? Well, it depends on *how* we teach the digital machine to perceive action. In one route, *proceduralization*, we specify for the machine all the features of action that we naturally attend to, and we tell the machine how to sense, determine, and compute aspects of these features to arrive at inferences about the action that are commensurate with our own inferences. Along the way, we become aware of what we ourselves attend to, even if our mind does not compute those features as does the machine. In another route, *machine classification*, we need not specify for the neural networks *how* we are perceiving and classifying action, but only the *results* of our judgments about the action (though, along the way, we might point out to the machine which *features* of action we are attending to in making our judgments). Hence, the machine learns which actions lead to which judgments by being shown numerous examples of actions and their resulting judgments. Just how it does so, nevertheless, may remain opaque for us, even as our tacit inference process may likewise remain obscure. In yet another route, *machine representation*, we provide the neural networks little to no framing, just the data itself, and let their deep-learning capacities predict actions and figure out the action "grammar."

We position these three approaches as important points along a spectrum that significantly oversimplifies the full realm of possibilities but that will allow us to examine the potential affordances conferred through each paradigm. In truth, researchers have creatively moved along a continuum that allows for repeated iteration between human intelligence and AI. Nonetheless, we invite the reader to settle into these three seemingly distinct approaches as we consider how this all connects to questions of EI.

Situating the Thesis in a Design-Based Educational Research Project

Our commentary on dialectics of human and Artificial Intelligence reflects on an ongoing empirical evaluation of OЯTHO, a novel science-hall mathematics exhibit (see figure 15.1). OЯTHO is an interactive body-scale digital tabletop designed for learning the basics of the Cartesian coordinate system through solving a dyadic motor-control problem of navigating a virtual ball along a designated path (vel Żabik et al., 2024). OЯTHO is similar in logic and task to popular physical games, such as mini golf, in the sense that there is an obligatory start point and then optional routes leading to an obligatory end point, with challenges along the way, such as impassable impediments, tight corners to negotiate, or narrow routes from which you might fall. However, key to OЯTHO's game mechanics is that one cannot *directly* control the ball, as one would with a hand or golf club; one can only control the *x* and *y* axial constituents of the ball's location in the Cartesian space (q.v., Etch-a-Sketch). Moreover, these two orthogonal axial controls are distributed over two players, where one player controls the ball's *x*-axis vectoral movement constituent and the other player controls its *y*-axis constituent. As such, if the path is diagonal, both players may have to operate their respective axial controls simultaneously and judiciously. For example, while a 45° path-leg would require equally paced *x* and *y* remote-controlling, a steeper-angled leg would require the *y*-player to move faster than the *x*-player. The first implementation of OЯTHO was in the LivingLab

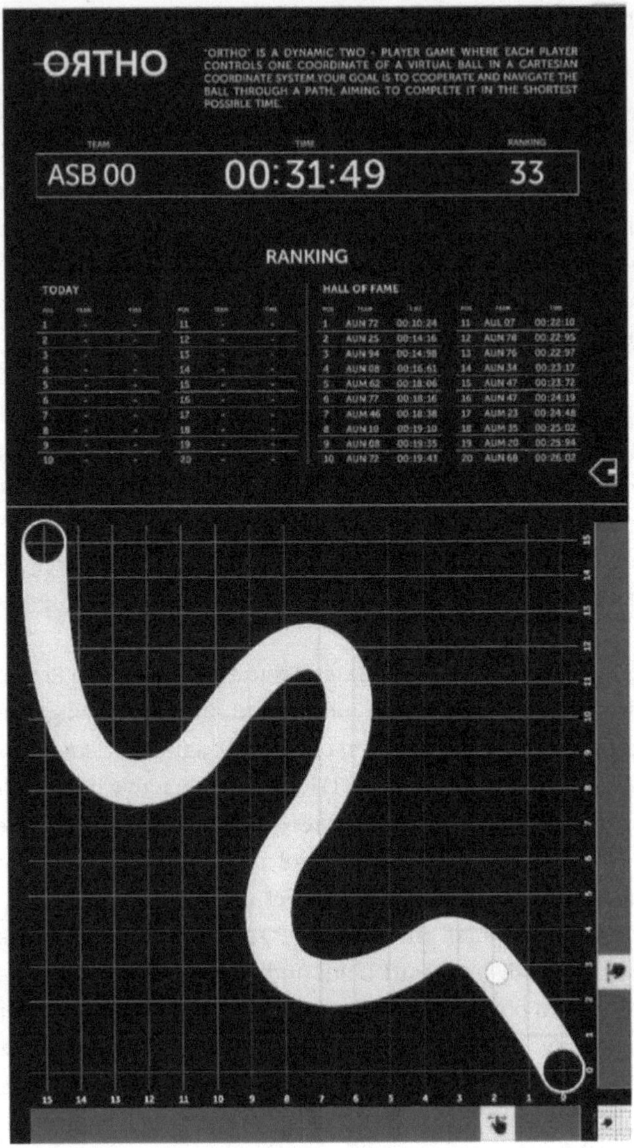

Figure 15.1

The x-player (little hand in green box to the right of the white ball) and y-player (little hand in green box below the ball) each control one of the ball's orthogonal coordinate locators along their respective axis. Negotiating tight passages might require simultaneous action, so as not to fall off the yellow path. Moreover, competitive movement—to get your names up on the wall of ranking fame—demands rapid and adroit simultaneous action. *Source*: the author.

wing of the Copernicus Hall of Science in Warsaw, Poland, where Abrahamson's original design was developed into a fully operable exhibit.

Drawing on heuristic principles of the embodied-design pedagogical framework (Abrahamson et al., 2023), the OЯTHO activity seeks to occasion opportunities for children to reinvent mathematical practice by collaboratively devising ad hoc discursive solutions to emergent problems inherent to coordinating the enactment of the joint action required for completing an engaging task. That is, a core learning objective—here, understanding the fundamental architecture of the Cartesian coordinate system as a two-dimensional space where points are construed as ordered number pairs (e.g., Abrahamson et al., 2023) referring to ratio scales on orthogonal axial lines—is not taught directly to students. Instead, this idea should occur to the students spontaneously through sensorimotor exploration, interpersonal movement coregulation, dyadic argumentation, and the adoption of interpolated semiotic resources as frames of reference for organizing and mobilizing their technologically mediated joint action. As design-based researchers operating from the epistemological perspectives of radical enactivist cognition and ecological dynamics, we look for students first to ground mathematical concepts in these new environmentally coupled perception-action loops—only later to formalize these new cognitive structures by utilizing available semiotic resources as auspicious means of enhancing, evaluating, and explaining their enactment strategies (Abrahamson et al., 2011).

Research Objectives: Learning about User Learning

While OЯTHO provides Copernicus museum visitors with an engaging interactive exhibit (supported by exit polls), it also serves as a valuable research environment for educational researchers to explore the complex interplay between digital technology, collaborative learning, and overall educational experiences. For example, we ask whether two children can learn mathematics together without any instructional guidance. If so, what would this learning look like? How might we theoretically characterize what changes in children's cognition as they figure out how to perform a collaborative task involving the joint manipulation of a digital object? And how would these changes in children's cognition, as they engage in the activity, relate to a formal understanding of the phenomenon they have thus coenacted? What might two people have in common, conceptually, once they are able to engage in this coordinated joint action? Does the need to collaborate stimulate learning and, if so, how?

To address these and other questions, the team needed to identify players' behavioral parameters indicating collaboration quality. But first, we needed an objective measure for the quality of the task performance itself—namely, how well the digital ball was moving through the tabletop path, regardless of how the pair were individually and jointly achieving this kinematic maneuver. That is, to say something about the players' *proximal* actions—what they were each doing manually with their own axial controller—we needed to gauge the players' compounded *distal* action as instantiated by the ball's movement. For our analyses, we had access to quantitative records of the *x* and *y* individual inputs with respect to the target track, which allowed us to reconstruct and understand what the two players were doing, thinking, and figuring out together as the game unfolded dynamically.

Still, how might we measure the quality of a movement-based task performance? For one, we can attend to summative performance markers, such as the duration of completing a track, and compare between dyads, perhaps acknowledging their age. But how about the dynamical attributes of playing the game, such as the fast-changing kinematic features of the unfolding path that the pair enacted through the game track?

Methodological Challenge: Scaling Up Qualitative Analysis of Movement Data

In order to attend to dynamical aspects of the players' performance, we needed to watch the game as though it were unfolding before our eyes in real time. Because the games were not encoded as dynamic actions, we first had to render the static numerical data into dynamic forms. We did this by reconstructing the digital scenes and reanimated the players' path along the track at the speed that it had originally been enacted. Once the inert symbols were thus resuscitated, we could witness the reinvoked historical performance directly. The data were now presented in a way that naturally aligned with our intuitive understanding, allowing us to leverage what Hutto et al. (2015) terms our "ecological intelligence." We could now form insightful judgments about the coordination dynamics of the players, even without access to the original video or audio recordings of the actual events.

We were surprised to experience empathetic resonance with the "ghost players," whose hesitations, decisions, and sentiments we could somehow intuitively conjure from their reanimated actions (cf. Heider & Simmel, 1944). We also were able to reconstruct apparent sociocognitive-socioemotional aspects of their experience in a manner far exceeding what we might glean from tables and graphs alone (cf. Sinclair, 2024). It was as though we *became* those anonymous players; we embodied the players through what felt as latent joint action. Thus, by examining the recorded action logs of past games at the museum, we were able to utilize our intuitive understanding of human interaction. This allowed us to easily recognize moments like smooth ball handling, skillful navigation of tight turns, or instances where one player was waiting for the other to act.

As we attempted to articulate the quality of the players' task performance, however, we became increasingly aware that our judgments were qualitative—we could at best say that the players' path looked excellent, good, fair, or poor. Yet, even if one were to concede such intuitive scoring, perhaps validating the judgments through inter-rater reliability tests and team deliberations, one would still encounter the challenge of needing to assess the quality of a much larger data set than could be feasibly managed by human evaluators. How do we capture these tacit impressions such that we can bring a machine *up to our speed* in making commensurate judgments and then bring it far *beyond our speed*? If we are to delegate our qualitative eyeballing to quantitative number-crunching, what instructions should our human intelligence give AI? How do you foster digital eyeballing?

It turned out we knew more about human movement than we knew to say, let alone formalize as procedural algorithms. Is that a bug of human intelligence, or perhaps a feature? Should we debug our minds, or do biological cognitive systems operate in inherently different ways than computers?

Theoretical Framing: On Articulating Tacit Knowledge

The epistemic tension between *knowing how* to do something and *knowing how to articulate* that knowledge is hardly new within the learning sciences. As philosopher Gilbert Ryle (1945) observed in his presidential address to the Aristotelian Society, practical competence often operates beneath the level of explicit rule-following: "In some ways the observance of rules and the using of criteria resemble the employment of spectacles. We look through them but not at them. And as a person who looks much at his spectacles betrays that he has difficulties in looking through them, so people who appeal much to principles show that they do not know how to act" (Ryle, 1945, p. 9).

Ryle's metaphor captures a foundational insight: skilled action is typically guided by tacit, embodied understanding rather than by conscious reference to formalized rules. Excessive reliance on explicit principles may therefore signal not mastery, but a lack of fluency in practice.

In our case, by juxtaposition, we can *see*, *feel*, and *know* how well OЯTHO players are manipulating the ball, but we may be hard-pressed to *say* how we are thus casting these judgments. We often overlook the tools we use every day, focusing on what we perceive through them rather than the tools themselves. As Dreyfus (2002) interprets Merleau-Ponty, we struggle to articulate our skilled actions because such expertise is rooted in perception, not language: "A phenomenology of skill acquisition confirms that, as one acquires expertise, the acquired know-how is experienced as finer and finer discriminations of situations paired with the appropriate response to each. Maximal grip names the body's tendency to refine its responses so as to bring the current situation closer to an optimal gestalt. Thus, successful learning and action do not require propositional mental representations. They do not require semantically interpretable brain representations either" (Dreyfus quoting Merleau-Ponty, p. 367). Polanyi (1958) resonates with this epistemic concept of a division between practical knowledge (know-how) and theoretical knowledge (know-that), suggesting that the former may surface as the latter through concerted introspection: "As observers or manipulators of experience, we're guided by experience and pass-through experience without experiencing it in itself. The conceptual framework by which we observe and manipulate things being present as a screen between ourselves and these things, their sights and sounds, and the smell and touch of them transpire but tenuously through this screen, which keeps us aloof from them. Contemplation dissolves the screen, stops our movement through experience and pours us straight into experience; we cease to handle things and become immersed in them" (Polanyi, 1958, p. 197).

The theme of articulating tacit knowledge was also pivotal for Heidegger (1927/1962), who explored the phenomenology of breakdown, when ready-to-hand tools become noticeable due to their conspicuousness, obstruction, resistance, or absence. In these moments, they transform into present-at-hand objects, forcing us to consciously acknowledge them and, as he puts it, "the environment announces itself afresh" (p. 105). For complementary perspectives from Dewey and Leont'ev on this idea, see Koschmann et al. (1998).

Yet how exactly might our tacit knowing become accessible to our conscious contemplation? From his semiotic-cultural theory of objectification, Radford (2013) argues that we make use of any available semiotic resource, such as a word, a gesture, a diagram, or a symbol, to objectify our presymbolic notions. Bamberger (1999) discusses computers as particularly auspicious semiotic partners for contemplating our tacit knowledge, whether musical or mathematical, as we attempt to build (digital) things. The exacting techno-epistemic constraints of information-computing machines require us to formulate our tacit skills in procedurally explicit and algorithmically specific forms, even if these logical forms do not quite mirror our epistemic architecture. And yet we can then dialogue with the computer, once it outputs the results of running our commands; And, in turn, we can evaluate this output against our intentions and then modify our input—"Oh, that's not quite what I had in mind, let me change this function."

As Papert (1980) maintained, by examining whether the computer built what we thought we'd told it to build, we debug our own thinking; that is, we learn. Along the way, we appropriate and instrumentalize cultural-historical conceptual systems that civilization has developed for instantiating, emulating, and perhaps augmenting tacit knowledge as techno-scientific application (Abrahamson & Bakker, 2016). Throwing a stone is not rocket science, but to launch a rocket we must rethink how we throw stones. We learn to think *with* or *through* artifacts; our cognition becomes culturally extended (Menary & Gillett, 2022).

Which brings us back to OЯTHO.

AI–Human Intelligence Partnership in Data Analysis

Our research objective is to develop computational methodologies that would scale up our investigations of dyadic user interactions from case-by-case human qualitative analysis to massive-case machine quantitative analysis. This process can take several quite distinct routes. Here, we survey three approaches that vary according to human agency in specifying data features that our AI research partner should foreground and compute: proceduralization (everything), machine classification (something), and representation learning (nothing). This distinction, in turn, bears implications for what we might learn about human intelligence.

Proceduralization

In one modus operandi, the method is to spell out our mental process. That is, we need to flow-chart in computational functions the tacit workings of our natural sensitivity to the social dynamics of coordinated joint action. That which just "comes to us" naturally must be articulated in the form of action parameters, input values, and output judgments. When we casually observe OЯTHO players' flowing behaviors, we just *see* that, for example, the *x*-player is hesitating amid a challenging passage. And yet to delegate this intuitive professional vision to a silicon "brain," we are obliged to highlight, code, and elaborate on each relevant feature, in a tight cascade of logical inferences (cf. Goodwin, 1994). Instrumenting ourselves with AI helps us figure out what we know that we don't know we know as we gaze at the humdrum actions of people doing things with stuff. By reflecting on our own naturalistic noticing, we become cognizant of our basic common sense about how things happen in the world. This is a skill that we've picked up along the evolutionary way—phylogenetically, ontogenetically—just by virtue of being human. We may never have stopped to think about this skill or even conceptualized it as, indeed, a skill.

As the OЯTHO research team sought to proceduralize our assessments of the dyads' actions, we became aware of *what* we were tacitly monitoring, such as changes in players' performance level, both as a pair and as individuals; inter-player agency dynamics, such as evidence of competence asymmetry and scaffolding; movement smoothness; and action efficiency. As such, our team's collaborative reflection was increasingly homing into aspects of the individual players' actions. We were unpacking the complex social dynamics of the social Gestalt of the kinematic Cartesian compound, attempting to identify its underlying psychological components. In the process, we became aware of our own analytical framework, much like Ryle's notion of "seeing the lens" through which we interpret the world.

As a first step, our team focused on deciphering the *synchronic* story within each individual frame—identifying *what* to observe and measure at each specific moment. This process naturally led us to consider the *diachronic* account of change across frames. It is one thing to say what we see now, now, and now; it is another thing to identify trends across now, now, and now; and it is yet another thing to *explain* those trends. As such, beyond articulating and measuring instances of performance aspects, we wondered *how* players become better: Do they jointly develop compatible anticipatory perceptions of things to come? Does their coordinative fluency—that is, their x-and-y interaxial spatial-temporal attunement—benefit from apparent figural patterns they project forward into their task flow? What would these patterns be? One such rudimentary pattern is a single path segment of constant gradient. One notch up in pattern complexity would be a *sequence* of segments, each of constant structural features, such as a zigzag whose zigs and zags are equally long and alternatingly angled.

We can then complexify our models by yet further considering relations between path features and action dynamics as mediated by the players' mutual microdevelopment of pattern anticipation: Are players slowing down amid impending troubled water, such as breaks in figural patterns, transition from linear to curvilinear path composition, path bifurcations, or narrowing pathways? Once they have negotiated a new feature, do we then witness a gradual increase in their speed? Are their actions smooth or jerky (Balasubramanian et al., 2015)? Do the players employ rhythmicity as a coordinative means (Farrera & Ramos-Fernández, 2022; Kugler & Turvey, 1987)? Does it take them less time to tackle a feature they had previously encountered (Abdu et al., 2023; Marwan et al., 2007)? And how does interplayer real-time coordination look? For example, will a more able player adjust the duration of their actions to communicate task-relevant information to a less able player (Abrahamson & Wilensky, 2005; Vesper et al., 2017)?

Thus, as we attempted to unpack and spell out our tacit knowledge, we realized both how much we know that we didn't know we know *and* that we are still unsure how we know it (Horsey, 2002). As such, even as we build systems that appear to output inferences similar to our own inarticulate know-how, we recognize that these systems are, at best, emulating, not simulating, how we think. This only underscores the gaps in our understanding of human cognition.

Machine Classification

In another methodology, we eschew introspection into our tacit inference processes. That is, we do not develop elaborate models of how we allegedly think. Rather, we may develop them but forgo presenting them to the machine. Instead, we let the AI figure out for itself, so to speak, what inputs are matched with which outputs. In practice, we feed the AI multiple cases representing each of our categories of interest, and we leave it to the machine to figure out for itself how to accordingly classify additional inputs that it deems as instances of the same category. The process ensues in four stages: data preprocessing, feature engineering, model training, and performance evaluation. For example, we might present to the AI incidences of what we view as players' well-coordinated actions and, separately, incidences of what we consider players' ill-coordinated actions, and then we evaluate whether, and at what precision rate, the machines can likewise sort out new cases into these two categories. Next, we can boost this process by performing feature extraction on the data in question, such as specifying recurring patterns or common themes, to generate a set of features that will serve as input variables for the machine learning (ML) algorithms. Feature extraction can either be conducted manually by the researchers or left to deep-learning algorithms. In the latter case, AI specifies distinctive features in the data enabling the classification. As such, we might learn what we ourselves are perceiving in the data.

Consider the case of dyad companions' age composition. We became aware that interpersonal dynamics of adult-child visitor dyads apparently differed from those of child-child or adult-adult pairs. For example, early qualitative analysis of available audio-video data captures parents scaffolding their children's actions, even while the game is running (a form of asymmetric dyadic interaction that Rosenbaum et al. (2020) call *participatory facilitation*). At the beginning of each gaming session, the digital system requires each player to enter their own basic demographic details, including their age, as well as their relation to the other player (family, friend, or stranger). We wondered whether running the quantitative data of players' digital actions through the AI neural networks might surface behavioral dynamics that differentiate the three dyadic age groups. That is, once the AI software has been fed three hefty batches of games—adult-adult, child-child, and adult-child—each with thousands of empirical data cases, could it perform above chance in guessing the dyadic composition when fed new random games?

In summary, this experiment used ML analyses to probe for significant differences among various data typifications, to both validate and generate hypotheses respecting correlations between user demographics, task parameters, and aspects of performance behavior (Proksch et al., 2023). These ML processes taught the system to categorize our data similarly to how we would. As such, the methodology proved successful.

However, for the purposes of our thesis, we note that conducting this process did not shed light on how we ourselves categorize the data along the parameters of interest. Moreover, just because the AI extracted certain features to emulate our classification, it does not necessarily mean that our human intelligence attended to the same features and their dynamic relations.

Representation Learning

A more modern focus of research on ML and its applications has been on *representation learning* using artificial neural networks (ANNs; i.e., deep learning). This is where human interpretation of the data (i.e., feature engineering) is bypassed in favor of ANNs learning a model of the structure of the domain space directly from the source data. This approach is most useful when the structure of the domain is too complex for a human-labeling approach to be tractable or when aspects of the domain may not be known, at least explicitly, to humans. A challenge with this approach is that representation learning requires a large amount of input and output data; however, in the case of a typical classification data set, there is far more input data (e.g., behavioral/telemetry data points) than there is output data (e.g., human labels of behavior). To solve this, modern AI methods model the input data itself. That is, the models predict what behavior is going to be observed next. Natural language has perhaps been the most notable and effective domain where representation learning has been applied. Modern large language models (LLMs), like those that power ChatGPT, do not learn language from domain experts explicitly enumerating the syntactic and semantic features of the language. Their capabilities are gained by learning to predict the next word in a sentence. In doing so, regularities of language are also learned (e.g., grammar [Mikolov et al., 2013]). This approach addresses a significant limitation to learning sciences research posed by machine classification.

Machine classification has been a mainstay of classical ML and has utility in *scaling human proceduralization*. As described above, it tries to improve the domain expert's explicit knowledge of the link between inputs and outputs by leveraging features given by the expert and augmenting those with its learned features or rules. Some of these learned procedures may represent human tacit knowledge and may or may not be interpretable, depending on the algorithm used. If it is sufficiently accurate in predicting the labels the domain expert has provided, then this function can be applied at scale. For example, if an interaction with a learner involves the subject matter expert classifying the learner's recent behavior and then conditionally suggesting the next activity, then this interaction could be scaled to any number of simultaneous learners interacting with a computer if AI has been taught how to perform this classification in a similar fashion to the expert. The shortcoming of this process is that, in the interest of advancing knowledge, merely mimicking the labeling ability of an expert is often not revelatory. Classification is limited to the labels already given to it by the expert and is therefore limited to deriving, at best, what the expert already tacitly knows. More interesting, however, from a scientific inquiry perspective is to advance our knowledge and explore, for example, if new labels are needed or if we should be thinking about the structure of the domain space in a completely different way.

In Pardos et al. (2021), representation learning was applied to the telemetry of learners interacting with a touch screen. Regularities in movements learned by the ANNs were visualized and then colored by the human labels associated with those movements. Interestingly, there were visually distinct groups of behaviors that had the same human-ascribed labels associated with them, perhaps suggesting that a new label, or expansion of the theoretical framework, may be in order.

Both human intelligence and AI develop tacit representations by observing behavior. Human reasoning, articulation of this representation, and experimental validation of conjectures go beyond what representation learning provides; however, it can serve as an alternative perspective on a domain—one gained from observing more data than a human can. By providing a novel, increasingly scrutable, and data-rich (though cognitively poor) perspective on a domain, our interaction as researchers with AI of this type may take on the complexion of a collaboration—understanding what it has learned about a domain and the implications of this on what we may come to know.

Conclusion

This chapter explored the profound implications of AI for educational research, particularly its potential to illuminate embodied learning and the intricate dance between perception, action, and cognition. The OЯTHO activity served as a prime example of embodied design, showcasing how